WATER CHLORINATION
Environmental Impact and Health Effects
Volume 1

WATER CHLORINATION
Environmental Impact and Health Effects
Volume 1

Edited by

Robert L. Jolley

Advanced Technology Section
Chemical Technology Division
Oak Ridge National Laboratory
Oak Ridge, Tennessee

Proceedings of the Conference on the
Environmental Impact of Water Chlorination
Oak Ridge National Laboratory
Oak Ridge, Tennessee
October 22-24, 1975

Sponsored by

Oak Ridge National Laboratory
Energy Research and Development Administration
U.S. Environmental Protection Agency

ANN ARBOR SCIENCE
PUBLISHERS INC
P.O. BOX 1425 • ANN ARBOR, MICH. 48106

Second Printing, 1978

Copyright © 1978 by Ann Arbor Science Publishers, Inc.
230 Collingwood, P. O. Box 1425, Ann Arbor, Michigan 48106

Library of Congress Catalog Card No. 77-92588
ISBN 0-250-40200-9

Manufactured in the United States of America
All Rights Reserved

FOREWORD AND ACKNOWLEDGMENTS

The formation of chlorine-containing organics during the chlorination of waters for drinking water sterilization, cooling water treatment, and sewage processing and disinfection is receiving much national attention. There exists a need for a thorough examination of the environmental effects of water chlorination. To meet this need the Conference on the Environmental Impact of Water Chlorination was conceived by Dr. Carl W. Gehrs and myself. Major objectives of the conference were to set down the present state of knowledge concerning aqueous chlorination with particular emphasis on chlorinated organic compounds and associated biomedical and environmental effects. The conference should establish a better understanding of both practical and theoretical aspects of water chlorination and perhaps set the stage for solving some of the concomitant problems.

I wish to express appreciation to my fellow concerned scientists and conference registrants for helping make this conference a success and to the speakers and authors of papers for their many hours of labor and preparation, for the quality of the conference is truly dependent on their efforts. I wish also to give credit to the capable conference planning committee: Dr. William A. Brungs, U.S. Environmental Protection Agency; Dr. Robert B. Cumming, Oak Ridge National Laboratory; Dr. Carl W. Gehrs, Oak Ridge National Laboratory; Dr. D. Heyward Hamilton, Jr., U.S. Energy Research and Development Administration; Dr. Sidney Katz, Oak Ridge National Laboratory; and Dr. W. Wilson Pitt, Jr., Oak Ridge National Laboratory.

This conference was sponsored by the Oak Ridge National Laboratory, the U.S. Energy Research and Development Administration (now the Department of Energy), and the U.S. Environmental Protection Agency. I especially wish to acknowledge the advice and assistance of Dr. Charles L. Osterberg, Dr. Robert L. Watters, and Dr. D. Heyward Hamilton, Jr., of the U.S. Energy Research and Development Administration, and Dr. Andrew J. McErlean and Dr. J. David Yount of the U.S. Environmental Protection Agency. Without their foresight concerning the need for this conference it would not have been held. I also wish to thank Dr. Herman Postma, Dr. Stanley I. Auerbach, Mr. Don E. Ferguson and Dr. Chester R. Richmond for their advice and securing the sponsorship of the Oak Ridge National Laboratory. A final thanks to Dr. Charles D. Scott, my supervisor, for permitting me to spend so very many hours on this challenging and rewarding undertaking.

Because of their patience and understanding, I dedicate this book to my wife, Doris, and my sons, Robert and Richard.

Robert L. Jolley
Conference Chairman
and Proceedings Editor

WELCOME

Herman Postma, Director
Oak Ridge National Laboratory
Oak Ridge, Tennessee 37830

Welcome to the Oak Ridge National Laboratory. With approximately 300 conference registrants, we have overloaded the system. If problems arise, be sure to let us know and we will try to resolve them.

The objectives of this conference, to see what is known, what is being done, and what should be done concerning the chlorination of various waters are very important because the real endpoint of this is to establish a base of data, models and understanding that will permit those who have the awesome responsibility of setting standards and regulations to do so based on facts.

This area of concern, the environmental impact of water chlorination, shares with so many other areas of modern technology in the last few years a high degree of visibility. It has the potential for being emotion-laden and provides a framework for those who prefer to be subjective and for those who consider personal gain to enter the battle. Therefore, it is very important that this conference be held to objectively assess where we are and what needs to be done. I hope you make the best of progress toward fulfilling the conference objectives because, in modern society and the technical areas of concern, the impact upon people and our way of life is vitally important.

CONFERENCE ON THE ENVIRONMENTAL IMPACT OF WATER CHLORINATION: PERSPECTIVE AND OBJECTIVES

Stanley I. Auerbach
Director
Environmental Sciences Division
Oak Ridge National Laboratory
Oak Ridge, Tennessee 37830

In a three-day symposium, bringing together scientists of diverse backgrounds to focus attention on the impact of one of man's activities, chlorination, on his environment, I believe it is appropriate to discuss ecology and the role it has come to occupy in our society.

In the last decade, ecology has become a household word, having different meanings for individuals of differing backgrounds. Although this general interest in ecology has been beneficial, it has also brought negative publicity to the science through the activities of nonecological instant experts and others who convey an aura of "leave everything like it is," and "no progress is good." Ecology, concerned with man's relationship to his environment, has made the general public aware of the unity of mankind and the biosphere. Even though ecology can trace its lineage to the writings of Hippocrates and Aristotle, it was not until early in the present century that it became recognized as a distinct field of science. Part of the reason for this lack of identity lay in the nature of the discipline. Ecology deals with organisms and their relationship to their abiotic environment.

It is also concerned with the unique relationships which exist between groups of different organisms (community ecology) and, ultimately, with the relationships, interactions and attributes of groups of organisms in and to their abiotic surroundings (ecosystem ecology). Because of this complexity of relationships and the essentially holistic approach of ecologists to their problems, in general parlance and in the public's view, ecology is regarded as the totality of the environment—a view which is somewhat misleading, but which does recognize that environmental science has physical as well as biological attributes and components.
In the mid-1970s it is refreshing to know that in our modern world we are beginning to realize that, as the popular advertisement goes, "we're all in this together." Or, more scientifically, all of man's activities in some manner affect his future home or environment.

This conference is called the "Environmental Impact of Water Chlorination." We have brought together distinguished scientists of diverse backgrounds; chemists, biomedical biologists, toxicologists, ecologists and engineers. We have come together to make an assessment of the available knowledge of the aqueous chemistry of chlorine, to examine effects of chlorine and chlorinated organic products on aquatic ecosystems and man, to assess the biological and ecological implications of chlorine usage for treatment of natural and process waters such as cooling waters and sewage effluents, and, through publication of the conference proceedings, to provide a permanent record and reference document containing the most recent relevant scientific data. This forum presents the opportunity for free interchange of information, both formally and informally, thereby permitting the development of ideas and concepts concerning areas of research needing emphasis. A principal result anticipated from the symposium is an increased understanding and better definition of the problems associated with chlorination.

We have gathered together through a common interest in the subject, but we must recognize that we do not all speak the same scientific language. This can, of course, cause some problems in communication. Important, however, is the realization that in coming to this symposium, each of us could learn from individuals of different scientific backgrounds. There are unique, valuable attributes in the scientific community apart from or in addition to those of our own specific disciplines.

Thus, engineers and chemists discuss the technology of water chlorination and many aspects of chlorine chemistry in a variety of waters of environmental concern. Others will discuss biomedical effects, the epidemiological evaluation of trace concentrations of chemicals, ecological transport and bioaccumulation of chemicals, toxicity studies, and predictive tools.

This conference was designed to cover most major aspects of water chlorination, particularly relating to the chlorination of organic constituents and possible biomedical and environmental effects resulting from the chloro-organic products. Each paper surveys its topic, presenting a summary of the known information in addition to individual author's research. I hope this conference will be a landmark in the field of environmental effects of water chlorination, and that the future results emanating from this symposium will be cognizant of the efforts presented here.

CONTENTS

1   Current Chlorination and Dechlorination Practices in the
    Treatment of Potable Water, Wastewater and Cooling Water. 1
    George Clifford White

SECTION I. AQUEOUS CHEMISTRY OF CHLORINE
Joseph E. Draley, Session Chairman

2   The Chemistry of Aqueous Chlorine in Relation to
    Water Chlorination 21
    J. Carrell Morris

3   Measurement and Persistence of Chlorine Residuals in
    Natural Waters................................. 37
    J. Donald Johnson

4   Organochemical Implications of Water Chlorination...... 65
    Robert M. Carlson and Ronald Caple

5   Chlorination of Organics in Drinking Water............ 77
    *Alan A. Stevens, Clois J. Slocum, Dennis R. Seeger and
    Gordon G. Robeck*

6   Chlorination of Organics in Cooling Waters and
    Process Effluents............................... 105
    *Robert L. Jolley, Guy Jones, W. Wilson Pitt and
    James E. Thompson*

xiii

7 Analysis of New Chlorinated Organic Compounds
   Formed by Chlorination of Municipal Wastewater 139
   William H. Glaze, James E. Henderson, IV and Garmon Smith

8 Chemistry of Halogens in Sea Water 161
   James H. Carpenter and Donald L. Macalady

9 Decision-Making in the Regulation of Chemicals........ 181
   Edward M. Brooks

SECTION II.
BIOMEDICAL EFFECTS OF CHLORO-ORGANICS
Robert B. Cumming, Session Chairman

10 Halogenated Organics in Tap Water:
   A Toxicological Evaluation........................ 195
   Robert G. Tardiff, Gary P. Carlson and Vincent Simmon

11 Origin, Classification and Distribution of Chemicals in
   Drinking Water with an Assessment of Their
   Carcinogenic Potential............................ 211
   Herman F. Kraybill

12 The Potential for Increased Mutagenic Risk to the Human
   Population Due to the Products of Water Chlorination ... 229
   Robert B. Cumming

13 The Epidemiologic Approach to the Evaluation of
   Water-Borne Carcinogens......................... 243
   Kenneth P. Cantor

SECTION III.
ENVIRONMENTAL TRANSPORT AND EFFECTS
William A. Brungs, Session Chairman

14 The Toxicity of Chlorine to Freshwater Organisms
    Under Varying Environmental Conditions............ 261
    Arthur S. Brooks and Gregory L. Seegert

15 A Revised Review of the Impact of Chlorination
    Processes Upon Marine Ecosystems: Update 1977...... 283
    William P. Davis and Douglas P. Middaugh

16 Chlorinated Compounds Found in Waste Treatment
    Effluents and Their Capacity to Bioaccumulate 311
    Herbert L. Kopperman, Douglas W. Kuehl and Gary E. Glass

17 Investigating the Effects of Chlorinated Organics 329
    Carl W. Gehrs and George R. Southworth

SECTION IV: MODELING AND PREDICTION
Carl W. Gehrs, Session Chairman

18 Modeling Residual Chlorine Levels: Closed-Cycle
    Cooling Systems 345
    Guy R. Nelson

19 A Kinetic Model for Predicting the Composition of
    Chlorinated Water Discharged from Power Plant
    Cooling Systems 367
    Milton H. Lietzke

20 Assessing Toxic Effects of Chlorinated Effluents on
    Aquatic Organisms: A Predictive Tool 379
    Jack S. Mattice

SECTION V. ROUNDTABLE DISCUSSION

21 Roundtable Discussion 401
    Carl W. Gehrs, Moderator. William A. Brungs, Robert B.
    Cumming, Joseph E. Draley, D. Heyward Hamilton,
    J. Carrell Morris and George Clifford White

INDEX 425

CURRENT CHLORINATION AND DECHLORINATION PRACTICES IN THE TREATMENT OF POTABLE WATER, WASTEWATER AND COOLING WATER

George Clifford White
  Consulting Engineer
  San Francisco, California 94118

ABSTRACT

The present annual production of chlorine in North America is estimated at 10.5 million tons. Only 3 to 4% of this is used for sanitary purposes such as potable water and wastewater treatment, swimming pools, household use, cooling water circuits and food packaging process water. The chemical industry accounts for at least 80% and the pulp and paper industry accounts for the rest, that is, about 15 to 16%. This is the perspective of chlorine in the environment.

The use of chlorine for sanitary purposes dates back to 1854 when it was used to deodorize London sewage. The first known use as a disinfectant was in 1879, also for sewage, and its first use as a potable water disinfectant on a continuous basis was 1903. Since that time the use of chlorine in potable water treatment has expanded to include (in addition to disinfection) several other important functions such as taste and odor removal; iron and manganese removal; hydrogen sulfide removal; color removal; prevention of water quality degradation in distribution systems; control of biofouling in long transmission systems, thereby preventing friction factor deterioration; prevention and control of biofouling in filter media; restoration of well capacity; and sterilization of water mains and reservoirs. In sewage treatment chlorine is being used primarily for disinfection, but it is also used very effectively for the prevention of septicity and control of hydrogen sulfide generation in collection systems and treatment plants. Chlorine is used in limited applications such as control of activated sludge, bulking, sludge thickening, the destruction of cyanides and foul air scrubbing. It is also used in the manufacture of ferric chloride, which is used as an effective coagulant for both potable water and wastewater treatment. There are many variations and combinations of chlorine application to potable water. These include chlorination followed by partial dechlorination or complete dechlorination followed by rechlorination, and ammoniation followed by chlorination and vice versa for specific odor control requirements. Chlorine dioxide generated on the site is used for the destruction of phenols. Chlorine is also used along with chlorine dioxide, ozone and activated carbon for the treatment of grossly polluted waters. The variations in selective treatment for potable water are many as is the amount of chlorine applied—for example, potable water, 5 to 100 lb/million gallon; wastewater, 50 to 400 lb/million gallon; cooling water, 20 to 200 lb/million gallon (intermittently). In the final analysis the required use of chlorine in potable water and wastewater depends on local regulatory requirements with an Environmental Protection Agency (EPA) override. The requirements for cooling water are for most economical heat transfer in the production of power.

Wastewater disinfection in California is being practiced on a total coliform destruction requirement for various receiving water conditions. Likewise the necessity to detoxify the chlorine residuals in these effluents is being accomplished by dechlorination with sulfur dioxide. Owing to the lack of necessary analytical equipment, proof of continuous dechlorination to zero chlorine residual has presented some problems.

Dechlorination of cooling water discharge is being taken care of by dilution of cooling water blowdown. This is accomplished in the basic design of the cooling water system.

Dechlorination of potable water is strictly for enhancement of the product.

INTRODUCTION

The perspective of chlorine and chlorine compounds in our environment is amply demonstrated by the percentage distribution of the annual chlorine production in North America, estimated at 95×10^{11} kilos (10.5 million tons) (Chlorine Institute, 1975). The chemical industry accounts for at least 80% of the chlorine use. Most of this amount is used for making plastics, pesticides, antifreeze fluids, synthetic fibers, gasoline additives, solvents and paint removers. The pulp and paper industry uses about 16% of the chlorine. The remaining 3 to 4% is used for "sanitary" purposes which includes but is not limited to potable water and wastewater treatment, swimming pools, household use, cooling water circuits and food packaging process water (White, 1972). The following discussion will be limited to potable water, wastewater and cooling water.

We do not know to what extent the housewife's use of sterilizing and bleaching compounds, encouraged by TV commercials and soap operas, reaches the environment via the sewage collection system. However, a quick survey of the detergent and soap manufacturers could give us a good idea of the amount of chlorine entering our environment by this route. It might be substantial.

POTABLE WATER

The most important use of chlorine for so-called sanitary purposes is for the disinfection of potable water. Because of chlorine's oxidizing powers, it has been found to serve other useful purposes in water treatment, such as taste and odor control; prevention of algae growths in water treatment structures; maintaining clean filter media; removal of iron and manganese; destruction of hydrogen sulfide; color removal by bleaching of certain organic colors; maintenance of distribution system water quality by controlling slime growths; restoration and preservation of pipeline capacity; restoration of well capacity; and sterilization of water mains and reservoirs.

Current chlorination practice of potable water varies in dosage from 1.0 mg/l to about 16 mg/l. This variation is directly related to the quality of the raw water supply, that is, degree of pollution, concentration of nutrients, temperature and pH. Winter ice cover on a water supply will almost automatically double the summertime chlorine demand. To combat the pollution and taste and odor problems, various combinations and dosages of chemicals are in use today in both the United States and Canada. These include combinations of chlorine and chlorine dioxide, ammonia, potassium permanganate and activated carbon, sometimes followed by dechlorination with either sulfur dioxide or activated carbon.

Prechlorination of low-quality water is most often the operator's salvation. It is one of the most important tools for maintaining the efficiency of a water treatment plant. In these situations, which are numerous, it would be virtually impossible to turn out an acceptable water if it were not for the unique ability of chlorine to maintain a persisting residual throughout the process. In these cases disinfection is just a side effect.

Two examples of low-quality water that require chlorine dosages up to 16 mg/l are the Grand River at Brantford, Ontario, Canada, and the Passaic River at Little Falls, New Jersey (Inhoffer and De Hooge, 1974; Williams, 1974). A notable foreign situation comparable to these two American cases is the Seine River at Paris, France. The Choisy-le-Roi treatment plant located on the outskirts of Paris has a filtration capacity of 900,000 m^3 per day [238 million gallon per day (mgd)]. The chemical treatment consists of a prechlorine dose of about 16 mg/l to reach a 1 to 1.5 mg/l free residual. Then chlorine dioxide at about 4 mg/l is added ahead of the filters. After the filters, ozone is added in accordance with its predetermined 10-min ozone demand. This is about 2 to 3 mg/l. Sometimes chlorine is added to the finished water in the event there is trouble in the distribution system.

In North America poor-quality water is treated by prechlorination sufficient to provide a substantial free residual in the flocculation or sedimentation basins. This may be followed by an intermediate dose of 2 to 5 mg/l chlorine or 0.5 to 4.5 mg/l chlorine dioxide to carry a residual through the filters. Posttreatment might consist of the addition of ammonia to convert the remaining residual to chloramines as the water enters the distribution system. Alternatively, sulfur dioxide might be used to trim the final residual. Ozone could be used in these situations for taste and odor control and/or color removal. Granular activated carbon may also be used for dechlorination of the finished water.

Now let us consider the case of a clean water where chlorine is used solely for disinfection. Two typical examples are the supplies for San Francisco and the East Bay (Alameda and

Contra Costa counties). These waters are derived directly from melted snow in separate runoff areas high in the Sierra Nevada Mountains. The San Francisco supply is transported by a 165-mile aqueduct and the East Bay supply travels about 90 miles to various local storage reservoirs. The chlorine dosage required for disinfecting these waters is from 0.8 to 1.2 mg/l. The San Francisco supply was plagued from the start by assorted difficulties. The most humiliating one occurred a few months after the system was put in operation. Part of the 165-mile aqueduct consists of a concrete-lined tunnel 25 miles long through the Coast Range Mountains. Although the concrete tunnel lining is up to 12 in. thick, many cracks developed in it. The ground water is laden with filamentous bacteria (*Crenothrix* sp.) which infiltrated the tunnel through the cracks, causing luxuriant growths on the tunnel lining. This created a debacle in the distribution system. Industries were forced to shut down because of this biofouling and the sloughing of the filamentous debris. To combat the problem, the city of San Francisco, after having investigated several treatment methods, concluded that a persisting chlorine residual was the only answer. Therefore, 2 mg/l of chlorine is applied at the beginning of the tunnel. This produces a 1 mg/l free residual at the end of the tunnel 25 miles away.

Now let us consider those areas not as fortunate as the San Francisco Bay area. St. Louis, which uses Mississippi River water, prechlorinates at an average 5 mg/l. This is supplemented by intermediate doses of chlorine to preserve the efficiency of the filter system. Kansas City, which uses Missouri River water, prechlorinates at about 12 mg/l. These chlorine applications are not for disinfection. This is to keep the water treatment process from deteriorating and, also, for taste and odor control.

The City of Chicago, which uses a cleaned-up Lake Michigan water, prechlorinates at only 1 to 1.2 mg/l. This is comparable to the melted snow waters of California. This prechlorination dose is designed to provide an absolute minimum of 0.25 mg/l free chlorine through the filters. It is supplemented by a post-chlorination dose of 0.25 to 0.36 mg/l to give a 0.75 mg/l free chlorine residual entering the distribution system. However, to exemplify the effect of pollution on chlorination, whenever there is an upset with the Chicago Ship Canal, which amounts to a reversal of flow and dumps the canal flow into Lake Michigan, the chlorine demand escalates to 10 mg/l. Even at this chlorine dosage there is so much ammonia-nitrogen in the raw water that a free chlorine residual is not attainable.

Surface waters in areas of Pennsylvania require prechlorination doses of 7 to 8 mg/l. The City of Philadelphia provides sufficient chlorination equipment capacity to doses as high as 30 mg/l at the raw water basin outlet (ice cover situation) and up to 4 mg/l chlorine dioxide for pretreatment.

It is significant to note the quality of the water provided by the billion dollar California Water Plan. This water is a mixture of several rivers, primarily the Sacramento and San Joaquin. This water requires only 2 to 4 mg/l prechlorination dose to produce a 0.7 mg/l free residual in the filter plant finished water without any intermediate chlorination. This is a tribute to the diligence and surveillance of both the California State Department of Health and the California Water Resources Quality Control Board for their programs designed to preserve the water quality of the receiving waters.

There are many other situations where the water responds to the various treatment processes but the quality subsequently deteriorates in the distribution system. These systems have no other choice but to maintain a residual as a matter of policy on the basis of preventive medicine. The following examples are cited to illustrate this practice: Brantford, Ontario, converts the remaining free residual to chloramine by postammoniation. Water leaving the plant carries a 0.4 to 0.6 mg/l combined residual in winter and 1.5 to 1.7 mg/l in the summer. Chicago water enters the distribution system at 0.75 mg/l free chlorine residual (Willey *et al.*, 1974). The Washington Suburban Sanitary Commission processes water from the Potomac and Patuxent rivers. The water leaving their treatment plants enters the distribution system with a 1.0 to 1.5 mg/l free residual. There are other systems that go as high as 2.0 mg/l.

In many ground waters the bacteriological quality is of no concern but the presence of a combination of iron and manganese causes dirty water in the distribution system. A California city afflicted with this problem receives water from some 50 wells with a manganese concentration varying from 0.1 to 0.5 mg/l. Each well is treated with chlorine and Calgon. The chlorine dose varies from 0.25 to 2.2 mg/l, depending on the well. This treatment provides control of the dirty water problem whereas other solutions have failed.

Other ground waters require the addition of chlorine as a significant part of the iron removal process. Chlorine dosages in these waters are typically 2 to 7 mg/l. The chlorine applied also controls the growth of iron-bearing filamentous organisms.

A large number of supplies have to contend with taste and odor control at the source—usually an impounded supply (*American Water Works Assoc.*, 1974). The city of Winnipeg, Canada, is an example. Chlorine is their main treatment for taste and odor control of the impounded water, which is separated from the treatment plant by a 90-mile-long aqueduct. Chlorine is applied as the water enters the aqueduct at a rate of 4 mg/l. Forty hours later the water arrives at the treatment plant with about 0.3 mg/l total chlorine residual. Many other systems use chlorine application to a transmission line solely for the purpose of maintaining carrying capacity.

A great many water systems in the British Isles resort to high free residual chlorination (1 to 3 mg/l) followed by residual controlled dechlorination with sulfur dioxide so that the residual entering the distribution system can be maintained at the desired level. In some cases the high free chlorine residual is used to compensate for short contact times. This procedure is beginning to be used in the United States (White, 1968).

There are some waters that require an artificially induced breakpoint to control taste and odors. This is accomplished by adding ammonia-nitrogen and subsequently chlorinating to a free residual.

One of the most important uses of chlorine in the purveying of potable water is main sterilization. This is also extended to sterilizing water reservoirs, ship tanks, etc. Dosages involved are usually 50 mg/l with a retention time of 24 hr. These parameters have evolved after a half century of experience by the water industry.

The recent publicity given to the formation of chloro-organics in potable water due to chlorination and to the possibility the chloro-organics may be detrimental to health has raised considerable interest in the use of ozone as an alternative to chlorination. This is not a practical approach because ozone cannot do all the things that chlorine can do. Besides, the equipment problems present practical considerations that always favor chlorine.

Ozone is an extremely reactive oxidant. It is a much better viricide than a bactericide. It reacts so quickly that the residual die-away is not much longer than 5 to 8 min. It is excellent in certain situations of color removal and taste and odor control. Most ozone installations are backed up by chlorination equipment (White, 1974).

Ozone must be applied on a demand basis. This means that ozone demands must be performed at frequent intervals. Operating ozone equipment from either a flow-proportional or a residual control signal is not an established fact. Ozone equipment is either on or off. Another disadvantage is the readout of ozone being applied. This is not a routine measurement as in a gas chlorinator because the gas leaving the ozonator is a mixture of air or oxygen and ozone. This has been the source of operating problems.

It is interesting to note that a recent installation at Strasburg, Pennsylvania involves a small domestic water supply consisting of 13 springs. This water could probably be handled by a chlorine dose of 0.8 mg/l. The flow range is 0.26 to 0.45 m^3/min (70-120 gpm). The installation cost is $23,000 not including the equipment, which is leased for $12,000/yr. The ozone generator has a capacity of 3.7 lb/hr. The dosage used is about 1.5 mg/l (Harris, 1975). An automatic hypochlorinator installation would have cost about $5,000 at the most.

The attributes of ozone should not be overlooked. I have and always will strongly advocate the exploitation of the properties of ozone to complement the chlorine compounds in the quest for a better-quality potable water. This is particularly significant in view of the virus problem and the water reuse consideration (White, 1975a; 1975b). We need more information relating to the chemistry of combinations of oxidants. It is folly to try and pit one against the other.

Finally there is the question of proof of disinfection. The existence of even a trace of chlorine residual, regardless of form, drastically reduces or eliminates total coliforms from the distribution system samples (White, 1975a). The presence of a chlorine residual as proof of disinfection would eliminate surveillance problems and eliminate risks for those small but numerous water systems; for example, ski and other resorts, roadside restaurants, bus stops, motorway rest areas, trailer camps and other similar water systems. This would not be possible with any disinfectant other than chlorine.

WASTEWATER TREATMENT

The first use of chlorine in wastewater treatment was for the control of odors. This dates back to 1854 when it was used to deodorize the London sewage. It was first used to disinfect sewage in North America in 1893 at Brewster, New York (White, 1973). This application was for the protection of the Croton watershed, which is a part of the New York City water supply. A review of the literature indicates a trend of active interest in wastewater disinfection beginning about 1945. Up to that time the main interest in chlorine was for odor control, hydrogen sulfide destruction and prevention of septicity. Most of the sewage treatment plants practicing disinfection during that time belonged to the U.S. Armed Forces. It was a matter of policy that sewage effluents had to be chlorinated at all army bases in the United States during World War II. Today, as a result of the 1970 Federal Water Pollution Control Act, all wastewater treatment plants in this country are subjected to some definitive disinfection requirement.

Chlorine also plays an important role in the treatment of cyanide wastes, which are highly toxic. Cyanide wastes must be treated before being discharged to either a sewage collection system or a receiving water. When discharging to a sewer the cyanides need only be oxidized to cyanates, but when discharging to a receiving water the cyanides must be completely broken down to elemental carbon and nitrogen. The new federal regulations for industrial discharges make treatment of cyanide wastes imperative. The state-of-the-art for onsite cyanide waste treatment has developed to a point where packaged systems are readily available for the individual discharger.

Today the emphasis is on disinfection of all effluents. The objectives of wastewater disinfection are: to prevent the spread of disease; to protect potable water supplies, bathing beaches and receiving waters used for boating and water contact sports; and to protect shellfish growing areas. So far, the most efficient way to accomplish these objectives is by chlorination. Over the last 25 to 30 years the various regulatory agencies—local, state and now the U.S. EPA—have been seeking a set of ground rules as proof of disinfection for various receiving water situations.

At one time it was thought that disinfection could be accomplished by showing a 0.5 to 0.75 mg/l orthotolidine residual at the end of 30 min contact time. Contact chambers were built to give a theoretical 30-min detention time at average flow based on the volume of the chamber. Effective mixing and contact chamber short-circuiting were never considered. The California State Department of Health, Bureau of Sanitary Engineering, made several intensive studies of wastewater disinfection. As a result of these studies the chlorine residual contact time concept was discarded. Some years ago California adopted the concept of disinfection evaluation based on compliance with a prescribed most probable number (MPN) of coliform organisms. The numbers now in effect are 80% of samples less than 1000/100 ml for coastal bathing waters (equivalent to a median of 240/100 ml), a median of 70/100 ml for shellfish growing areas and a median of 23/100 ml for confined waters used for bathing or other water contact sports, assuming the dilution is at least 100 to 1.

An essentially coliform-free effluent (*i.e.*, a median MPN not greater than 2.2/100 ml) is the requirement for discharge into ephemeral streams, negative estuaries or other areas where the public is exposed to effluents receiving little dilution. There is a subtle implication of the necessity for good operation and adequate (secondary) treatment to achieve the 23/100 ml requirement. The severe effluent standard of 2.2/100 ml implies the necessity of some type of advanced treatment—for example, filtration or nitrification. Such a water quality coliform requirement suggests some virus removal or destruction capability for the system beyond that which normally occurs.

The EPA requirements for disinfection are in general less stringent than the requirements of California and other states. The EPA has temporarily committed itself to a 200/100 ml to 400/100 ml fecal coliform requirement as consistent with adequate disinfection, depending upon the classification of the receiving waters. This figure is comparable to a total coliform concentration of about 2000/100 ml to 4000/100 ml (White, 1975).

The stringent requirements for disinfection as set by California focused much needed attention on the proper design of chlorination facilities for this purpose. The result of this attention has been well documented in the literature from 1970-1975. I have concluded that if a chlorination facility is to be an optimum design it should consist of: (1) good mixing of chlorine with wastewater (1 to 3 sec); (2) a minimum of 30 min contact time at peak flow in a contact chamber with superior plug-flow characteristics; (3) a good chlorine control system which should include flow-proportional plus residual control followed by effluent residual monitoring; and (4) competent operators who understand the process chemistry and instrumentation.

On the basis of my own research and a personal investigation of some 60 plants over the past five years, I have made the following observations: A good secondary effluent, dosed in the range of 10 to 15 mg/l, can achieve a total coliform count of 23.2/100 ml on a consistent basis. If the same effluent is filtered this same dosage will achieve an MPN of 2.2/100 ml. If this same effluent is nitrified, but not filtered, it is possible to achieve a 2.2/100 ml MPN by virtue of the presence of a free chlorine residual due to the nitrification process. It follows that if the secondary effluent is nitrified, coagulated and filtered prior to disinfection, virus removal can be achieved. The residuals from such optimum systems will be on the order of 2 to 4 mg/l.

Quite a few plants are operating in this range of dosage and residual. A surprising number are achieving a total coliform MPN less than 9/100 ml. There are also a number of plants whose disinfection systems are inefficient and consequently require dosages up to 25 mg/l, resulting in residuals as high as

8 mg/l. Then there are the days when the best of plants suffer a biological upset. These periods require higher doses of chlorine, resulting in higher residuals.

It has been convincingly demonstrated that any chlorine residual greater than 0.05 mg/l is toxic to a considerable variety of aquatic life. This has brought about a requirement for dechlorination of these effluents at the end of the contact chamber. This is now an accomplished fact in many California plants.
The use of sulfur dioxide as a dechlorinating agent has proved satisfactory. It is easy to handle and utilizes the same equipment and instrumentation as that used for chlorine. This simplifies the engineer's design problems. Each 1 mg/l of chlorine residual requires 1 mg/l of sulfur dioxide. Mixing is not as critical as for chlorine and, since the reaction is almost instantaneous, contact chambers are not required.

There are many ramifications in wastewater treatment which have a profound effect on the disinfection process. Industrial wastes are the worst offenders. They usually cause an increase in chlorine demand or render the wastewater septic or both. The latter causes all kinds of problems. The biological oxygen demand (BOD) and suspended solids do not have as great an effect on the disinfection process as does the reduction of coliforms in the process before disinfection. The lower the concentration of coliforms, the easier the wastewater can be disinfected.

COOLING WATER

Chlorination of cooling water was first tried in the United States in 1924 at the Commonwealth Edison Company, Chicago, Illinois. Up to that time power generating stations were plagued with biofouling of the heat exchangers. This caused plant shutdowns for cleaning the slime and debris in the condenser tubes. This is a costly and messy business. Now, almost 50 years later, the power industry experience has proven that the chlorination of cooling water is the most efficient and economical method for the control of biofouling in the water-cooled heat transfer equipment (Cole, 1975).

All but 10% of the electric generating plants chlorinate the cooling water on a programmed basis. The few who do not are in areas where there is excessive scouring by sand in the cooling water. The total amount of chlorine used in this sector is about 90 x 10^6 kilos per year (98,915 tons per year).

The most important factor in cooling water chlorination is the consideration that the cooling water is the vehicle for transporting the disinfectant to the condenser tubes. Therefore current practice consists of intermittent doses of chlorine—for example, 1 to 2 mg/l for 20 to 30 min for two or three times each 24 hr. The magnitude and frequency of dosage varies with local conditions. There are two types of systems. The most common is the once-through cooling water flow (as from receiving water and back again on a continuous basis). These systems chlorinate on an intermittent basis. The EPA complains about two things: (1) the chlorine residual in the effluent which may persist in a plume for 2 to 3 hr is objectionable, and; (2) during the period of chlorination the chlorine is destroying the biota in the cooling water taken from the source.

Once-through systems now under design are arranging chlorination so that the effluent from a group of condensers is diluted with the effluent from the rest of the condensers in order that the residual in the chlorinated condenser effluent is diluted sufficiently to disappear by the time it reaches the receiving water.

The other system of cooling water use is the closed recirculation system using atmospheric or forced-draft cooling towers. There are many advantages to this system from a conservation point of view. However, a large amount of organic matter does develop on the tower and in the sludge accumulation of the tower basins. This not only increases the chlorine demand but prevents the formation of a free chlorine residual. Therefore practically all tower systems operate on the less efficient chloramine residual. These tower systems do have a distinct advantage over the once-through systems insofar as protecting the environment is concerned. This is because the blowdown from the tower can be turned off during periods of chlorination, or it can be diverted to the ash sluice system or to a lagoon, or the blowdown can be dechlorinated. The blowdown system on many towers would only have to dechlorinate the 60-min residual, which would be less than 0.2 mg/l in most situations.

SUMMARY AND CONCLUSIONS

It is demonstrably clear that there is no alternative to chlorine as a disinfectant and chemical tool in the treatment of potable water, wastewater and cooling water. While there appear to be some disadvantages, such as the formation of some undesirable chloro-organics, no other oxidant can combine all the attributes of chlorine. The popularity of chlorine is deserved because of its potency and wide range of effectiveness as a germicide, and it is easy to handle, apply, measure and control. Most important of all it can be applied to effect a predictable persisting residual over long distances and extended periods of time. This turns out to be somewhat of a disadvantage in wastewater treatment. The other halogens such as iodine, bromine and bromine chloride are so much more expensive and cumbersome to handle that they could only be considered for special situations. Ozone has some distinct advantages over chlorine, not the least of which is its superior viricidal efficiency. However, it is such a rapid-acting oxidant that it is impossible to maintain a persisting residual. Any remaining residual has a very short half-life.

Instead of looking for an alternative for chlorine we should be investigating the chemical attributes of combinations. For example, in Europe combinations of chlorine, chlorine dioxide, ozone and activated carbon are used, and in Great Britain chlorination is followed by dechlorination and sometimes this treatment is followed by rechlorination and ammoniation.

We should investigate the difference in efficiency of various strengths of applied chlorine solutions and at different pH levels.

We should investigate the use of preformed chloramines for special situations.

We should investigate various combinations of dechlorination with both sulfur dioxide and activated carbon.

We should investigate all possible combinations of sequences using chlorine, chlorine-dioxide and ozone.

Lastly, we should investigate the merit of certain types of aeration sequences and dechlorination followed by rechlorination with and without postammoniation.

REFERENCES

American Water Works Association (AWWA). 1974. Special Taste and Odor Research Committee Meeting. Chicago, Illinois. November 12-13.
Chlorine Institute. 1975. Chlorine–Alkali Production in North America, Pamphlet No. 10. New York.
Cole, S. A. 1975. "Chlorination for the Control of Biofouling in Thermal Power Plant Cooling Water Systems," Presented at Biofouling Workshop, Johns Hopkins University, Baltimore, Maryland. June 16-17.
Harris, W. C. 1975. "Ozone Disinfection of the Strasburg, Pennsylvania Water Supply," Presented at Pennsylvania Section Meeting, Am. Water Works Assoc., Champion, Pennsylvania. April 27-29.
Inhoffer, W. R. and F. J. De Hooge. 1974. "Free Residual Chlorination of Passaic River Water at Little Falls, New Jersey," Presented at Am. Water Works Assoc. Annual Conference. Boston, Massachusetts. June 16-21.
White, G. C. 1972. *Handbook of Chlorination.* Van Nostrand Reinhold, New York.
White, G. C. 1974. Unpublished notes from a field survey of water supplies in France and England.
White, G. C. 1975a. "Disinfection: the Last Line of Defense for Potable Water." *Am. Water Works Assoc.* 67:410-413.
White, G. C. 1975b. Disinfection Committee Report. Presented at Am. Water Works Assoc. Annual Conference. Minneapolis, Minnesota. June 8-13.
Willey, B. F., C. M. Duke and J. Rasho. 1974. "Chicago's Switch to Free Chlorine Residuals," Presented at Am. Water Works Assoc. Annual Conference. Boston, Massachusetts. June 16-21.
Williams, D. B. 1974. Presented at Am. Water Works Assoc. Research Committee on Taste and Odor Control. Chicago, Illinois. November 12-13.

DISCUSSION

John R. J. Sorenson, Quad Corporation. As I understand it, the chlorine used in water purification is essentially the dregs of the chlorine production process. Is that correct? Also, as I understand it, these dregs contain many different chlorinated aliphatic and aromatic hydrocarbons including methylene chloride, chloroform, carbon tetrachloride, hexachloroethane, hexachlorobenzene, etc., which are being measured in chlorinated water. Could we be mistaken about their production and could they result from impurities in the chlorine itself? I also have another question concerning the alternative water purification process, ozonization. Shouldn't there be some concern with regard to ozonides and epoxides produced in this process and their potential carcinogenicity?

White. Do you mean are these impurities in the chlorine itself? Oh, no. Somebody has done a chemical balance study and determined that none of these chloro-organics are necessarily from the impurities in the chlorine. So we can't indict the impurities in chlorine, not yet anyway. Incidentally, you notice that I mentioned the amount of chlorine that is used in the manufacture of gasoline. Now, since you brought this subject up, I'll just expand a little on something that is interesting. If any of you were in Minneapolis, you may have heard Dr. Rook from Amsterdam. Somebody asked him a question about this catalytic action of the bromide ion or bromine. The question was, "How does the bromine get into the environment?" He answered that it was the runoff from the highways because of the bromine that is used in gasoline additives. Now bromine was found to be the best lead scavenger in gasoline, and so this is emitted in automobile discharges. To confirm this idea even further, the people at East Bay have a catchment high up in the Sierras, very near a freeway, and they're detecting trihalomethanes in the water before it's even treated. So they suspect that the bromide possibly comes from gasoline additives.

Albert Dietz, Jones Chemical Company. Have you had experience with the problem of regrowth of bacterial organisms after dechlorination?

White. Yes. There is regrowth after dechlorination. This is, of course, due to the nutrients that are in the effluent. The health department's posture on this particular question is that the pathogens have already been killed. I don't know how valid that is. We have to play by the rules that are laid down for us.

Arthur S. Brooks, University of Wisconsin-Milwaukee. You have advocated dechlorination in various water treatment processes. Is dechlorination 100% effective in removing active chlorine species? I ask the question because biological studies have indicated that chlorine levels on the order of a few ppb will produce adverse biological effects.

White. So far as I know, it is as complete as will be registered when you make an amperometric titration. In other words, you cannot titrate a residual if you are dechlorinating properly and in the stoichiometric ratio. There is a little loss in the system but the reaction is very rapid and complete. This is with sulphur dioxide. Now there are a lot of things we don't know about activated carbon. You can't say the same for that and I haven't had enough experience with it.

CURRENT PRACTICES 17

Richard F. Unz, Pennsylvania State University. You stated that the nonchlorine halogens were too expensive and cumbersome for practical application in water and wastewater treatment. What is the basis for this conclusion?

White. Yes. Let's take bromine as an example. The only case of bromine use in water treatment, that I know of, is in the little town of Irvington, California, which was very near a West Virginia Chemical Company's plant. I guess they had bromine to throw away at one time. The Irvington Water Company was experiencing a lot of troubles with their distributor system. Their water was supplied by wells. So the water company made atmospheric brominators using liquid bromine. Two things happened that were devastating. First, it produced the foulest tasting water that you could believe, a medicinal taste and odor, and it couldn't do the job because it reacted so fast with the slime growths in the distributor system that the residual wouldn't persist more than 800 ft from the well. Then, on top of that, one of the operators burned himself with the bromine. With the combination of the two, it just didn't get off to a good start. There has been a lot of experimentation done by Dow Chemical with bromine chloride. I think that there are other ways to handle bromine although it is very expensive. I think that there are some possible applications. I'm working with some people in France that are patenting a device where bromine and chlorine are used together. It looks like it has some promise for water reuse because the die-away of the residual in that combination is very rapid. Iodine is just too expensive.

Unz. Lack of interest in exploring the utility of halogens other than chlorine for disinfection purposes reflects the perpetuation of early taboos having little scientific credibility. I hope the conference will not result in an admiration society for chlorine in matters of pollution control. Iodine, for example, possesses interesting properties which make consideration of its use in certain applications highly attractive. It is a poor algicide. However, we have been working on the discovery of a compatible algicide and have found two which work well in the presence of iodine and which provide satisfactory combined bacterial-algal control in test swimming pools.

White. Well, my experience with iodine is not the same. I don't have that same feeling. For example, if you want to learn about a chemical as a disinfectant and as an algicide, get yourself a swimming pool and operate it. Now, Dr. Marks patented the Iodine Bank System whereby you put potassium iodide in a swimming pool and you activate it by releasing the iodine with intermittant applications of chlorine. It sounded swell on paper and it was supposed to work. But two things happened, namely:

(1) it colored the water and (2) it wasn't a good algicide. They just had to give up on it. Several pools in California tried it. Now that is not true with bromine. They have bromine sticks which made it easy to apply. But if you talk to the people in Illinois, I think you'll find that they say that it is limited to small pools. I think we first need to learn more about chlorine before we go to bromine and iodine. That is my feeling. I agree not to shut my eyes at any of these things.

C. Sengupta, Public Service Electric and Gas Company. I want to comment on bromine. Our company did some experiments with bromine chloride. Our experiments were abandoned for other reasons, but one problem was that of handling the bromine chloride. Cost was an important factor, but handling it was not easy, although it can be done.

White. Yes. There are ways that you can handle bromine that are very interesting. Now, one of the things that I didn't mention is this, that when you are chlorinating a cooling water condenser with sea water and you're putting the sea water through the chlorinator injector, you're going to release the 60 mg/l of bromide ion that is in the sea water and get a hotter residual than if you just used fresh water.

SECTION I.
AQUEOUS CHEMISTRY OF CHLORINE

Joseph E. Draley, Session Chairman
    Assistant Director
    Argonne National Laboratory
    Argonne, Illinois 60439

    I am impressed by how many people are here who want to hear about the aqueous chemistry of chlorine. I think for the purpose of the conference it would be worthwhile saying that we are interested in the chemistry of chlorine for two reasons. One is to be certain that the chlorine will do the good things that we would like to have it do. This includes assuring potability of water supplies, disinfection of wastewaters and the defouling of surfaces. You have heard some things about those already. The other thing we would like to assure is that, when we discharge from systems, the chemistry of the products is known enough so that we are able to deal with it. We need to know what the reactions of chlorine are with the water and with the impurities in the water. I use the word impurities loosely to mean everything that is in the water, because they will influence the possibilities of the reactions that you generally want to occur. We will need to know the reactions with the system. These have to do with the walls and the components, and the substances that form on the walls and components. Oftentimes those are the things that will interfere with what we want and, at other times, they will be the very reactions that we want. Again, with respect to discharging safely and acceptably, we will need to know toxicities of substances that formed because chlorine was used. I have chosen those words in an effort to avoid saying chlorinated compounds. In other words, I think that the subject is very broad and I do not want to use any specific language. So we need to know toxicities and persistencies of these substances. It is really to that kind of question, to those questions, that we address our attention in today's split session on chemistry.

2

THE CHEMISTRY OF AQUEOUS CHLORINE IN RELATION TO WATER CHLORINATION

J. Carrell Morris

>Division of Engineering and Applied Physics
>Harvard University
>Cambridge, Massachusetts 02138

ABSTRACT

When chlorine is dispersed in water, a variety of molecular and ionic species is produced, including H_2OCl^+, $HOCl$ and OCl^- as well as Cl_2. Of these, the reactively dominant species for most aqueous chlorination reactions is $HOCl$. Other species generally are present in too small a concentration or have specific reactivities too low to be significant.

The $HOCl$ may act as an electrophilic reagent at either the oxygen or the chlorine atom. When reaction is at oxygen, chloride ion is formed by displacement, as has been shown for certain inorganic oxidations. For reactions at amine-N or at carbon the electrophilic attack is by the chlorine atom which acts as Cl^+. Reactions of chloramination, of chlorophenol formation or other aromatic substitution, of addition to double bonds and of haloform formation are all examples of this form of electrophilic attack.

INTRODUCTION

It is well known to almost everyone, I presume, that the term aqueous chlorine is a misnomer when it is applied to the usual conditions of water and wastewater treatment. For despite the statements in numerous elementary chemistry textbooks that elemental chlorine (Cl_2) is just partially or only slightly hydrolyzed in water, already at a formal concentration of 10^{-3} M in pure water Cl_2 is hydrolyzed more than 99%. Moreover, the hydrolysis increases with decreasing concentration or with neutralization of librated H^+ to greater pH values.

The hydrolysis is in accord with the equation

$$Cl_2 + H_2O = HOCl + H^+ + Cl^- \qquad (1)$$

with hydrolysis constants ranging from 1.5 to 4.0 x 10^{-4} (mol/liter)^2 for the temperature range from 0 to 25°C (Jakowkin, 1899; Connick and Chia, 1959). Also, the hydrolysis is rapid, with equilibrium conditions being established within a few seconds or less (Shilov and Solodushenkov, 1945; Eigen and Kustin, 1962). So the major oxidizing species present in dilute acidic "aqueous chlorine" solutions is hypochlorous acid, HOCl.

Hypochlorous acid is a weak acid that dissociates in accordance with the equation

$$HOCl = H^+ + OCl^- \qquad (2)$$

with a dissociation constant ranging from 1.6 to 3.2 x 10^{-8} for the temperature range 0 to 25°C (Morris, 1966). At pH values in excess of 7.8 to 7.5 for 0 to 25°C, hypochlorite ion becomes the dominant species in aqueous chlorine solutions.

Table I shows the distribution of principal oxidizing species for aqueous chlorine solutions at even pH values between 5 and 9 for a temperature of 15°C with a chloride content equal to 350 mg/l (10^{-2} M). The fraction present as Cl_2 is only in the range of parts per million (ppm) for most of the conditions tabulated, as can be seen. Even the presence of 0.5 M chloride in sea water, as shown in the last line of the table, increases the fraction of Cl_2 only to about 10 ppm of total oxidizing chlorine.

Other transient species that have been considered to be formed in some circumstances in aqueous chlorine solutions are H_2OCl^+, Cl^+ and Cl_3^-. Only the first of these, H_2OCl^+, seems likely to be of significance in the dilute aqueous media characteristic of water and wastewater treatment. This H_2OCl^+, however, is the species probably responsible for acid catalysis of many reactions of HOCl.

Table I. Distribution of Aqueous Chlorine Species, 15°C

pH	pCl	Fraction of Cl_2 (x10^6)	Oxidizing HOCl	Chlorine as OCl^-
5	2	360	0.997	0.003
6	2	36	0.975	0.025
7	2	2.9	0.797	0.203
8	2	0.10	0.280	0.720
9	2	0.001	0.038	0.962
7.8	0.3	11	0.382	0.618

The issue of the major reactive pathway or species, however, is not determined solely on the basis of the constituent that is present in predominant concentration. There is also the question of the specific reactivity of each of the forms, for generally it is the product of concentration times specific reactivity that determines the contribution of a given mechanism or pathway to the overall reaction.

Define and express the reactivity of a given reaction pathway by the equation

$$\text{reactivity} = R_i = r_i C_i \quad (3)$$

where C_i is the concentration of a particular species and r_i is its specific reactivity toward the substrate under consideration. The r_i are, of course, equivalent to specific reaction rates with the inclusion of concentration of substrates other than the C_i.

The overall reaction rate in a given situation is then the sum of the reactivities of the different pathways, that is

$$\text{rate} = \Sigma R_i = \Sigma r_i C_i \quad (4)$$

Now, if this overall rate is divided by the total stoichiometric concentration of reactant, the quotient, which may be termed the observed specific reactivity, r, exhibits the relationships

$$r = (1/C_T)\Sigma R_i = \Sigma r_i f_i \quad (5)$$

where the f_i are the fractions of the total stoichiometric reactant concentration in specific molecular forms. Variations in r with pH, ion content and solution conditions other than temperature and pressure may then be associated with changes in the f_i, so long as the substrate remains unchanged.

Unfortunately the r_i are not universal constants; their absolute and even their relative values depend upon the substrate with which the reaction is occurring. So, no widely applicable tables of absolute or relative specific reactivities can be constructed. Nonetheless, when similar types of reaction are under consideration, relative specific reactivities often maintain themselves sufficiently fixed that at least approximate extrapolation from one substrate to another is feasible.

Table II presents a sample evaluation of relative reactivities for several forms of oxidizing chlorine in dilute aqueous solution at pH 7. The tabulated specific reactivities are relative to $r_{HOCl} = 1$; they are estimated values for exemplification, based principally on the reactivities of forms of aqueous chlorine toward nitrogenous compounds, and are not to be regarded seriously or universally in any quantitative sense. However, the order of reactivities is likely to remain the same for many reactions in which electrophilic or oxidizing properties of the forms of aqueous chlorine are involved.

Table II. Estimated Net Reactivities of Forms of Active Chlorine; pH 7, 15°C

Species	Estimated Specific Reactivity	Fraction of Total Cl	Net Relative Reactivity
Cl_2	10^3	3×10^{-6}	0.003
HOCl	1	0.80	0.80
OCl^-	10^{-4}	0.20	0.00002
H_2OCl^+	10^5	10^{-8}	0.001

It is clear from a table such as this that HOCl must be regarded as the major reactive species for most oxidizing reactions of aqueous chlorine in dilute solutions ($<10^{-3}$ M) at pH values between 5 and 9. This is very different from the situation in typical nonaqueous solvents where very little hydrolysis occurs and where the dominant reactive form is Cl_2. It is also very different from strongly acidic solutions where Cl_2, H_2OCl^+ or even Cl^+ may dominate and from strongly basic solutions where essentially only OCl^- is present.

So, the types of reactions with organic substances found for chlorine in organic solvents, where Cl_2 is dominant and free radical reaction mechanisms are common, are not to be expected in dilute aqueous solution unless photochemical generation of radicals occurs. Instead the reactions and reaction patterns occasioned by the electrophilicity of HOCl are to be anticipated.

REACTIVITY OF HOCl WITH INORGANIC ANIONS

Either the chlorine atom or the oxygen atom of HOCl may act as a center of reaction. It appears that the Cl is the more electropositive atom, so that electrophilic processes proceed by way of the chlorine atom. On the other hand, the attraction of the Cl for electrons is so great that it may split from the molecule directly as chloride ion.

An instance is the reaction of nitrite with aqueous chlorine according to the elementary reaction

$$NO_2^- + HOCl \rightarrow NO_2OH + Cl^- \qquad (6)$$

with the nitrite displacing Cl⁻ (Anbar and Taube, 1958; Lister, 1961). The oxidation of sulfite may proceed similarly, at least in part (Halperin and Taube, 1952).

Somewhat similar transfers, with displacement of either Cl⁻ or OH⁻, may occur as initial steps in the oxidations of Fe^{++} and Mn^{++} and in the reaction with H_2O_2. The situation with regard to the exact elementary step is less clear in these latter instances, however.

HOCl AS ELECTROPHILE

The usual reactant behavior of HOCl with organic carbon and with amino nitrogen is as an electrophilic agent in which the chlorine atom takes on partially the characteristics of Cl^+ and combines with an electron pair in the substrate. Simultaneously or subsequently the hydroxyl ion is split off, often with the assistance of H^+ from the solvent or of reactive centers from other parts of the substrate.

This type of behavior serves to account fundamentally for the reactions of dilute aqueous chlorine with ammonia and amines, with phenolic and other aromatic substances, and in the formation of chloroform from organic substrates. Each of these classes of reaction will be considered in some detail in the following sections.

Clearly, Cl^+ itself is the strongest electrophilic species related to HOCl, but it occurs only when the solvent has dehydrating properties. The ion, H_2OCl^+, is also a stronger electrophile than HOCl, partly because of its positive charge, and partly because of the easier release of H_2O as compared with OH^-.

Reaction of electrophilic Cl^+ can be viewed analogously to reactions of H^+. Naturally, transfer of Cl^+, like that of H^+, is facilitated by negative charge, basicity or nucleophilicity of the recipient atom.

REACTIONS WITH AMMONIA AND AMINES

Reaction of HOCl with the nitrogen atom of ammonia and amines follows the pattern that has just been described. The electrophilic chlorine atom can be visualized as attaching itself to the bare electron pair of the nitrogen atom with concurrent release of H^+ from the ammonia or amine and of OH^- from the HOCl. These latter ions would appear to react forming water with the participation of solvent water as a part of the overall pattern of the single elementary process.

Conversely, the reaction may be viewed as a displacement reaction in which the nucleophilic ammonia or amine displaces OH^- from HOCl. The nucleophilicity of NR_2^- as compared with OH^- is then of importance.

In accord with either of these models of the reaction, the specific rates of chloramination have been found to vary directly with the basic strength (or nucleophilicity) of the nitrogenous substrate (Morris, 1967). Catalysis by H^+, indicating participation of the more electrophilic H_2OCl^+, has been observed only when the basicity of the receptor compound is very low.

Although amides appear also to accord with the previous reaction pattern with regard to basicity and overall rate of reaction, there is another possible mechanism available to these latter compounds. Since acidic dissociation of amides occurs relatively readily to give anionic structures resembling those of enolates, it is possible that chlorination of amides proceeds by a pattern similar to that of the haloform reaction.

REACTIONS WITH THE AROMATIC RING

It has been known for a long time that dilute aqueous chlorine reacts with phenolic compounds to form chlorinated derivatives. Already by 1926 Soper and Smith had shown that the mechanism of the reaction is electrophilic attack of HOCl on the phenoxide ion. Here, as in other attacks by electrophilic agents, reaction is facilitated by the presence of a negative charge on the nucleophilic substrate.

Benzene itself and many of its derivatives require a more active electrophile than HOCl, such as Cl^+ or H_2OCl^+, for measurable chlorination at ambient temperature. Only when the ring is "activated" by substituents such as the oxide ion does chlorine or other halogen substitution occur easily. It then may proceed at the activated ortho and para positions until these are occupied fully by halogen atoms; following this, rupture of the aromatic ring may occur by mechanisms that are not as yet fully understood.

The accepted detailed mechanism involves, first of all, an addition of Cl^+ to the aromatic ring to give a transitory intermediate which, in the formation of p-chlorophenol, for example, can be represented by the following formula:

This then becomes $ClC_6H_4O^-$ by loss of H^+ from the para position.

When there is more than one activating group in meta positions on the aromatic ring, then reaction with HOCl should be much accelerated. There is evidence in the very recent work of Rook that such accentuated reactivity is observed with resorcinol, m-dihydroxybenzene, and that the reaction may lead ultimately to the production of chloroform after ring rupture. He found, for example, that millimolar (mM) resorcinol treated for 4 hr at pH 7.5 and 10°C with 8 mM aqueous chlorine gave 0.38 mol of $CHCl_3$ per mole of resorcinol (Rook, 1975).

Heterocyclic aromatic rings may be either activated or inactivated toward electrophilic attack as compared with benzene. Pyridine, for example, is greatly deactivated, so that reaction in the ring with HOCl in dilute aqueous solutions is not to be expected. On the other hand the α-hydrogens of pyrrole are greatly activated, so that electrophilic chlorine substitution by HOCl in dilute aqueous solutions is quite likely (Sykes, 1961).

ADDITION TO DOUBLE BONDS

Addition of halogens or hypohalous acids to double bonds of organic compounds in polar solvents like water is another example of electrophilic attack. In this case attack by the hypochlorous acid is more vigorous than that by Cl_2, for the chlorine in the former is more electropositive.

The initial rate-determining step is believed to be transfer of Cl^+ to the double bond to give a chloronium ion, as shown by

$$\underset{/}{\overset{\backslash}{C}} = \underset{\backslash}{\overset{/}{C}} + ClOH \rightarrow \underset{/}{\overset{\backslash}{C}} \underset{Cl^+}{-} \underset{\backslash}{\overset{/}{C}} + OH^- \qquad (7)$$

Reaction is completed by addition of OH^- or other anion from the solvent at one or the other of the chloronium bonds.

Reactions of this sort appear to be too slow in general to be of significance in water or wastewater chlorination unless the double bond is strongly activated by substituent groups. The dynamics and products of HOCl reactions of this sort with conjugated molecules like terpenes, carotenoids and xanthophylls need investigation.

THE HALOFORM REACTION

Reaction of aqueous hypohalites with methyl ketones or compounds oxidizable methyl ketones to yield haloform is a classic reaction of organic chemistry that has been known since 1822 (Fuson and Bell, 1934). Two forms of the reaction are known— acid-catalyzed and base-catalyzed. The base-catalyzed reaction pattern is the one that is predominant for reactions in dilute aqueous solution at pH>5.

The pattern of the reaction is the successive replacement of hydrogen by chlorine on carbon alpha to a carbonyl group followed by eventual hydrolysis to produce CHX_3 and, generally, a carboxylate. The mechanism is believed to be an initial proton dissociation from the α-carbon, giving an enolate carbanion which is then subject to electrophilic attack by HOCl or OCl^-. The rate of proton dissociation is ordinarily the slow, rate-determining step for the conditions under which the reactions have been studied extensively. It should be noted that most investigations have either been at pH>11 or with quite strong buffer solutions. There are few results in simple aqueous media.

Figure 1 gives a detailed diagrammatic representation of the course of the haloform reaction with hypochlorous acid and acetone. The electrophilic attack of HOCl on the enolate ion is very similar to the attack on the phenoxide ion. Indeed, as was pointed out by Bartlett and Vincent (1935), "It would seem, then, that the chlorination of aliphatic enols and that of phenols are essentially alike, differing only in the magnitude of some of the constants."

It seems peculiar, in view of this statement and the well known formation of chlorophenols in water chlorination, that no one until Rook (1974) should have looked for or found the haloform reaction occurring in connection with water chlorination.

Actually, the reaction of aqueous chlorine with acetone or other simple methyl ketones is too slow at 10^{-4} M concentration and pH values of 5 to 9 to account for any of the observed chloroform formation in water chlorination. Moreover, the oxidation of simple secondary alcohols to methyl ketones is also slow for these concentrations and pH values.

More highly activated structures than those of simple methyl ketones are required to account for the formation of chloroform in water chlorination. These must involve more acidic carbons

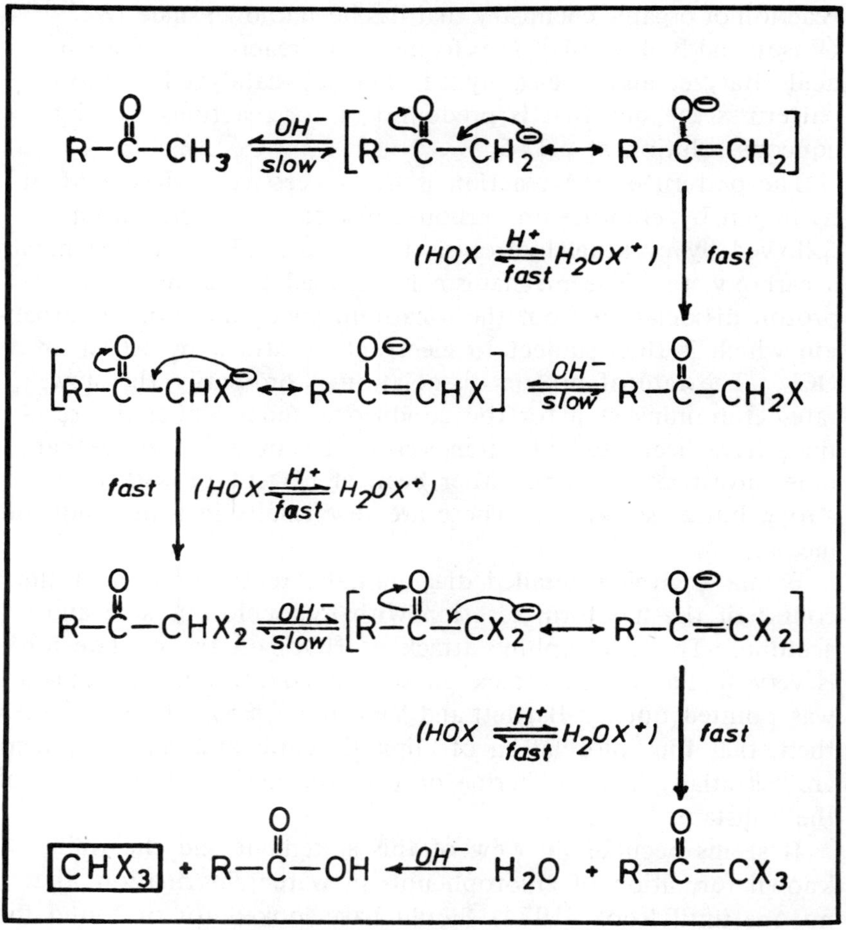

$$R-\overset{O}{\underset{\|}{C}}-CH_3 \underset{slow}{\overset{OH^-}{\rightleftharpoons}} \left[R-\overset{O}{\underset{\|}{C}}-CH_2^{\ominus} \longleftrightarrow R-\overset{O^{\ominus}}{\underset{\|}{C}}=CH_2\right]$$

$$(HOX \underset{fast}{\overset{H^+}{\rightleftharpoons}} H_2OX^+) \quad \Big| \text{ fast}$$

$$\left[R-\overset{O}{\underset{\|}{C}}-CHX^{\ominus} \longleftrightarrow R-\overset{O^{\ominus}}{\underset{\|}{C}}=CHX\right] \underset{slow}{\overset{OH^-}{\rightleftharpoons}} R-\overset{O}{\underset{\|}{C}}-CH_2X$$

$$\text{fast } \Big| \ (HOX \underset{fast}{\overset{H^+}{\rightleftharpoons}} H_2OX^+)$$

$$R-\overset{O}{\underset{\|}{C}}-CHX_2 \underset{slow}{\overset{OH^-}{\rightleftharpoons}} \left[R-\overset{O}{\underset{\|}{C}}-CX_2^{\ominus} \longleftrightarrow R-\overset{O^{\ominus}}{\underset{\|}{C}}=CX_2\right]$$

$$(HOX \underset{fast}{\overset{H^+}{\rightleftharpoons}} H_2OX^+) \quad \Big| \text{ fast}$$

$$\boxed{CHX_3} + R-\overset{O}{\underset{\|}{C}}-OH \overset{OH^-}{\longleftarrow} H_2O + R-\overset{O}{\underset{\|}{C}}-CX_3$$

Figure 1. The haloform reaction.

than those alpha to a single carbonyl group, ones for which dissociation of a proton is several orders of magnitude faster than from a simple alpha carbon.

Acidic carbons like this are found in methylene groups between two carbonyl groups, such as that in acetyl acetone, $CH_3COCH_2COCH_3$. Rook (1975) has investigated three such compounds: indanedione, cyclohexanedione-1,3, and 5,5-dimethylcyclohexanedione-1,3. All of them produced substantial chloroform, 0.38 to 0.52 mol per mol of substrate, when allowed to react with 6 to 8 mM aqueous chlorine at pH 7.8 and 10°C for 4 hr. The mechanism of rupture to yield chloroform after chlorine substitution is still obscure.

Other acidic carbons that yield carbanions without forming an enolate structure are also subject to electrophilic attack by HOCl. An instance is chloroform itself, which is a relatively strong carbon acid. Following loss of proton, the CCl_3^- carbanion is susceptible to electrophilic attack by HOCl to yield CCl_4. However, it appears that the rate of this reaction is too slow to be of significance in water chlorination.

EFFECTS OF BROMIDE ON AQUEOUS CHLORINE REACTIONS

It has become apparent in the past few years that bromide is a common constitutent of natural waters at concentrations in the range of fractions of a part per million. The importance of these minute concentrations of Br^- is great, however, especially in connection with the reactions of aqueous chlorine.

Whenever chlorine or hypochlorite is added to a water containing Br^-, there is rapid formation of HOBr according to the reaction

$$Br^- + HOCl \rightarrow HOBr + Cl^- \tag{8}$$

The resulting HOBr is also an electrophilic agent, but one that tends generally to react much more rapidly than HOCl. Much of the HOBr that reacts will be reduced to the bromide ion, to be reoxidized to HOBr by residual aqueous chlorine in the water. The overall result is an enhanced reactivity exhibited by aqueous chlorine in the presence of an original concentration of bromide ion.

It is quite likely, therefore, that there will be apparent inconsistencies in the reactions of different water supplies and wastewaters with aqueous chlorine, depending on the bromide content of the waters. Already it appears that the rate or pattern of the breakpoint reaction, in which halogenation of ammonia is involved, may vary for different water supplies depending on the concentration of bromide. Effects on reactions with organic carbon may also be expected. It may be noted that, although the absolute rates of haloform formation depend on the rates of carbanion formation, the relative amounts of bromine and chlorine substituted into the haloforms depend on the relative rates of the halogenation reactions.

There are two types of needed information. First, there should be efforts to determine the bromide content of numerous waters to provide a background for assessing observed differences in reactivity with aqueous chlorine. Second, there should be examination of the reactivity and reaction patterns of several waters when spiked with small concentrations of bromide to determine the practical catalytic effect of this ion on the reactions of aqueous chlorine.

CONCLUSION

The most important conclusion to be drawn from this paper is the observation that the reactions of aqueous chlorine in water chlorination are not indiscriminate and unpredictable, but rather that they follow quite well defined pathways in accord with general principles of organic reaction mechanisms.

So, even when the exact composition of the organic material in a water or wastewater is not known, it is still possible to predict something of the nature and extent of the reactions with aqueous chlorine to be anticipated.

REFERENCES

Anbar, M. and H. Taube. 1958. "The Exchange of Hypochlorite and Hypobromite Ions With Water." *J. Am. Chem. Soc.* 80:1073-1079.

Bartlett, P. D. and J. R. Vincent. 1935. "The Rate of the Alkaline Chlorination of Ketones." *J. Am. Chem. Soc.* 57:1596-1600.

Connick, R. E. and Y. Chia. 1959. "The Hydrolysis of Chlorine and Its Variation with Temperature." *J. Am. Chem. Soc.* 81:1280-1285.

Eigen, M. and K. Kustin. 1962. "The Kinetics of Hydrolysis of Cl_2, Br_2 and I_2." *J. Am. Chem. Soc.* 84:1355-1358.

Fuson, R. C. and B. A. Bull. 1934. "The Haloform Reaction." *Chem. Rev.* 15:278-309.

Halperin, J. and H. Taube. 1952. "The Reaction of Halogenates with Sulfite in Aqueous Solution." *J. Am. Chem. Soc.* 74:375-379.

Jakowkin, A. A. 1899. "On the Hydrolysis of Chlorine." *Z. physik. Chem.* 29:613-657.

Lister, M. W. and P. Rosenblum. 1961. "Kinetics of the Reaction of Hypochlorite and Nitrite." *Can. J. Chem.* 39:1645-1653.

Morris, J. C. 1966. "The Acid Ionization Constant of HOCl from 5 to 35°C." *J. Phys. Chem.* 70:3798-3802.

Morris, J. C. 1967. "Kinetics of Reactions Between Aqueous Chlorine and Nitrogenous Compounds," in *Principles and Applications of Water Chemistry*, S. D. Faust and J. V. Hunter, Eds. (New York: John Wiley and Sons, Inc.), pp. 23-53.

Rook, J. J. 1974. "Formation of Haloforms During Chlorination of Natural Waters." *Proc. Soc. Water Treatment and Exam.* 23:234-243.

Rook, J. J. 1975. "Formation of Haloforms During Chlorination," presented at Am. Water Works Assoc. Annual Conference, Minneapolis, Minnesota, June 9-13.

Shilov, E. A. and S. M. Solodushenkov. 1945. "The Velocity of Hydrolysis of Chlorine." *J. Phys. Chem.* (U.S.S.R.) 19:405-407.

Soper, F. G. and G. F. Smith. 1926. "The Halogenation of Phenols." *J. Chem. Soc.* 1926:1582-1590.

Sykes, P. 1961. *A Guidebook to Mechanism in Organic Chemistry*. John Wiley and Sons, Inc., New York. 247 pp.

DISCUSSION

David H. Rosenblatt, U.S. Army Medical Bioengineering Research and Development Laboratory. I wonder, with all this interest now about chlorination of water producing small chlorinated hydrocarbons and plural brominated hydrocarbons, whether anybody has given thought to what might happen when you brominate the same type of water. Would you in fact get bromination in the absence of chlorine? Also, things might be a little more complicated than you show here. You're going entirely by two electron reactions. The question in my mind is whether there aren't some free radical paths that would require the presence of HOCl or other active chlorine species in order to introduce bromine into the molecule?

Morris. So far as I can tell, when one is operating in these dilute aqueous media, the occurence of free radical mechanisms is relatively remote. The tendency is almost always for a split with these hypohalous acid molecules, with one being positive and one negative.

Robert B. Dean, U.S. Environmental Protection Agency. I think it's worthwhile considering that in real chlorination reactions, adding chlorine to a little bit of water produces a low pH, a lot of local acidity. Now are the reactions at the acid site fast enough to be useful for substitution—

in other words, to attack the organic compounds via the acid flow before that chlorine gets neutralized by the alkalinity of the water?

Morris. I think that depends on whether you have used Cliff White as your engineer in designing your chlorination facilities or not. If you designed it to mix the chlorine with the water very rapidly, within a period of 1 to 3 sec, then there is not time for any extensive reactions to occur. But, if you simply bleed the chlorine in and let it drift in a plume downstream, almost anything could happen.

Herbert S. Posner, National Institute of Environmental Health Sciences. You mentioned three reactions that would yield chloroform: (1) the degradation of resorcinol; (2) the successive chlorination of methyl ketone; and (3) substitution into a molecule like benzylacetone or decenylacetone. Does this exhaust the possibilities or are there other possibilities?

Morris. It does not exhaust the possibilities, except when I generalized and said I thought that any time you got a carbanion you could begin to get the substitution. Then, whether or not this will eventually result in chloroform depends on the structure of the rest of the molecule. So you would have to look at the individual molecule to see whether it could eventually end up as chloroform or not.

George Clifford White, Consulting Engineer. Dr. Morris, I am very interested in this oxidation of the bromide ion situation. I have done a lot of reading about it. You say that the bromide ion is readily oxidized by hypochlorous acid. I am under the impression that a stoichiometric reaction of oxidizing bromide ion with hypochlorous acid only occurs if the pH is below 5, and that you have to have an excess of bromide ion to make the reaction go at all if the pH is near neutral. Is that correct?

Morris. When we were doing some studies on the bromamines, the experimental and theoretical work was carried out which indicated that this reaction of hypochlorous acid with bromide increased with decreasing pH, or increased linearly with the hydrogen ion concentration of the solution and became more rapid than the reaction of hypochlorous acid with ammonia at a pH of about 7.8. So on the acid side of pH 7.8, you would expect the reaction to be more rapid than the reaction with ammonia. Thus if you put hypochlorous acid into, say, equal concentrations of bromide and ammonium ion, you would form hypobromous acid rather than forming chloramine in such a situation.

White. But how about low concentrations of bromide, say 10 to 15 mg/l, in a similar or equal concentration of chlorine?

Morris. The absolute concentration would make no difference. It's only the relative concentration that would make a difference. You would have to go a pH unit lower to get predominance if you had had 1/10 as much bromide as ammonia. Every power of 10 would reduce you that much. But otherwise it's very rapid.

Walter J. Blogoslawski, National Marine Fisheries Service. In view of the presence of about 50 to 55 ppm bromide in sea water, with a pH of about 7.8 to 8.0, would you feel that there is a significant reaction of hypochlorous with that bromide in forming a hypobromide compound?

Morris. Yes, I would.

Blogoslawski. Would you feel that would be significant from a power plant which might be considering chlorinating large volumes of sea water in effluent?

Morris. Yes, indeed I would.

3

MEASUREMENT AND PERSISTENCE OF CHLORINE RESIDUALS IN NATURAL WATERS

J. Donald Johnson

    Department of Environmental Sciences
      and Engineering
    School of Public Health
    The University of North Carolina
    Chapel Hill, North Carolina 27514

ABSTRACT

Selective measurement of only the good disinfectant chemical species of chlorine will minimize the concentration of chlorine which must be added to produce a microbiologically safe water. The lower dosage and better control permitted by selective analytical methods minimizes the formation and persistence of toxic chlorination products, while destroying pathogenic microorganisms.

Chlorine residual, as normally measured, is the concentration of all oxidizing agents produced by chlorination of natural water and remaining after some time. These oxidizing agent products of chlorination may contain no chlorine but are still measured and referred to as chlorine residual. Of those which do retain chlorine in the plus one oxidation state, all have radically different ability to disinfect. The really good disinfectant, hypochlorous acid (HOCl), is never measured selectively.

New methods are needed to distinguish HOCl from the poor disinfectant hypochlorite (OCl^-) in drinking water. Monochloramine (NH_2Cl) is the good disinfectant in wastewater and poor-quality cooling waters. N-chloroorganics are poor disinfectants measured along with NH_2Cl. The formation of toxic residuals and their persistence results from the excess chlorine required for reliable disinfection because our analytical methods are not selective.

The nature and persistence of chlorination products classed as chlorine residuals are discussed by type of compound. Actual measurements of chlorine residual decay for free and combined chlorine in water, wastewater and estuarine water are discussed. The decay of chlorine residuals by apparent first-order specific rate constants of 0.03 hr^{-1} to 7.5 hr^{-1} are related to residual type, demand, volatility and photochemical effects.

Present field, laboratory and continuous methods for free and combined chlorine residual are compared for specificity, reagent stability, accuracy and simplicity. Included in this discussion are the acid orthotolidine methods, DPD, SNORT, LCV, FACTS, amperometric titration, continuous amperometric cells and the NBS flux monitor.

A new analytical method specific for HOCl in the presence of the poor disinfectants OCl^-, organic chloramines and other interferences will be presented. The advantages and disadvantages of this method will be discussed in terms of selectivity, sensitivity and other effects in disinfection efficiency measurements.

INTRODUCTION

Chlorine residual measurements are made to determine the efficiency of disinfection—the objective of chlorination. If only the good disinfectant species of chlorine are measured, the disinfection process can be done with small concentrations of chlorine. Low dosages, resulting from careful control of the disinfection process, will minimize the formation of toxic chlorination by-products. For this purpose, selective measurement of the good disinfectant hypochlorous acid (HOCl) is needed to the exclusion of hypochlorite ion (OCl^-) and other poor disinfectants commonly measured.

Residual measurements are also needed for toxic chlorine compounds–particularly monochloramine (NH_2Cl). Monochloramine is important because of its toxicity, rapid formation rate, stability and widespread appearance due to the ubiquity of ammonia in natural water.

Monochloramine is also the principal chlorine disinfectant in any water containing as much as 0.1 mg/l NH_3-N and less than 1.0 mg/l Cl_2. Disinfection control in water of this type requires selective measurement of NH_2Cl to distinguish it from other chlorine oxidizing agents formed in natural water which are less persistent and are poor disinfectants.

False chlorine residuals are commonly measured, especially in the methods available for monochloramine. These methods are universally based on the reaction of monochloramine with iodide to give iodine. The selectivity of this approach is particularly poor for low concentration levels at which monochloramine is toxic to fish and, also, in the low-quality waters in which this measurement is important for determination of environmental toxicity. It is this problem of poor selectivity in monochloramine measurements which makes interpretation of much of our fish toxicity data difficult. Lack of selectivity and sensitivity makes control of a maximum combined chlorine residual for the purpose of limiting toxicity to fish nearly impossible.

This paper reviews measurements of persistence and decay for free and combined chlorine residual in both estuarine and fresh water and, also, in the dark and sunlight. The selectivity of analytical methods for chlorine to determine disinfection efficiency and environmental toxicity is discussed. The meaning of chlorine residual is presented in terms of what we should be measuring as contrasted with what we are currently measuring. Common interferences will be discussed by interference type and effect of procedure. The basis of the iodide procedure for measurement of various combined chlorine forms is described and new data presented. Electrode methods, both laboratory and field procedures, are reviewed and a new membrane electrode method of obtaining analytical selectivity is described for $HOCl$ and NH_2Cl amperometric measurement.

FREE "AVAILABLE" CHLORINE

When added to water, chlorine gas reacts rapidly and practically completely with water to form a mixture of $HOCl$ and OCl^- depending on the pH, temperature and dissolved solids or chlorinity (Morris, 1966; Sugam and Helz, 1975). This mixture is referred to as free available chlorine (FAC) as shown below:

$$FAC = HOCl + OCl^- \tag{1}$$

Free available chlorine is hypochlorous acid plus hypochlorite ion. These two forms of free available chlorine are produced when either chlorine gas (Cl_2) or hypochlorite solution (NaOCl) or solid $Ca(OCl)_2$ are added to water:

$$Cl_2 + H_2O \rightarrow HOCl + H^+ + Cl^- \tag{2}$$

$$OCl^- + H_2O \leftrightharpoons HOCl + OH^- \tag{3}$$

A small amount of acid, H^+, is produced in dissolving chlorine gas (Equation 2) and a small amount of alkali, OH^-, comes from adding hypochlorite to water (Equation 3). However, the quantities are small compared to the HCO_3^- concentration of natural water. These small amounts of acid and alkali in Equations 2 and 3 are neutralized by HCO_3^-. The pH of the solution determines the proportion of the FAC present as HOCl or OCl⁻ as shown by Equilibrium (3) or (4):

$$HOCl \leftrightharpoons OCl^- + H^+ \tag{4}$$

The FAC is half HOCl and OCl⁻ at pH 7.5 and 25°C, while at 0°C the 50:50 value is at pH 7.9. At higher pH values the FAC shifts to more OCl⁻ and at lower pH to more HOCl for the same concentration of FAC.

HOCl, REAL FREE CHLORINE, THE ONLY GOOD DISINFECTANT

Figure 1 shows the minimum safe bactericidal free chlorine residual necessary after 10 min at 20 to 25°C or 70 to 77°F (*Am. Water Works Assoc.*, 1973). The basis of these data is the work of Butterfield at the U.S. Public Health Service (USPHS) as reinterpreted by the National Academy of Sciences. Note that disinfection requires twice as much chlorine at pH 8 as it does at pH 7, while it takes four times as much at pH 9. The reason higher concentrations of FAC are required is the shift from HOCl to OCl⁻ at higher pH values, making it necessary to increase the level of FAC to maintain a constant level of the only good and effective disinfectant—hypochlorous acid, real free chlorine. The other form of FAC, hypochlorite ion, is not a good disinfectant and is not "available" to kill micro-

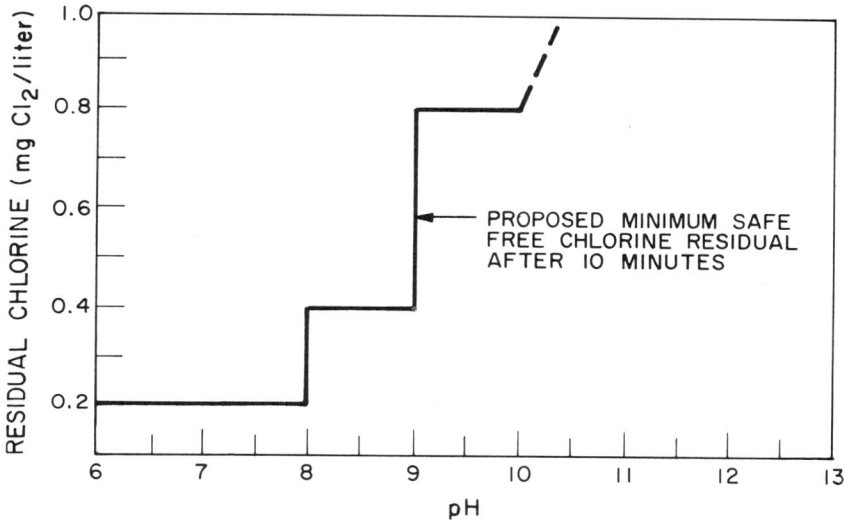

Figure 1. Minimum free bactericidal chlorine residuals.

organisms. But it is as strong an oxidizing agent and so is available to the analytical reagents for measuring FAC.

Figure 2 shows the increasing concentration, C_T, of FAC, as $HOCl + OCl^-$, required to maintain a constant level of hypochlorous acid, the effective disinfectant form of FAC, as pH increases (McClanahan, 1975). The lines are calculated from the equilibrium shown in Equations 3 and 4 so as to keep the HOCl level constant, independent of pH. As more OCl^- is formed at the higher pH values, as shown by the positive slope of the concentration of OCl^- line above pH 6, more free chlorine, C_T, is required to maintain a constant level of HOCl. The data points plotted on the diagram are the actual FAC measurements of residual necessary to produce 99.6% virus inactivation, 99.999% coliform kill and 99.999% cyst disinfection. This comparison shows that disinfection is not pH-dependent in the neutral range but that we have been measuring the wrong chemical or rather sum of chemicals, $HOCl + OCl^-$. If we could measure only real free chlorine, HOCl, and do not confuse it with hypochlorite in a FAC measurement, we would not need to change the concentration of residual HOCl required as a function of pH.

Table I summarizes the primary factors affecting disinfection efficiency. As we normally measure the sum of $HOCl + OCl^-$

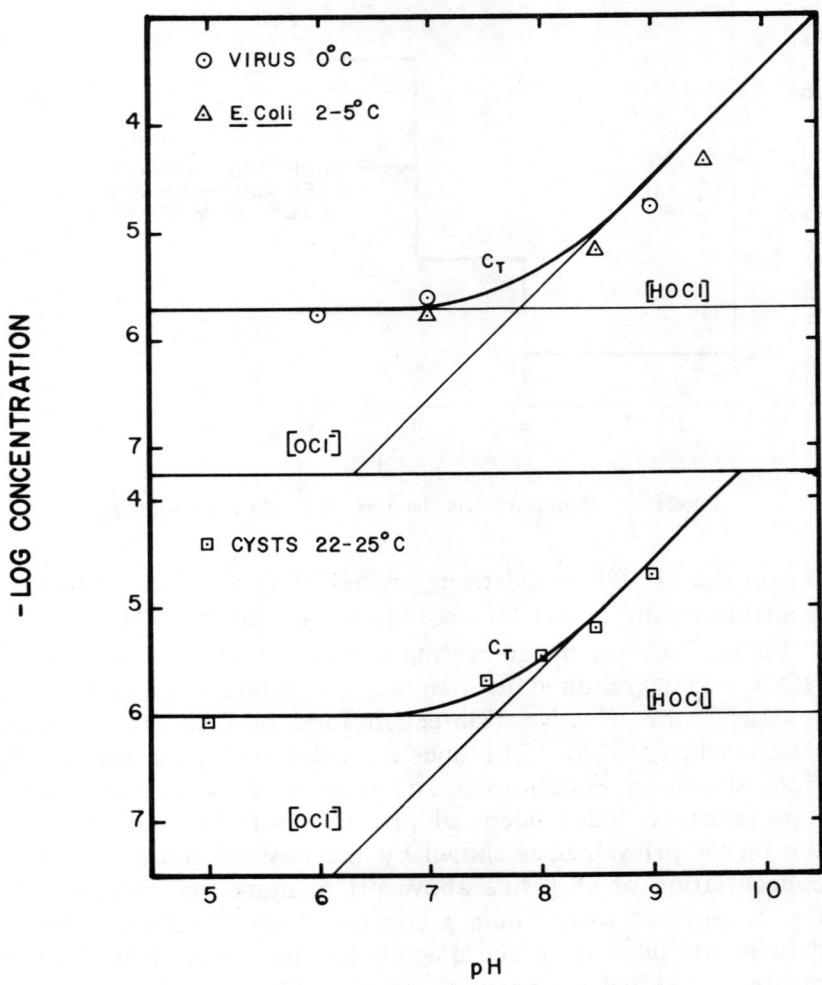

Figure 2. Log diagram for the system HOCl-H$_2$O with the concentration of HOCl constant. Data points show effect of pH on the concentration required to inactivate virus (99.6%), *E. Coli* (99.999%) and *E. histolytica* cysts (99.999%).

as FAC, free available chlorine, higher quantities of chlorine residual are required at higher pH and lower temperature. The effect of temperature is a combination of the shift in the HOCl

Table I. Primary Factors Affecting Disinfection Efficiency

1. Residual FAC measured
   pH effect
   Temperature effect—complex
2. Residual HOCl measured
   No pH effect
   Temperature effect—simple

—OCl⁻ equilibrium constant for Equation 3 with temperature and the increase in the quantity of HOCl required because disinfection is slower at low temperature. These two effects are in opposite directions. As temperature decreases the pK for HOCl ionization increases from 7.5 to 7.9, increasing the HOCl fraction at a given pH. At the same time the lower temperature markedly increases the concentration of disinfectant required. This produces a complex total effect when FAC is measured. If HOCl, real free chlorine, is measured, the temperature effect is simply the slower rate of kill at lower temperature, requiring higher concentration.

COMBINED AVAILABLE CHLORINE, CAC

In wastewater, and even in many drinking and river waters, sufficient ammonia is present to make it difficult to add enough chlorine to produce FAC. Combined available chlorine (CAC) forms rapidly from pH 7 to 10 to give NH_2Cl, monochloramine:

$$HOCl + NH_3 \rightarrow NH_2Cl + H_2O \qquad (5)$$

Reaction 5 may proceed further to replace another H^+ from the nitrogen of ammonia to give dichloramine ($NHCl_2$) and trichloramine (NCl_3). These latter compounds are formed more slowly, at higher concentrations of chlorine relative to the ammonia present and at lower pH. The slowness of their formation and the instability of $NHCl_2$ make NCl_3 concentrations generally very small. The instability of $NHCl_2$ via oxidation to nitrogen gas and nitrate produces the effect known as the breakpoint. This breakpoint reaction is so fast above pH 7.5 that $NHCl_2$ is not found because it is oxidized as rapidly as it is produced.

Figure 3 shows a typical curve of chlorine residual as a function of chlorine added or dosage in the neutral to alkaline pH range. The first bit of chorine added to any water is reduced to chloride or unavailable chlorine by its reaction with chlorine demand. By definition, chlorine demand is the sum of the reducing agents producing this reaction with chlorine in a given time.

This initial loss of disinfectant shown in Figure 3 is due to easily oxidized organic coumpounds, ferrous ion, sulfide, nitrite and other reducing agents. After the demand is satisfied, combined available chlorine is formed that is primarily monochloramine (NH_2Cl). As additional chlorine is added beyond that needed to react with ammonia-nitrogen (5 mg Cl_2/1 for each 1 mg NH_3-N/1), dichloramine ($NHCl_2$) begins to form and

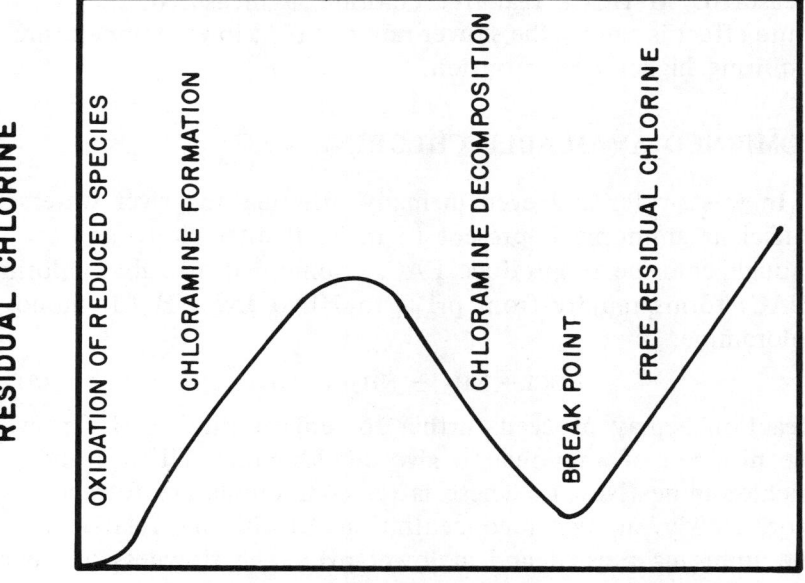

Figure 3. Relationship between chlorine dosage and residual chlorine for breakpoint chlorination.

decompose. This decomposition results in the loss of residual CAC, until the breakpoint occurs (near 8 mg Cl_2/l/mg N) as the ammonia present in the CAC is oxidized to N_2 and NO_3^- and the chlorine becomes chloride (Cl^-). Beyond the breakpoint excess FAC is added and remains in solution except as it is consumed by difficult to oxidize organic and inorganic compounds. Seldom does the breakpoint actually return to the zero residual chlorine level since CAC includes difficult to oxidize N-chloro-organic compounds ($RNCl_z$, where R represents many possible organic compounds). Figure 4 shows monochloroglycine, curve B, is measured as monochloramine. CAC is thus a complex mixture of chloramines of ammonia (NH_xCl_y)

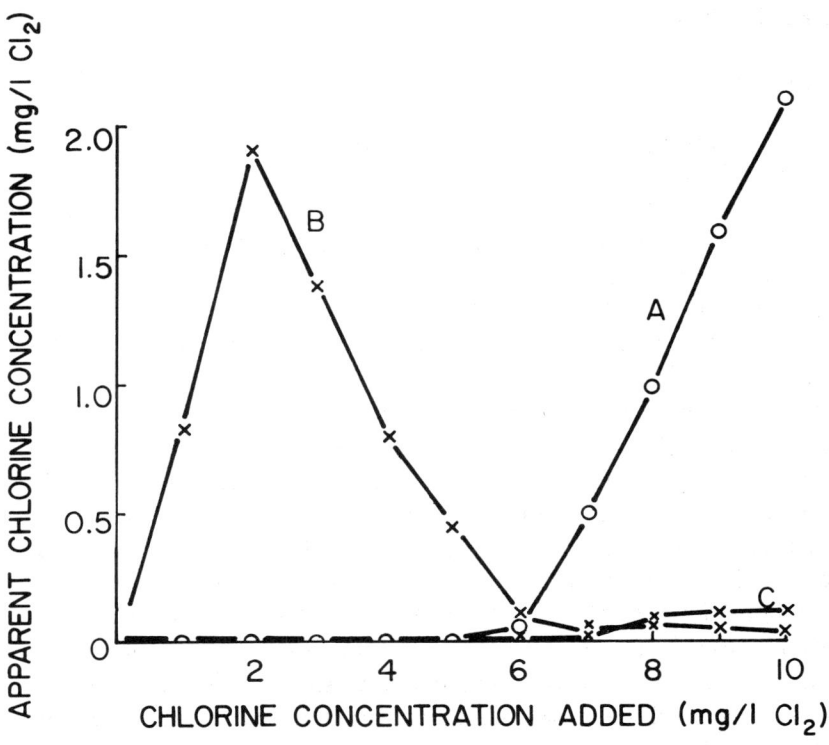

Figure 4. Breakpoint reaction with glycine (2 mg/l Cl_2) at pH 7.0 after 20 hr. A—free chlorine; B—monochloramine; C—dichloramine.

and organic chloramines ($RNCl_z$, where x, y and z are small whole numbers between 0 and 3):

$$CAC = NH_xCl_y + RNCl_z \qquad (6)$$

By far the most important of these chloramines is monochloramine (NH_2Cl). In general the ammonia chloramines are good disinfectants compared to the organic chloramines but are poor compared to HOCl (Table II). In wastewater, however, monochloramine is generally relied upon as the desired disinfectant. The high stability and fish toxicity of monochloramine is a problem. Because of the much greater disinfection efficiency of HOCl than NH_2Cl (Table II), the measurement of the latter as an interference in the free available chlorine measurement has been a major problem.

PERSISTENCE

Free Chlorine

The stability of free chlorine in natural water is very low, expecially below pH 7, because it is a strong oxidizing agent; for example, $E°$ is 1.49 V at pH 0 or 1.28 V at pH 7. Hypochlorous acid rapidly oxidizes inorganic compounds such as Br^- and I^- in sea water. The rate of reaction with iodine is too fast to follow. The rate of formation of OBr^- from Br^- is also fast but has been measured (Farkas, Lewin and Black, 1949). From the $E°$ value it is concluded that hypochlorous acid should be capable of oxidizing many compounds which actually can reduce chlorine to chloride only slowly. The oxidation of water to O_2 is an obvious example. This reaction does not occur at a significant rate unless the solution is irradiated with ultraviolet light (Hancil and Smith, 1971). Other compounds, especially organics, react only slowly with chlorine and at rates strongly dependent on the pH and chlorine concentration. $Manganese^{++}$, $iron^{++}$ and CN^- are important inorganic compounds in natural water which react slowly. Some of these reactions produce new oxidizing agents that are mistaken by even the best analytical method for free chlorine.

Table II. Relative Germicidal Activity of N-Chloro
Compound Compared to HOCl (Marks and Strandskov, 1950)

Compound	Relative Activity
Monochloramine	1/36
N-chlorosuccinimide	1/13
N-chloropiperidine	1/300
N-chloro-p-toluenesulfonamide	1/27,000

Oxidation of organic compounds with chlorine is generally slow, especially at the low concentrations of chlorine used for disinfection, except in the presence of strong ultraviolet light (Hancil and Smith, 1971). The reactions of chlorine and oxychlorine species in natural water have recently been reviewed by Rosenblatt (1975). Atkinson and Palin (1972) have reviewed chemical oxidation in water treatment.

The wide variety of rates of loss of free available chlorine reacting as an oxidizing agent in demand reactions make generalizations difficult. Recent measurements by Snoeyink and Markus (1973; 1974) give decay rates for free chlorine measured in nitrified secondary effluents at pH values of 8.2 to 8.5. These rates depended on sunlight, depth of stream, turbulence, temperature and type of residual. Samples exposed to prevailing winds and daylight, and dosed with 3.12 mg/l Cl_2, gave first-order decay rates from 2.1 to 7.4 hr^{-1} with half-lives of 8 to 28 min. The chlorine residual in the samples was predominantly (80%) OCl^- measured with the DPD, FAS titration (Palin, 1957). For samples kept indoors, without the stirring and ultraviolet light present in the outdoor samples, a tenfold greater persistence was measured. The range of first-order decay constants were 0.19 to 0.77 hr^{-1} for OCl^- after an initial reduction of half of the added initial dose at a rate similar to the outdoor samples. The first half-life required only 10 to 30 min while the second half-life was 1.3 to 5 hr.

The initial rapid reduction of chlorine or "demand" is due to the easily oxidized reducing agents present. These might be iron^{++}, sulfide and the low concentrations of easily oxidized organic compounds contained in these well nitrified effluents. The presence of sunlight, copper, bromide or other redox catalysts may be effective in increasing the rates of these demand reactions. Sunlight promotes the oxidation of organic compounds and even water to oxygen by free chlorine. Volatility of chlorine would not be expected to be large for the OCl^- predominant in these alkaline samples studied by Snoeyink and Markus (1973; 1974). In the more acid pH range where HOCl predominates, volatility losses would also be expected.

Combined Chlorine

Ammonia at 1 mg N to 8 mg or greater Cl_2 weight ratios acts as a demand or reducing agent with the classical breakpoint reaction shown in Figure 3. Beyond the breakpoint where chloramine decomposition has occurred, ammonia has acted simply as a reduced species consuming chlorine. At higher concentrations of ammonia than 1 mg N to 5 mg Cl_2, monochloramine is formed as a stable combined chlorine. Monochloramine is a weaker oxidizing agent than free chlorine. The E° value of monochloramine recently calculated by Rosenblatt (1975) is 1.16 V. At pH 7 the E_H decreases to 0.95 V. These values are both approximately 0.25 V more positive than the values reported by White (1972), but both sets of values show NH_2Cl is 0.25 to 0.33 V less positive than HOCl.

The much lower oxidation potential of NH_2Cl compared to HOCl predicts that NH_2Cl would be less able to participate as an oxidant in chlorine-demand reactions. This is found in the measured first-order decay rates of 0.03 to 0.075 hr^{-1} for NH_2Cl determined indoors by Snoeyink and Markus (1973). They found higher rates of 0.28 to 0.31 hr^{-1} for these same high-ammonia trickling filter effluents when subjected to sunlight and turbulence, but these rates are still a factor of 10 slower than the much less persistent free chlorine. Monochloramine would be expected to persist for hours to days compared to the minutes to hours persistence expected for free chlorine.

In estuarine water Bender *et al.* (1975) found decay rates for free chlorine in the York river of 0.046 to 0.052 hr^{-1} where the pH was 7.5 to 8.0, ammonia concentrations were 0.1 to 0.5 mg N/l, salinity was 18 to 20 parts per thousand and temperatures were 17 to 28°C. Although hypochlorite solutions were added in the free chlorine test, the pH and concentrations of ammonia were sufficient to form monochloramine. They also prepared solutions of monochloramine from chlorine and ammonia. When this monochloramine solution was added to York River water the linear portion of the first-order decay curve gave a k of 0.052 hr^{-1}, the same rate seen for free chlorine addition. In both cases monochloramine may be responsible for the observed persistence. Their results are shown in Figure 5.

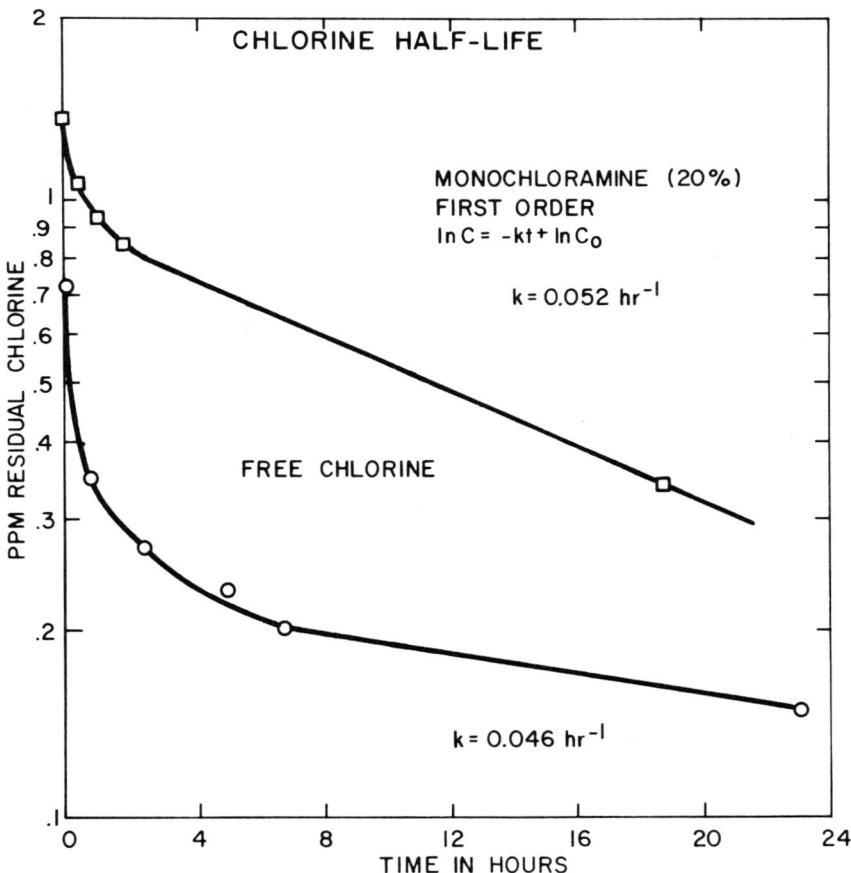

Figure 5. Persistence of chlorine in York River water (after Bender et al., 1975).

The slow monochloramine decay rates with 20-hr half-lives were reached after an initial rapid decay in the first hour of contact. The major difference in the two curves is the more rapid initial loss from free chlorine before it formed combined chlorine. In sea water, with its significant concentration of bromide, it is possible that bromine and bromamines may also be formed. These compounds, although slightly weaker as oxidizing agents than their chlorine analogs, are likely to react even quicker than chlorine and are even less persistent.

A similar pattern of a rapid initial decay followed by a persistent residual was measured by Baker (1970) in Passaic River water at Little Falls, New Jersey. When dosed with 20 mg Cl_2/l this water gave 77% of its total demand after 2 hr contact in the first 4 min. This sample contained about 1.5 mg N/l as ammonia. A majority of this ammonia should be completely oxidized by this high concentration of chlorine. In spite of this, the amperometric titration measurements showed a persistent 0.5 mg/l dichloramine fraction. Baker points out that this dichloramine label is "likely inaccurate since it could include other available chlorine compounds besides dichloramine."

ANALYTICAL METHODS

Several compounds are commonly present or formed on chlorination which are falsely measured by one or more of the usual methods as chlorine. These compounds are then incorrectly interpreted as chlorine residual in either the free or combined available chlorine fractions. The most common interference in free chlorine measurements is manganese dioxide formed by chlorine oxidation of soluble manganous ion. Even though manganese dioxide is a solid, it interferes stoichiometrically in almost all methods, including the continuous amperometric monitors. Only the free chlorine amperometric titration is free of manganese dioxide interference. Even this method, however, measures manganese dioxide in the combined available chlorine test. Other strong oxidizing agents such as ozone, peroxide, bromine, iodine and chromium[6]$^+$ or dichromate universally interfere in chlorine (disinfection) measurements.

The purpose of this portion of the paper is to describe the major limitations of each of the common methods for chlorine measurement. Finally, a new instrumental method developed in my laboratory will be described.

Common Methods For Chlorine Measurement

A. Acid orthotolodine (OT, OTA, flash, drop dilution)

This method is the simplest but least accurate of all the methods. It is being removed from *Standard Methods* (AWWA, WPCF, APHA, 1971) in its next edition. The yellow color fades rapidly. The free chlorine test has a combined chlorine (NH_2Cl) interference of 3%/sec at 20°C or 77°F. Nitrite and $iron^{3+}$ interfere as well as MnO_2, which interferes stoichiometrically.

B. Diethylparaphenylenediamine (DPD) (Palin, 1957)

This method is available in colorimetric and ferrous titration procedures. Its major disadvantage is the instability of its reagents. The interference from combined chlorine is small compared to the OT methods but much larger than the SNORT, LCV and FACTS methods discussed below. Combined chlorine interference in the free chlorine measurement is 1%/min at 25°C. Manganese dioxide also interferes stoichiometrically in the free chlorine measurement. It is the best developed method with many modifications and is readily available in kit form. It is a standard method.

C. Leuco crystal violet (LCV) (Black and Whittle, 1967)

This method is a colorimetric procedure with low (less than 0.1% per minute) NH_2Cl interference in the free chlorine test. It is a complex procedure requiring that reagents be added down the side of the tube. The reagent stains containers. Manganese dioxide interferes stoichiometrically in the free chlorine measurement. It is available in kit form from Taylor Chemical and Hach. It is a standard method.

D. *Stabilized neutral orthotolodine (SNORT) (Johnson and Overby, 1969)*

This is a colorimetric method with low (0.1%/min) NH_2Cl interference at 25°C in the free chlorine test. Figure 6 compares the NH_2Cl interference at 35°C to the DPD method. The blue color fades slowly. Manganese dioxide interferes stoichiometrically in the free chlorine measurement. It requires two liquid reagents which are quite stable. It is available from La Motte in kit form (#7846). It also is a standard method.

E. *Syringaldazine (FACTS) (Sorber, Cooper and Meier, 1975)*

This method is a new method. The color fades slightly and the reagents are colored, which makes interpretation difficult at low concentration, especially using reagents capable of measuring above 1 mg/l. It has little if any NH_2Cl or MnO_2 interference. It is not yet a standard method but will be published as a tentative standard in the next edition. It is not available yet in kit form.

F. *Amperometric titration*

This method is generally considered to be the laboratory standard. Its main problems are its complexity and the operator experience required. To obtain good results, blank corrections should be made. Because of the fast stirring rate used with commerical instrumentation, volatility and reduction losses can be significant unless the majority of the titrant is added quickly. It has low (0.1% per minute) NH_2Cl interference and no MnO_2 interference in the free chlorine measurement. Electrode poisons and films are a problem and careful cleaning to remove iodide between titrations is important. Equipment is available from Fisher and Porter, Wallace and Tiernan and others.

G. *Continuous amperometric cells; bare electrode*

These instruments are on-line amperometric measurements. They should not be confused with the amperometric titration. They use no titrant but do consume buffer. They are claimed

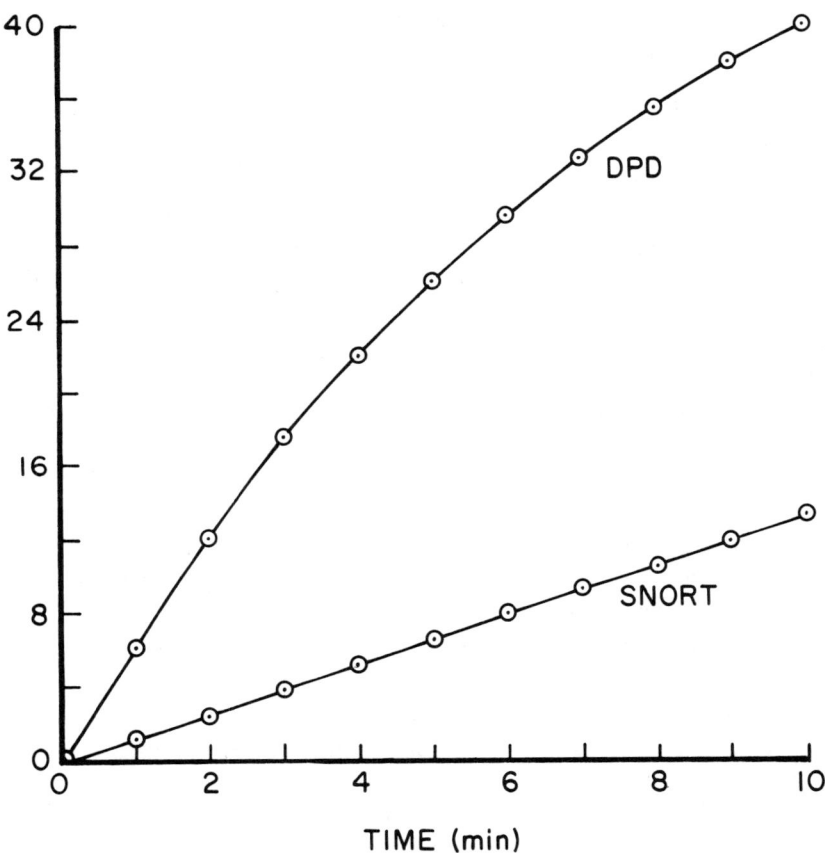

Figure 6. Relative interference error from monochloramine (4.9 mg/l Cl_2) in free chlorine measurement at 35°C.

to measure in either a free chlorine mode with acetate buffer or in a combined chlorine test with iodide added to the buffer. These methods have stoichiometric NH_2Cl and MnO_2 interference even in the free chlorine mode (Morrow and Roop, 1975). These methods are also subject to electrode films and poisons. Equipment is available from Wallace and Tiernan, Fisher and Porter, Capital Controls, Honeywell and others.

A modification on this procedure which also uses an amperometric cell is the National Bureau of Standards Flux Monitor

(Marinenko, Huggett and Friend, 1975). It has the novel feature of a built-in coulometric standardization but also has the same interference problems as the regular procedure.

H. Summary

The selective free chlorine methods such as SNORT, DPD, LCV, FACTS or amperometric titration have only interferences from MnO_2 and other strong oxidizing agents. For combined chlorine, these methods, with added iodide or the iodometric back titration and the continuous amperometric monitors or the NBS Flux Monitor, can be used. Some determination and corrections for interferences can and should be done especially in the combined chlorine methods. This can be done by adding arsenite to the sample which will react with free and combined chlorine but leave manganese dioxide, nitrite and other compounds whose reaction or the blank reading can be subtracted. Use fresh reagents, especially if they contain color or growths. Iodide solutions should also be fresh each week when measuring low concentrations.

None of the methods above are capable of measuring HOCl in the presence of OCl^-. Cyanuric acid and other acidic nitrogen functions such as those found in amides and peptide bonds when they form N-chloro derivatives have an acidic and easily hydrolyzed chlorine (O'Brien, Morris and Butler, 1974). These compounds are in equilibrium with HOCl but have lower ability and probably no inherent ability in themselves to act as disinfectants (Table II). Because of this equilibrium, any analytical method which removes or reacts with the HOCl from these compounds will measure them stoichiometrically as free chlorine (FAC). For this reason, even the measurement of free chlorine with the best of current methods must be interpreted with care with respect to disinfection.

The measurement of combined chlorine (CAC) and its interpretation in terms of fish toxicity or disinfection from NH_2Cl are difficult in natural water and impossible in wastewater. The CAC methods are generally based on the use of iodide and/or acid to give an equivalent of I_2 from the chlorine associated with nitrogen. In general, organic chloramines are measured by this method (Palin, 1950) with iodine generated in the same way as with NH_2Cl. As shown in Table III, methods which employ acidification are especially poor because of chloramine decomposition produced by acidification. Thus the acid orthotolidine method (OT) gave much lower readings in all the samples. After only 30 sec at pH 1.5 the chlorine residual determined by amperometric titration in a chlorinated sewage dropped from 4.75 to 0.58 ppm. Table III also shows N-chloropiperidine is measured as if it were NH_2Cl in the amperometric titration. Table II shows, however, that this compound is only 1/9 as effective as NH_2Cl as a germicide.

Table III. Organic Chloramines Measured as Residual Chlorine 15 Minute Contact at pH 7.0 (Marks and Joiner, 1948)

Base	Chlorine Dose (ppm)	Amperometric Titration (ppm)	OT (ppm)
1-Cystine, 25 ppm	15	1.85	0.04
Piperidine, 61 ppm	6	5.60	0.20
Above piperidine + SO_2	- 0.56	5.20	0.17
Raw sewage, pH 7	10.5	3.95	0.90
Above sewage + SO_2	- 0.84	3.15	0.50
Raw sewage, pH 7	12	4.75	0.10
Titration of above sample after 30 sec at pH 1.5	12	0.58	

The HOCl Membrane Electrode

An amperometric membrane electrode similar to the oxygen probe has been perfected for chlorine, bromine and iodine (Johnson and Edwards, 1975). The response of this electrode is in proportion to the decreasing effectiveness of the halogens in their various chemical forms X_2, HOX and OX^-. Figure 7 shows the cell design developed for the halogen electrode. This cell is similar to the amperometric oxygen electrode. It uses a platinum or gold cathode separated from the solution to be measured by a microporous film that permits the transport of X_2, HOX and OX^- species with decreasing sensitivities. Several membranes were screened for use in the system. A microporous polypropylene film and two microporous fluorocarbon films were found to be the most useful. The microporous polypropylene and one of the fluorocarbon films used were hydrophobic in character. The other fluorocarbon film, an ion exchange membrane, absorbed approximately 18% water, had a microporous structure of 10-mil thickness and was used in the hydrogen form.

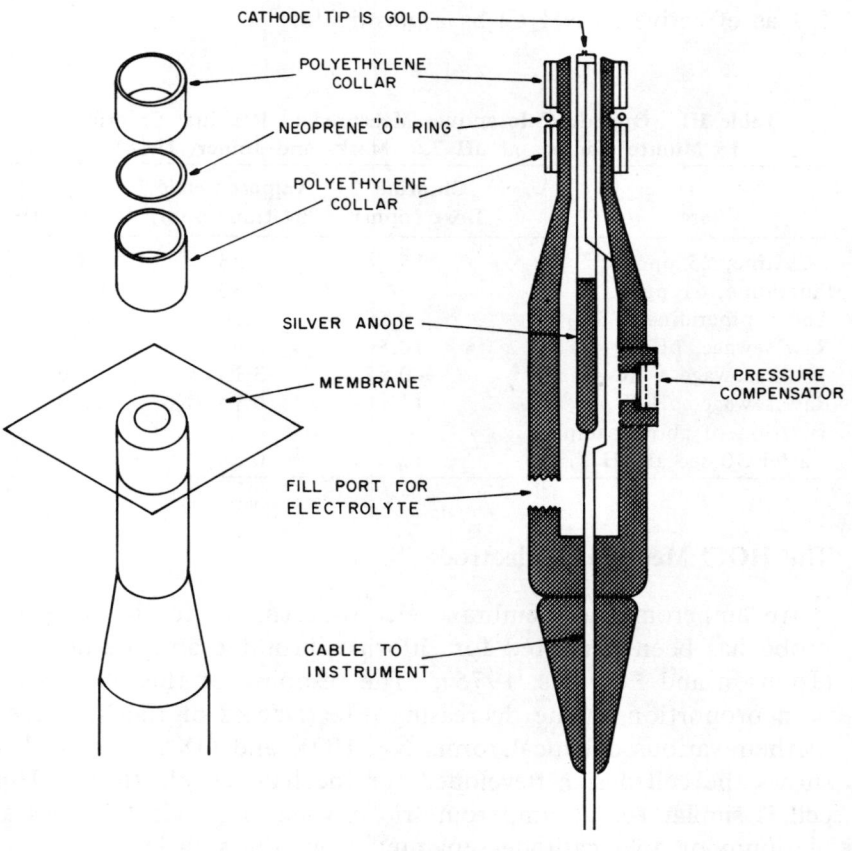

Figure 7. The HOCl membrane electrode.

Chlorine, bromine and iodine all exist in several chemical forms as determined by the following equilibria:

$$X_2 + X^- = X_3^- \tag{7}$$

$$X_2 + H_2O = HOX + H^+ + X^- \tag{8}$$

$$HOX = OX^- + H^+ \tag{9}$$

Of these forms, the X_2 form is generally the best disinfectant, while HOX is also highly effective, depending on the biological system being disinfected. For chlorine, only the HOCl form is important. The OX^- and X_3^- species are relatively poor disinfectants compared to the HOX and X_2 species. The halide ion itself, X^-, is completely ineffective as a disinfectant.

Figure 8 shows calibration curves for HOCl taken with several electrodes over a 6-month period. This curve shows the response of the electrode system is linear as a function of concentration and therefore gives a current signal in direct proportion to the concentration of the sample, as indicated by Equation 10:

$$I = SC \tag{10}$$

where S is the sensitivity in microamperes per ppm and C is the concentration in ppm or mg/l.

Table IV gives the sensitivities in microamperes per ppm determined from calibration curves like those in Figure 8.

Table IV. Sensitivities for Halogen Species with Ion Exchange Fluorocarbon and Polypropylene Membranes

	Sensitivities in Microamperes/ppm	
Species	Ion Exchange Fluorocarbon	Polypropylene
Cl_2	4.27	13.64
HOCl	2.98	0.41
OCl^-	1.43	-
Br_2	1.22	4.13
HOBr	0.98	0.15
OBr^-	0.40	-
I_2	-	0.37

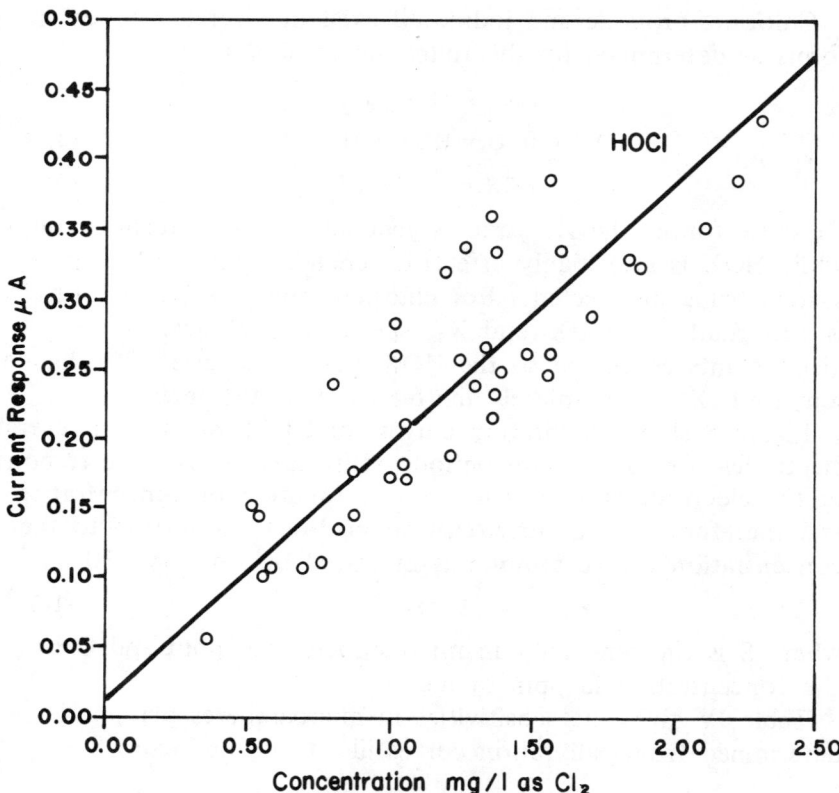

Figure 8. Composite of 9 months of HOCl measurements at pH 5 and 337 mV of applied potential using the microporous hydrophobic membrane at 20°C and at an ionic strength of 0.02.

The ion exchange fluorocarbon electrode response to the HOX species has approximately twice the sensitivity it shows to the OX⁻ species, while the X_2 species gives twice as much response as HOX. Using the hydrophobic fluorocarbon or polypropylene, the halogen electrode response is approximately 30 times greater for the X_2 species than for the HOX species and is completely insensitive to the OX⁻ species.

The temperature dependence for the response of the electrode system obeys Equation 11:

$$\log \frac{S_2}{S_1} = \frac{-E_p}{2.3R}\left(\frac{1}{T_2} - \frac{1}{T_1}\right) \tag{11}$$

where E_p is the energy of permeation, R is the gas constant in cal/mol-degree and T is the absolute temperature in degrees Kelvin. For the ion exchange fluorocarbon film E_p averages 2.36 kcal/mol with a standard deviation of 0.45 for the six chlorine and bromine species given in Table IV. The value of E_p averages 4.26 kcal/mol with a standard deviation of 0.90 for the microporous polypropylene with the five chlorine, bromine and iodine species which can be measured. This compares with 17 kcal/mol reported for oxygen diffusion through homogenous polyethylene. Thus the oxygen electrode has approximately 3 to 7 times greater temperature sensitivity than the halogen electrode. This gives the halogen electrode a sensitivity to temperature as Q_{10} at 25°C of approximately 1.26 for the polypropylene film and 1.14 for the ion exchange fluorocarbon membrane. This is slightly less temperature sensitivity than has been reported for the disinfection of bacteria by halogens.

The halogen electrode, like the oxygen electrode, can be placed directly in the solution or sample stream and the current calibrated in terms of effective chlorine residual or disinfection efficiency. There is an important difference between the halogen electrode response and typical concentration measurement systems. The electrodes respond with sensitivities for various halogen species in proportion to their efficiency as disinfectants. The current measurement, then, gives not a measurement of the sum of the total halogen concentration with complete disregard of the lower effectiveness of OCl^- as a disinfectant, for instance, but gives a measurement of effective residual halogen concentration. Thus the response of the electrode does not suffer from the OCl^- interference typical of total concentration measurements, but reads current, which is in proportion to the efficiency of the chemical species as water disinfectants.

If desired, FAC concentrations can be determined by calibrating at the pH of the sample. Calibration factors would then be used which were appropriate to the chemical form desired as a function of pH. The FAC concentration measurement is not the measurement which is really needed. The measurement needed is the determination of residual effective disinfectant. On this basis FAC concentration measurements determining the sum of HOCl and OCl$^-$ are in error to the extent that they measure OCl$^-$ as if it were as effective as HOCl. This electrode gives us for the first time a method capable of measuring only HOCl, *real* free chlorine.

The electrode is not sensitive to manganese dioxide, which is a common interference in all current analytical methods for chlorine. The electrode can be placed directly in the inplant flow downstream from the point of application of chlorine to determine chlorine residual at any contact time without the necessity for the time delay and sample loss typical of remote systems. The electrode is not as subject to electrode fouling, which is a common problem with other amperometric electrode systems. Depending on the applied voltage, it will measure either HOCl alone for use in drinking and cooling waters or HOCl plus NH$_2$Cl for use in distribution systems and wastewater. Table V summarizes the electrode characteristics. The electrode is available commercially from Delta Scientific Corp., Lindenhurst, N.Y.

Table V. Characteristics of the Halogen Membrane Electrode

1. Measures only (HOCl) independent of pH, 0.4 µA/ppm
2. Insignificant interferences from poor disinfectants:
   Monochloramine—1% NH$_2$Cl
   Hypochlorite ion—1% OCl$^-$
   Nitrite ion—1% NO$_3^-$
   Manganese^{4+}—0% MnO$_2$
   Iron^{3+}—0% Fe(OH)$_3$
   Color, organics, oxygen—0%
3. Temperature dependence—2 to 6% per °C; 1 to 3% per °F
4. Sensitive to 0.03 mg/l `
5. Accurate, ϕ = 0.40 to 0.42 µA/ppm at 95% C.I.
6. Independent of pH (FAC depends on HOCl/OCl$^-$ ratio)

ACKNOWLEDGMENT

This work was supported in part by the U.S. Army, Medical Research and Development Command, Contract No. DADA 17-72-C-2053.

REFERENCES

American Water Works Association. 1973. "Water Chlorination Principles and Practices." AWWA Manual M20, New York.

Atkinson, J. W. and A. T. Palin. 1972. "Chemical Oxidation in Water Treatment, p. E1-E9." In Proceedings of International Water Supply Conference, Special Subject No. 5, New York.

AWWA, WPCF, APHA. 1971. *Standard Methods for the Examination of Water and Wastewater*, 13 ed., American Public Health Association.

Baker, R. J. 1970. "Engineering Considerations in Disinfection," p. 685-697. In Am. Soc. Civil Eng., *Proceedings of the National Specialty Conference on Disinfection*, Amherst, Massachusetts.

Bender, M. E., M. H. Roberts, R. Diaz and R. J. Huggett. 1975. "Proceedings Technology and Ecological Effects of Biofouling," Maryland Power Plant Siting Program, Baltimore, Maryland.

Black, A. P. and G. P. Whittle. 1967. "New Methods for the Colorimetric Determination of Halogen Residuals. Part II: Free and Total Chlorine." *J. Am. Water Works Assoc.* 59:607.

Farkas, L., M. Lewin and R. Black. 1949. "The Reaction Between Hypochlorite and Bromides." *J. Am. Chem. Soc.* 71:1987.

Hancil, V. and J. M. Smith. 1971. "Chlorine-Sensitized Photochemical Oxidation of Soluble Organics in Municipal Wastewater." *I&EC Process Design and Dev.* 10:515-523.

Johnson, J. D. and J. W. Edwards. 1975. "A Halogen Membrane Electrode." Presented to Annual Conference, American Water Works Association, Minneapolis, Minnesota.

Johnson, J. D. and R. Overby. 1969. "Stabilized Neutral Orthotolodine Method for Chlorine." *Anal. Chem.* 41:1974.

Marinenko, G., R. J. Huggett and D. G. Friend. 1976. "An Instrument with Internal Calibration for Monitoring Chlorine Residuals in Natural Waters." *J. Fish. Res. Board Can.* 33:822-826.

Marks, H. C. and R. R. Joiner. 1948. "Determination of Residual Chlorine in Sewage." *Anal. Chem.* 20:1197.

Marks, H. C. and F. B. Strandskov. 1950. "Halogens and Their Mode of Action." *Ann. N.Y. Acad. Sci.* 53:163-171.

McClanahan, M. 1975. "Recycle—What Disinfectant for Safe Water Then?" pp. 49-66. In J. D. Johnson (Ed.), *Disinfection: Water and Wastewater.* Ann Arbor Science, Ann Arbor, Michigan.

Morris, J. C. 1966. "The Acid Ionization Constant of HOCl from 5 to 35°C." *J. Phys. Chem.* 70:3798.

Morrow, J. C. and R. N. Roop. 1975. "Advances in Chlorine Residual Analysis." *J. Am. Water Works Assoc.* 67:184.

O'Brien, J. E., J. C. Morris and J. N. Butler. 1974. "Equilibria in Aqueous Solution of Chlorinated Isocyanate," p. 333. In A. J. Rubin (Ed.), *Chemistry of Water Supply Treatment and Distribution.* Ann Arbor Science, Ann Arbor, Michigan.

Palin, A. T. 1950. "A Study of the Chloro Derivatives of Ammonia and Related Compounds with Special Reference to Their Formation in the Chlorination of Natural and Polluted Waters." *Water Water Eng.* 1950:248.

Palin, A. T. 1957. "The Determination of Free and Combined Chlorine in Water by the Use of Diethyl-p-phenylenediamine." *J. Am. Water Works Assoc.* 49:873-880.

Rosenblatt, D. H. 1975. "Chlorine and Oxychlorine Species Reactivity with Organic Substances," pp. 249-276. In J. D. Johnson (Ed.), *Disinfection: Water and Wastewater.* Ann Arbor Science, Ann Arbor, Michigan.

Snoeyink, V. L. and F. I. Markus. 1973. "Chlorine Residuals in Treated Effluents." Report prepared for Illinois Institute for Environmental Quality, Urbana, Illinois.

Snoeyink, V. L. and F. I. Markus. 1974. "Chlorine Residuals in Treated Effluents." *Water Sew. Works* 121:35-38.

Sorber, C., W. Cooper and E. Meier. 1975. "Selection of a Field Method for Free Available Chlorine," p. 91. In J. D. Johnson (Ed.), *Disinfection: Water and Wastewater.* Ann Arbor Science, Ann Arbor, Michigan.

Sugam, R. and G. R. Helz. 1975. "Apparent Ionization Constant of Hypochlorous Acid in Seawater." *Environ. Sci. Technol.* 10:384-386.

White, G. C. 1972. *Handbook of Chlorination.* Van Nostrand Reinhold, New York: p. 212.

DISCUSSION

George R. Helz, University of Maryland. On your chart you show essentially 2 ppb as the sensitivity of your electrode. On your previous slide the calibration curve was essentially 1 to 10 ppm. Can you clarify that? Can you really measure at the ppb level?

Johnson. No, the electrode will not measure 3 ppb; 30 maybe, and then only if you're willing to accept 100% error. That is, it measures plus or minus 0.03 ppm or 30 ppb. It's not as sensitive as I'd like it, that's for sure. The NBS flux monitor is nice and sensitive but then you don't get selectivity for chlorine. So you're sort of between the selectivity devil and sensitivity devil.

George Clifford White, Consulting Engineer. Don, you sound very confident about being able to selectively measure monochloramine. This is very exciting news because this would be very helpful in wastewater treatment, obviously. Now, have you done anything measuring monochloramine in the presence of all the rest of the garbage that normally interferes in monochloramine measurements in wastewater?

Johnson. I just made an experiment last week with N-chloroglycine. We ran a curve with a constant concentration of glycine, adding chlorine up to where we have just N-chloroglycine without cleaving off the amine and doing oxidative deamination. So we have this compound present and it measures with the standard analytical methods; for example, SNORT, DPD, amperometric titration. It measures like monochloramine. Then we put the electrode in and we measure with the lower voltage where the electrode has some response to monochloramine. We don't get any response. So the electrode does respond to monochloramine under proper conditions but it does not measure the chloraminoacid, N-chloroglycine.

4

ORGANOCHEMICAL IMPLICATIONS OF WATER CHLORINATION

Robert M. Carlson and Ronald Caple
Department of Chemistry
University of Minnesota
Duluth, Minnesota 55812

ABSTRACT

The desire for structural information on the specific organic compounds present in a given sample of renovated water makes it imperative that there be a basic understanding of those principles of mechanistic organic chemistry that apply to the situation in question. The process will be illustrated for aqueous chlorination, where the observed chloro-organics can be readily explained on the basis of commonly recognized reactive intermediates and those stereochemical and electronic features associated with the organic moiety. The relationships of these mechanistic processes to pH, product distribution, BOD, oxidative capacity and chloramine formation are considered.

INTRODUCTION

The common thread that has woven its way throughout the entire fabric of research into the chemical and biological implications of water renovation processes is a desire for structural information regarding the specific "parent" and "second-order" compounds under investigation. This awareness of potential structural implications may be as subtle as the negative feeling synonymous with polychlorinated organics or as direct as the development of new gas chromatography mass-spectral techniques. Although most of the "awareness" is necessarily "after the fact," it should be reaffirmed that a major goal of this effort is to provide a predictive capability. That is, what structural features —and thereby ultimately what biological response—should be anticipated under a given set of circumstances? If indeed this goal is to be met, it is imperative that a greater number of those individuals concerned with water renovation be aware of those basic elements of mechanistic organic chemistry that apply to their particular situation.

In the case of aqueous chlorination, we should be concerned with two distinct electron-deficient species of chlorine—the "chloronium ion" and the chlorine radical. As might be expected from the charged nature of the chloronium ion, it is the major contributor to reactions occurring in aqueous media. On the other hand, significant product formation resulting from the chlorine radical should be anticipated in the presence of light and the absence of a significant polar reaction pathway involving the loss of the parent chlorine molecule.

$Cl^+ \equiv$ *Chloronium Ion*	$Cl\cdot \equiv$ *Chlorine Radical*
Outer shell has 6 e$^-$	Outer shell has 7 e$^-$
Electron-deficient	Electron-deficient
Positively charged	Neutral

In predicting the mode of action of these reactive electrophilic intermediates we should examine those potential electron sources that would satisfy this inherent electron deficiency. Indeed, such electron-rich centers as the unshared electrons of the nitrogeneous bases and the π electron systems of olefins and aromatic compounds appear particularly vulnerable.

Amines represent a major class of compounds recognized to be present in waters subjected to renovation with the result that chemistry of chloramine formation has received considerable attention. However, such chemistry remains as a topic for the current discussion only in that within the chloramine there remains a potential that has yet to be determined for the *subsequent* formation of carbon-bound chlorine.

RESULTS AND DISCUSSION

The reaction of chlorine with aromatic systems results in "electrophilic aromatic substitution" (Figure 1). In this process the general principles of relative substrate reactivity and the orientation of substitution are documented to the extent that the elements of predictability are particularly attractive. This was confirmed in a limited study conducted in our laboratories on the chlorination of aromatic compounds possessing substituents of diverse electronic properties. Under the conditions of dilute aqueous chlorination, phenol was found to be the most reactive, with the other compounds decreasing in chlorine incorporation with the increasing electronegative nature of the substituents (see Table I).

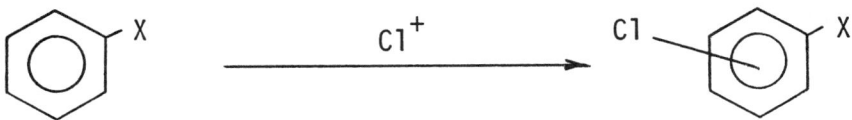

Figure 1. Electrophilic aromatic substitution.

Table I. Electrophilic Aromatic Substitution: Chlorine Incorporation in Selected Aromatic Compounds

Compound ($9.5 \pm 0.6 \times 10^{-4} M$)	Percent Chlorine Remaining After Reaction Time[a]		
	pH 3	pH 7	pH 10
phenol	2.2 ± 0.1	2.4 ± 0.1	2.4 ± 0.2
anisole	19.3 ± 0.2	88.6 ± 0.4	97.2 ± 0.3
acetanilide	44.7 ± 0.5	98.6 ± 0.2	-
toluene	88.9 ± 0.1	97.1 ± 0.4	-
benzyl alcohol	97.7 ± 0.2	-	-
benzonitrile	97.9 ± 0.2	-	-
nitrobenzene	98.2 ± 0.1	-	-
chlorobenzene	98.2 ± 0.1	-	-
methylbenzoate	98.2 ± 0.2	-	-
benzene	98.5 ± 0.1	-	-

[a] Chlorine (7.0×10^{-4} M); 20 min; 25°C.

It should also be noted that in three cases (anisole, acetanilide and toluene) the extent of chlorine incorporation increased with decreasing pH. This observation clearly appears to be related to an existing equilibrium in the potential chlorinating species. This result also closely parallels the varying disinfecting capability of chlorine with pH, where low pH values promote effective disinfection.

$$HOCl \rightleftharpoons H^+ + OCl^-, pKa \cong 7.5$$

A related point of interest is illustrated by a detailed examination of biphenyl chlorination conducted in our laboratories and those of Monsanto (Dr. Harold Weingarten and co-workers) where the major products were those containing limited numbers of chlorines incorporated into the biphenyl nucleus (Figure 2). Although this observation can be partially ascribed to the dilute nature of the reaction medium, it also illustrates the decreasing reactivity of the aromatic nucleus with increasing chlorine substitution.

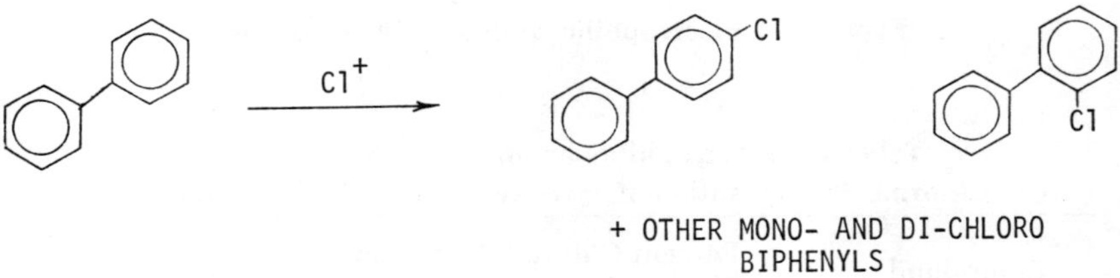

Figure 2. Chlorination of biphenyl.

The relationship of aromatic chlorination to such commonly used parameters as BOD was ascertained by determining the BOD of a given molar amount of parent phenol relative to the mono- and dichloro progeny. The results indicate that at low concentrations the chlorophenols are not significantly metabolized over the test period and that at high concentrations the chlorophenols are actually adversely affecting the microbial population (Table II).

AQUEOUS CHEMISTRY OF CHLORINE 69

Table II. Summary of Phenol Results (BOD)

Test Number	Concentration	Phenol 5 (day)	10	20	o-Chlorophenol 5 (day)	10	20	p-Chlorophenol 5 (day)	10	20	(2,4-) Dichlorophenol 5 (day)	10	20
2	1.06×10^{-5} M (1 ppm)	312											
3	7.78×10^{-6} M (1 ppm)				191								
	7.78×10^{-5} M (10 ppm)				195								
	7.78×10^{-4} M (100 ppm)				157								
4	1×10^{-5} M (1.29 ppm)							226					
	1×10^{-4} M (12.9 ppm)							225					
	1×10^{-3} M (129 ppm)							94	140				
5	2×10^{-6} M (0.19 ppm)	168	215	284									
	6×10^{-6} M (0.56 ppm)	214	306	290									
	1×10^{-5} M (0.94 ppm)	229	335	335									
6	1×10^{-5} M (1.29 ppm)				165	207	229						
	1×10^{-3} M (129 ppm)				187	223	245						
7	2×10^{-6} M (0.19 ppm)	183	233	269									
	6×10^{-6} M (0.56 ppm)	227	290	290									
	1×10^{-5} M (0.94 ppm)	243	278	273									
8	2×10^{-6} M (0.33 ppm)										135	159	166
	6×10^{-6} M (0.99 ppm)										154	156	173
	1×10^{-5} M (1.65 ppm)										151	149	147

These results have the obvious environmental implications that: (1) chlorinations at low pH (*e.g.*, certain industrial wastes) have a greater probability of generating "second-order" chloroorganics; (2) *poly*chlorinated aromatics isolated from a given effluent are most likely *not* derived via a dilute aqueous chlorination process; (3) an *a priori* estimate of the product type and distribution can be made if the parent aromatic organic content of a waste is known; and (4) qualitative criteria such as reduction of BOD upon chlorination must be viewed with some suspicion.

Olefins and acetylenes represent an additional group of compounds that are potentially vulnerable to attack. However, in this situation each molecule has its own complex set of electronic and stereochemical features that dictate the ultimate set of products. Although this complexity makes a detailed analysis of these diversified groups of compounds more difficult, several principles are apparent from the examination of some representative olefinic materials.

For example, the aqueous chlorination of oleic acid provides a mixture of chlorohydrins (Figure 3) – a result anticipated on the basis of the participation of the solvent (*i.e.*, water) in the second stage of the addition sequence and the "isolated" nature of the disubstituted double bond. This successful competition of water with chloride should have been expected considering the dilute nature of the reaction but also raises the question of the possible implications of using chlorination processes in substantially different media such as sea water, which contains significant amounts of "impurities" (*e.g.*, Cl^-).

In another instance, the anticipated stereochemistry of the intermediates in the aqueous chlorination of the environmentally ubiquitous α-terpineol assisted in the determination of the products and their variation with pH (Figure 4). For example, it should be expected that epoxides would be observed at higher pH values where the nucleophilic substitution of the alkoxide (RO^-) on the carbon bearing the chlorine would be promoted by the trans-anti parallel arrangement of the chlorine relative to the hydroxyl in chlorohydrin IV that is not present in compound II (*i.e.*, II → III at high pH). Likewise, the presence of the bridged system can be explained by the involvement

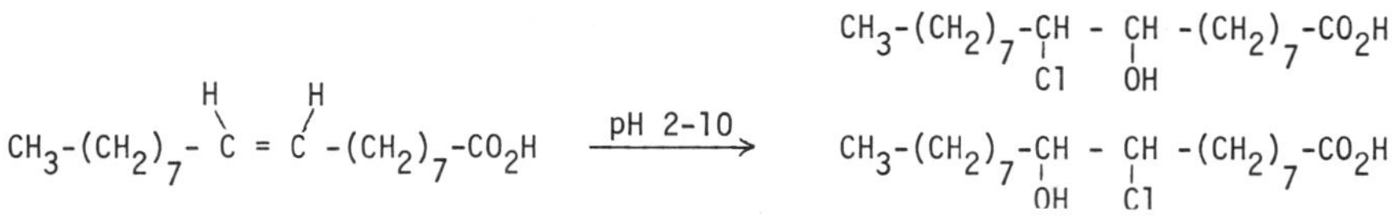

$$CH_3-(CH_2)_7-\underset{H}{\overset{H}{C}}=\underset{}{\overset{}{C}}-(CH_2)_7-CO_2H \xrightarrow{pH\ 2-10} \begin{array}{c} CH_3-(CH_2)_7-\underset{Cl}{CH}-\underset{OH}{CH}-(CH_2)_7-CO_2H \\ \\ CH_3-(CH_2)_7-\underset{OH}{CH}-\underset{Cl}{CH}-(CH_2)_7-CO_2H \end{array}$$

Figure 3. Chlorination of unsaturated fatty acid.

of the 3°–OH in an intramolecular attack on the "down" epoxide (Figure 5). The other observed products can similarly be derived by examining alternative reaction pathways from these or related intermediates (Figure 4).

The search for those products derived from "free-radical" processes should begin where the molecule in question can provide a relatively stable odd electron system. An example of such a system would be that obtained from the abstraction of a hydrogen atom from a benzylic system where the resulting radical owes its stability to the delocalization of the odd electron throughout the molecule by resonance.

Examples of such "free radical" processes appear to have been observed among the chlorination products of the resin acid, dehydroabietic acid (Figure 6).

CONCLUSIONS

The investigation of the dilute aqueous chlorination of typical compounds known to be present in water subjected to chlorine renovation indicates that chlorine is readily incorporated into the carbon framework by a pathway that is predominantly ionic. Furthermore, the use of the basic principles of mechanistic organic chemistry is helpful in elucidating the structure and/or evaluating the distribution of the aqueous chlorination products. A major result of such a capability is thereby the provision for an apparent element of predictability of the environmental impact of variations that might occur during the chlorination process.

72 THE ENVIRONMENTAL IMPACT OF WATER CHLORINATION

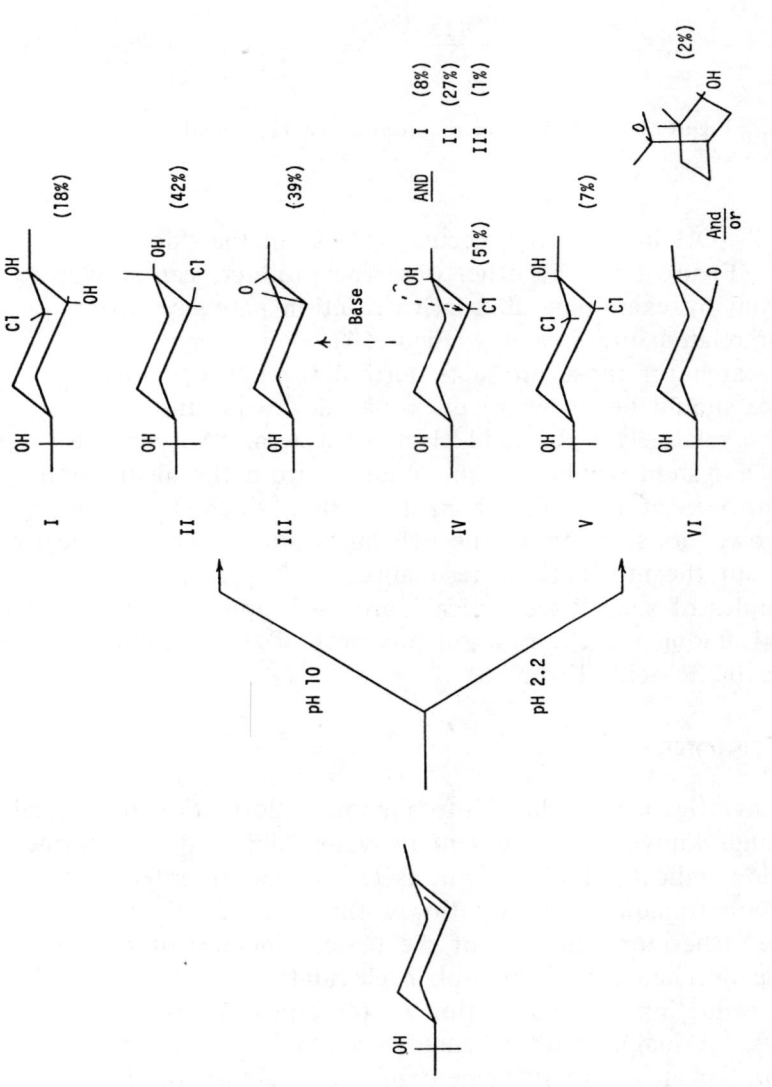

Figure 4. Chlorination of α-terpineol.

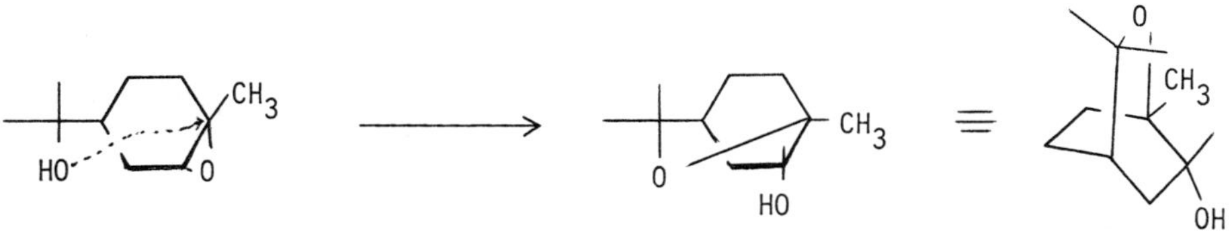

Figure 5. Formation of bridged system by intramolecular rearrangement.

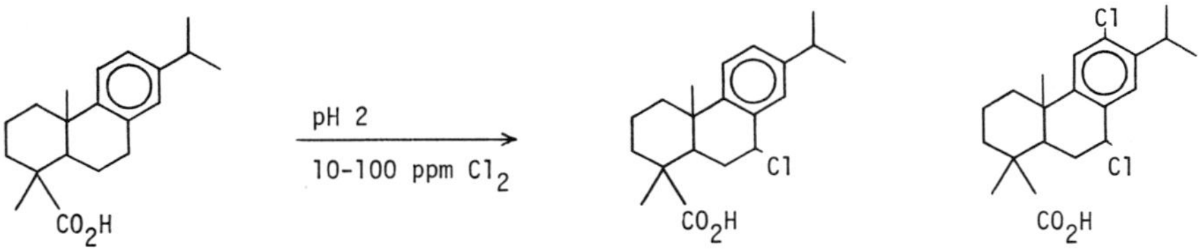

Figure 6. Chlorination of resin acids (abietic and dehydroabietic acid).

In addition, the current study has generated several major questions that require immediate attention:

1. How much of this structural information can be translated to a "real world" situation where the presence of ammonia and a complex combination of organic and inorganic materials are involved?
2. Are the present requirements (*i.e.*, standards) for chlorination meaningful, especially where the environmental "trade-off" would mean the production of significant quantities of degradatively resistant chloro-organics?
3. Is the present effort to discredit the use of chlorine to be matched by a concurrent investigation of the potential consequences of using an alternative procedure?

REFERENCES

American Water Works Association, Inc. 1971. *Water Quality and Treatment*. McGraw-Hill, New York.

Carlson, R. M. and R. Caple. 1977. "Chemical/Biological Implications of Using Chlorine and Ozone for Disinfection." Ecological Research Series, EPA-600/3-77-066.

Carlson, R. M., R. E. Carlson, H. L. Kopperman and R. Caple. 1975. "The Facile Incorporation of Chlorine into Aromatic Systems During Aqueous Chlorination Processes." *Environ. Sci. Technol.* 1975:674.

De La Mare, P. B. and J. H. Ridd. 1959. *Aromatic Substitution: Nitration and Halogenation.* Academic Press, New York.

Johnson, J. D. 1975. *Disinfection: Water and Wastewater.* Ann Arbor Science, Ann Arbor, Michigan.

Jolley, R. L. 1973. "Chlorination Effects on Organic Constituents in Effluents from Domestic Sanitary Sewage Treatment Plants." Ph.D. Thesis, U. of Tennessee, Oak Ridge National Laboratory, TM-4290.

Kopperman, H. L., R. E. Carlson, R. Caple and R. M. Carlson. 1974. "Structure-Toxicity Correlations of Phenolic Compounds to *Daphnia magna*." *Chem. Biol. Interactions* 9:245.

Kopperman, H. L., R. C. Hallcher, A. Riehl, R. M. Carlson and R. Caple. 1976. "Aqueous Chlorination of α-Terpineol." *Tetrahedron* 32:1621-1626.

Liberles, A. 1968. *Introduction to Theoretical Organic Chemistry.* Macmillan, New York.

Manufacturing Chemists' Association. 1973. "The Effect of Chlorination on Selected Organic Chemicals." EPA Water Pollution Control Research Series Project #12020EXG.

Morris, J. C. 1975. "Formation of Halogenated Organics by Chlorination of Water Supplies." Environmental Health Effects Research Series, EPA-600/1-75-002.

White, G. C. 1972. *Handbook of Chlorination.* Van Nostrand-Reinhold, New York.

DISCUSSION

George R. Helz, University of Maryland. You stated that multiple substitution in aromatic compounds is unlikely. Won't this depend on the initial conditions, especially the chlorine-to-aromatic ratio? You might get one answer at the sewage treatment plant and another answer in drinking water treatment. Have you investigated this?

Carlson. Yes. We have in a sense and we still are convinced that unless your aromatic molecule is highly reactive you are still not going to incorporate a lot of chlorine into that molecule. The only way we could get a highly chlorinated biphenyl was to go to 250 ppm at a pH of 2. That's pretty high. I think that relates to your question.

AQUEOUS CHEMISTRY OF CHLORINE 75

Carl W. Gehrs, Oak Ridge National Laboratory. Bob, I'm not sure I followed you correctly. What pH were you working with? Were they strictly pH 2 and 10?

Carlson. We went through the whole series. We looked at the extremes in that one case. But we studied a variety of reaction conditions and, depending on the conditions, there were varying combinations of all those different products.

Gehrs. You did go across the pH range?

Carlson. Right.

J. Donald Johnson, University of North Carolina-Chapel Hill. Did you look at the effect of UV catalysis on any of your reactions?

Carlson. No. That would be very interesting to do. I think you would expect on the basis of what is known to see a lot more free radical reactions and probably a lot more chlorine incorporation. I think a good example of the use of these mechanistic principles in product prediction is in this business of polychlorinated or halogenated methanes. I have been convinced by looking at the problem and talking with people about polychlorinated methane, that we're going to find that the incorporation of chlorine is going to be enhanced quite readily by having multiple carbonyl groups present. We've talked about the iodoform reaction and the ability to have multiple carbonyl groups. Activating that carbon is going to provide a very facile mode of incorporation of chlorine. Any molecule that can generate an oxygenation pattern in a 1-3 relationship is going to give these halogenated methanes. I'm convinced of it.

5

CHLORINATION OF ORGANICS IN DRINKING WATER

Alan A. Stevens, Clois J. Slocum, Dennis R. Seeger
and Gordon G. Robeck
> Water Supply Research Division
> Municipal Environmental Research Laboratory
> U.S. Environmental Protection Agency
> Cincinnati, Ohio 45268

ABSTRACT

Halogenation of organic compounds occurs during chlorination of drinking water. The major known products of these reactions are trihalomethanes. Recognition of the potential health significance of these compounds has led to a search for alternatives to present treatment practice for potable water supplies. Some factors influencing trihalomethane production are precursor compound concentration, pH, type of disinfectant used (*e.g.*, free vs combined chlorine) and temperature.

Appropriate control of these factors reduces concentrations of trihalomethanes in the finished water. The influence of these factors as determined by bench- and pilot-scale experiments is demonstrated and application of some appropriate control measures at a full-scale treatment plant is discussed.

INTRODUCTION

Recently there has been great interest in the study of organic compounds in drinking water—interest that stems largely from the results of (and the publicity that followed) a 1974 study of New Orleans drinking water (U.S. Environmental Protection Agency, 1974). About the same time, two studies (Rook, 1974; Bellar et al., 1974) called attention to the presence in finished drinking water of some trihalomethanes (mostly chloroform) which were not found in the respective raw waters at the locations of study. Conclusions were drawn in both reports that the trihalomethanes were formed during the chlorination step of the water treatment process.

Because of the interest in the various organic contaminants and the concern as to their significance, the U.S. Environmental Protection Agency (EPA) undertook a survey of 80 selected cities to measure the concentration of six halogenated compounds in raw and finished water. Those six included four trihalomethanes (chloroform, bromodichloromethane, dibromochloromethane, bromoform) suspected of being formed on chlorination, plus carbon tetrachloride and 1,2-dichloroethane, known contaminants at New Orleans, but not necessarily formed on chlorination. During this National Organics Reconnaissance Survey (NORS) a more comprehensive organic analysis was also performed in five of the 80 cities and has just been completed in another five.

The occurrence of trihalomethanes in finished drinking water was demonstrated to be widespread and a direct result of the chlorination practice. No hard evidence was found in this regard with respect to 1,2-dichloroethane or carbon tetrachloride.

Based on the survey results, a theoretical finished water with the median concentration of each compound would contain about 21 $\mu g/l$ of chloroform, 6 $\mu g/l$ of bromodichloromethane, 1.2 $\mu g/l$ of dibromochloromethane, and an amount less than the detection limit for the method used (Symons et al., 1975) of bromoform (Figure 1). Although most of the finished waters tested demonstrated this decreasing order of concentration, this was not always the case. The finished water at one location had a chloroform concentration of only 12 $\mu g/l$, but a bromoform concentration of 92 $\mu g/l$. It was speculated that this concentration reflected a relatively high bromide concentration in the raw water, with oxidation of bromide to hypobromite by hypochlorite, and subsequent reaction of hypobromite with

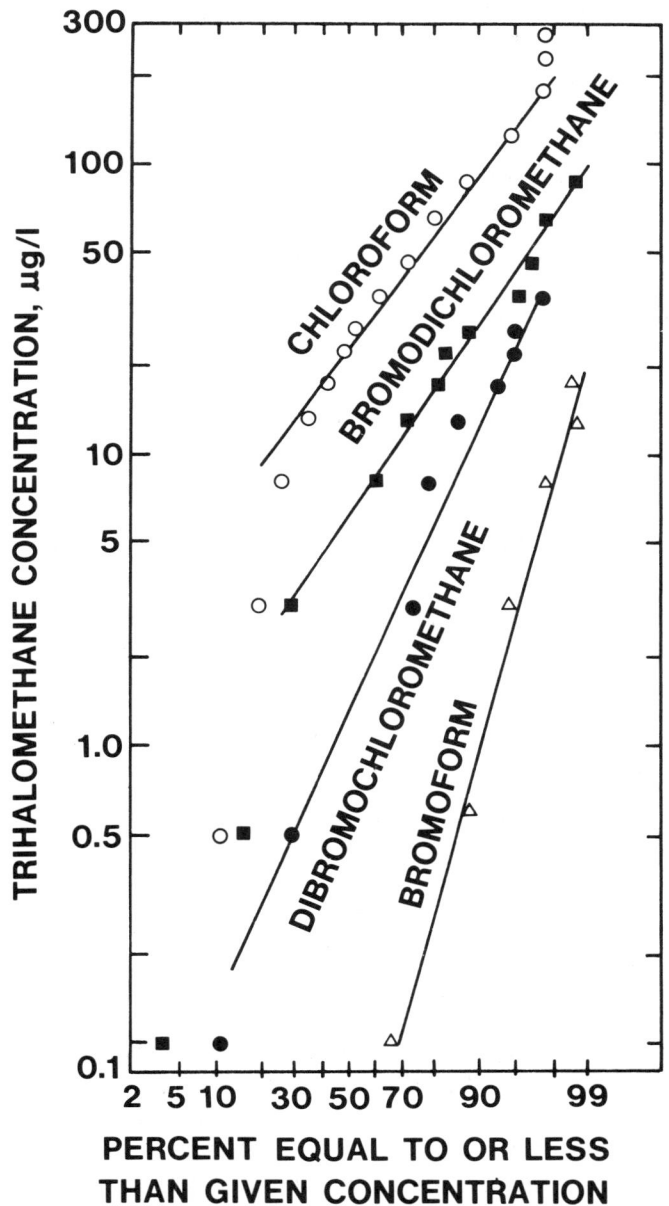

Figure 1. Frequency distribution of trihalomethane data (NORS).

precursor compounds to form the bromine-substituted trihalomethanes. Recently workers at another EPA laboratory (Bunn et al., 1975) have adequately demonstrated this effect by experimentally adding the halides fluoride, bromide and iodide in the form of salts to Missouri River water and subsequently chlorinating that water. The detected reaction products included all 10 possible nonfluorine mixed and single halogen-containing trihalomethanes. Finally, the range of chloroform concentrations was less than 0.1 to 311 μg/l; bromodichloromethane, none found (NF) to 116 μg/l; dibromochloromethane, NF to 100 μg/l; and bromoform, NF to 92 μg/l.

While the health significance of trihalomethanes produced during chlorination of drinking water has not been completely evaluated at this time (1975), understanding the factors affecting the ultimate formation of the trihalomethanes was considered prudent. The goal was then to develop general conclusions applicable to rational modification of water treatment processes if removal of trihalomethanes was finally deemed important for public health reasons. Basic approaches to affect finished water trihalomethane concentrations considered for study were: reducing precursor compound(s) concentration, changing disinfectant (e.g., ozone, chlorine dioxide, etc.) and removing trihalomethanes after formation. The last of these is being studied as an alternative and has been discussed elsewhere (Love et al., 1975). Changing disinfectant without an intense research input to study other public health ramifications could be a catastrophic step. Therefore, because a chlorine residual must be maintained within the distribution system, removing of precursor compounds or controlling their reactions with chlorine was considered the most logical approach.

The foremost consideration in adjusting a series of water treatment processes to remove an organic precursor is identifying the compound(s). Bellar et al. (1974) proposed ethanol as the compound–with oxidation by hypochlorite to acetaldehyde, or acetaldehyde itself, followed by the classical haloform reaction– to be the mechanism of trihalomethane production. Organic chemistry texts typically cite acetone as the simplest example of a methyl ketone that undergoes the haloform reaction. Indeed, Fairless et al. (1975) have investigated the reactivity of simple methyl ketones in water supplies and consider them to play a major role in trihalomethane production. Glaze and Henderson (Glaze, 1975) have identified chlorinated acetone derivatives that could be haloform reaction intermediates in superchlorinated sewage effluents. These theories are attractive because the precursor compounds mentioned have been qualitatively identified during gas chromatographic-mass spectrometric (GCMS) analysis of Ohio River water that contains the unknown precursors which react to form trihalomethanes upon chlorination.

In December 1974 Rook proposed that natural humic substances were responsible. Later Rook (1975) discussed the probable role of the fulvic acid fraction in trihalomethane production, elaborating with examples of very reactive *m*-dihydroxy aromatic compounds suspected to be basic building blocks of the humic (fulvic) acid structure.

Some clarification of the relative roles played by these two groups of compounds (humic materials vs acetyl derivatives of low molecular weight) is paramount to the eventual understanding needed to predict the success of any water treatment process change designed to bring about a reduction in the ultimate trihalomethane concentrations. The roles of other treatment parameters such as NH_3 addition with chlorine (free vs combined chlorine), pH and temperature should also be clarified.

METHODS

Reagents

Chlorine was obtained from Union Carbide Corporation (high-purity grade, Ohio Valley Sales, Cincinnati, Ohio). Stock solutions were prepared by passing the pure gas through nitrogen-purged distilled water. Freshly prepared stock solutions were standardized by amperometric titration as described in *Standard Methods* (American Public Health Association, 1971). Experimental mixtures were prepared by appropriate volumetric dilution of the stock solutions in the test media.

Water for the various experiments was obtained from the pilot plant facility of the Water Supply Research Division (WSRD), Municipal Environmental Research Laboratory (MERL),

U.S. EPA, Cincinnati, Ohio. This plant has previously been described in detail (Love *et al.*, 1975). "Raw" water was that obtained directly from the Ohio River intake at the Cincinnati Water Treatment Plant, Cincinnati, Ohio. This water was used as an untreated source water in all pilot plant work. "Settled" water was that obtained from the pilot plant after alum coagulation and sedimentation. "Dual-media filtered" water was the "settled" water after anthracite-sand filtration. "Activated carbon filtered" water was the same settled water after passage through 1.5 m of Filtrasorb 200 (Calgon Corporation, Pittsburgh, Pennsylvania) granular activated carbon (GAC). Filtration rates through this plant were similar to those found in a conventional water treatment plant: 2 to 2.5 gpm/ft^2 (5 to 6.25 m/hr).

Blank water for analytical purposes was obtained by exhaustively purging laboratory distilled water with helium gas.

The test precursor substances [humic acid (Pfaltz and Bauer, Flushing, New York, or Aldrich Chemical Company, Milwaukee, Wisconsin), acetone (Mallinckrodt, "Nanograde," St. Louis, Missouri), acetaldehyde (Aldrich), acetophenone (Fisher Scientific, Fairlawn, New Jersey)] were used as obtained from the suppliers.

Standard analytical solutions of chloroform (Fisher Scientific, "Spectro-analyzed"), bromodichloromethane (Aldrich), dibromochloromethane (Columbia Chemical Company, Columbia, South Carolina), and bromoform (Fisher Scientific) were prepared as described in the NORS 80-city report (Symons *et al.*, 1975).

Procedures

Analyses for the trihalomethanes were performed by a modification of the "volatile" organic gas chromatographic technique described by Bellar and Lichtenberg (1974) using specific halogen electrolytic conductivity detection (Stevens and Symons, 1975) as described in the NORS 80-city report (Symons *et al.*, 1975).

Nonvolatile total organic carbon (NVTOC) was measured using the method and apparatus described in the NORS 80-city report (Symons et al., 1975). Samples were acidified with nitric acid and purged with carbon-free air for about 10 min to remove carbon dioxide before the actual analysis. Some volatile organic materials were lost during this step. NVTOC was defined as that organic carbon remaining in the sample after this treatment.

Briefly, the experimental procedure was as follows: All reactions described were carried out in the presence of phosphate buffers. Reaction solutions were made up at pH 7 and adjusted to the desired pH with the addition of either hydrochloric acid or sodium hydroxide. Reaction mixtures were prepared with the appropriate source water, buffer was added and pH was adjusted. The mixtures were then spiked with the test compounds, and chlorine stock solution was added. The reaction mixtures were typically 1 to 2 liters. Immediately after mixing, zero-time samples were taken by pouring from the larger vessel into a 50-ml serum vial containing an appropriate quantity of 0.1 N sodium thiosulfate (Fisher Scientific) to halt the reaction by removing chlorine. Samples for storage (extended reaction time) were taken in a similar manner without sodium thiosulfate. All vials were sealed headspace-free with Teflon*-faced septa immediately after filling as described in the NORS 80-city report (Symons et al., 1975). The sealed samples were stored at the indicated temperature in either a water bath or incubator controlled at ±0.5°C. At the appropriate time, the vials were opened, and aliquots were quickly transferred to a 30-ml vial containing sodium thiosulfate. The smaller vial (headspace-free) was then sealed as described above. All preserved samples were then stored under refrigeration until analysis.

RESULTS AND DISCUSSION

Precursor at pH 7

General

Trihalomethanes must result from a reaction or series of reactions of chlorine with a precursor material. Simple methyl

*Registered trademark of E. I. du Pont de Nemours and Company, Inc., Wilmington, Delaware.

ketones react through the classical haloform reaction mechanism. More complex substances, such as humic materials, also react by this mechanism or by some other mechanism that includes an oxidative cleavage step. Because control of trihalomethane production by precursor removal or control of precursor reaction rate was considered the best approach, some knowledge of precursor identity was required. Suggestions, as mentioned above, as to identity of precursor, varied from complex humic materials to simple methyl ketones or simple compounds with the acetyl moiety.

The earliest work at this laboratory with precursor removal was simply an experiment to determine whether GAC adsorption had any effect on precursor concentration. In this work, samples of water taken from the pilot plant were chlorinated at a dose of 8 mg/l—that used at that time by the Cincinnati Water Treatment Plant on the same raw water to satisfy chlorine demand and maintain a free residual in the distribution system. In this experiment, not only were settled and activated carbon filtered water samples chlorinated to determine the effect of the carbon, but dual-media filtered and raw water samples were also chlorinated at the same concentration for comparison. All four samples were buffered at pH 7. The results in Figure 2 show that when the result of chlorination of fresh GAC-filtered water was compared with the result of chlorinating the settled water, removal of precursor was indicated. The effectiveness of GAC filtration, however, was shown later to be relatively short-lived—a matter of only a few weeks under conditions of pilot plant operation (Love *et al.*, 1975). The other important aspect of this experiment was the observed dramatic change in rate of chloroform formation when the results of raw and settled water chlorination were compared. Conventional alum coagulation and sedimentation caused the removal of most of the precursor material from the raw water.

Particulates

The above experimental results indicated that precursors are one or more of the following: some sort of particulate; a substance associated with the particulates; a substance reacting in association with the particulates; or possibly a substance that could be complexed with the alum and precipitated with the floc. The nature of the role of the particulates was, therefore, further investigated. A simple vacuum filtration of raw water through filter paper (Whatman No. 1) was carried out. The filtrate, particulates with filter resuspended in GAC-filtered water, original raw water, and GAC filter effluent with and without filter paper were each chlorinated and subsequently analyzed for trihalomethane content after varying periods of storage.

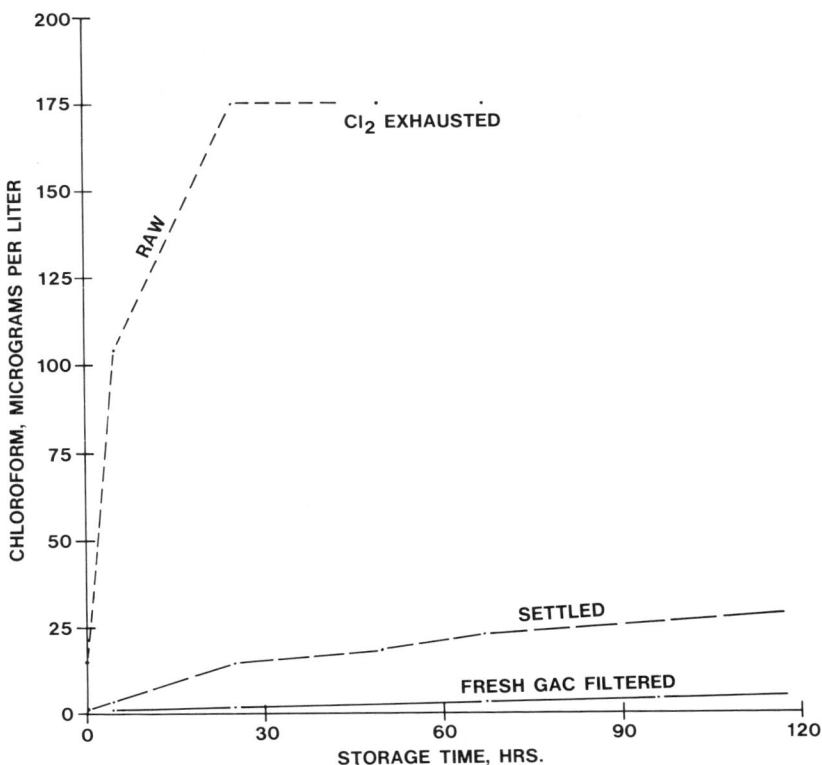

Figure 2. Effect of treatment on chloroform production. Conditions: chlorine dose, 8 mg/l; 25°C; pH 7.

Comparison of the reaction rate curves for raw and filtered raw water shown in Figure 3 illustrates a reduction of the rate of trihalomethane production caused by removal of particulates. The rate curve for GAC filter effluent with resuspended filter paper and particulates from the raw water indicates that essentially all of the difference between the raw and filtered raw rate curves can be accounted for by the substances trapped on the resuspended filter paper. The curves for GAC filter effluent and GAC filter effluent plus filter paper are simply the appropriate controls and are nearly identical. They indicate essentially no reaction interference or enhancement by the filter paper itself. According to these results, simple filtration either removed some trihalomethane precursor from the raw water or the removal of some of the particulate matter reduced the

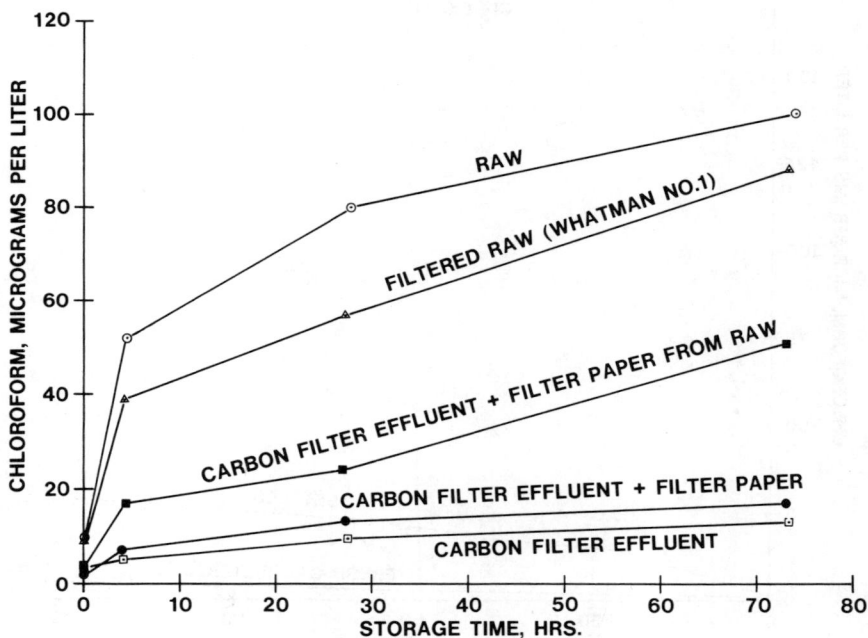

Figure 3. Effect of simple filtration on trihalomethane production. Conditions: chlorine dose, 10 mg/l; pH 7; 25°C.

reaction rate of dissolved precursor. The particulate matter, therefore, played some direct role in trihalomethane production when Ohio River water was chlorinated.

To determine which of these mechanisms was important, the effect of potentially active surfaces was investigated by spiking two sets of GAC-filtered water samples with simple acetyl derivatives and then suspending Bentonite clay in one set and powdered activated carbon in the other set. Neither of the two added particulates caused any detectable increase in rate of trihalomethane formation. Therefore, active surface effects were not considered significant but particulate matter or substances strongly sorbed on the particulate matter were found to be important precursors to trihalomethane production at pH 7.

Humic Acid

Because humic substances are more likely to be found in natural waters as small particulates or sorbed on clay particles (Schnitzer and Khan, 1972) than are soluble simple methyl ketones, a direct test of Rook's (1974) hypothesis was attempted using commercially available humic acid, both suspended at pH 7 and dissolved at a higher pH but later readjusted to pH 7. At concentrations of humic acid representing a NVTOC concentration similar to that found for Ohio River water (approximately 3 mg/l of NVTOC), the rate curve for formation of trihalomethanes was observed to be very similar to that seen for chlorination of the natural water (Figure 4). In addition, a filtration experiment (0.2-μm pore filter) similar to that carried out on the raw water described above was conducted on suspensions and solutions of humic acid. The results (Figure 5) observed were similar to those reported for the raw water filtration experiment (Figure 3). Thus, in terms of rate of trihalomethane formation on chlorination, the physical and chemical characteristics of humic acid in suspension and solution at these concentrations were found to be similar at pH 7 to those of the unknown precursor substances present in the Ohio River.

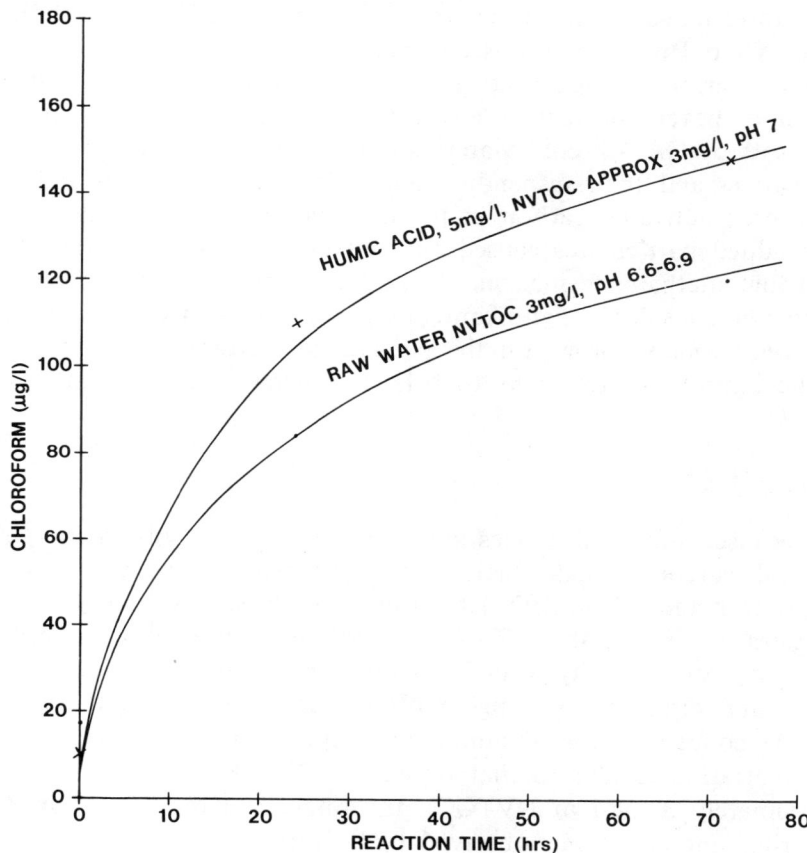

Figure 4. Comparison of humic acid, raw water reaction rates at similar NVTOC concentrations. Chlorine dose, 10 mg/l.

Finally, attempts to react chlorine at pH 7 with simple acetyl compounds (acetone, acetaldehyde and acetophenone), when these compounds were spiked at 5 μmol/1 into raw and GAC-filtered water, failed to produce trihalomethanes at rates significantly higher than those observed for the blank samples (Figure 6). Therefore, for chlorination of natural waters at pH values near 7, the humic acid precursor hypothesis of Rook seemed the most valid.

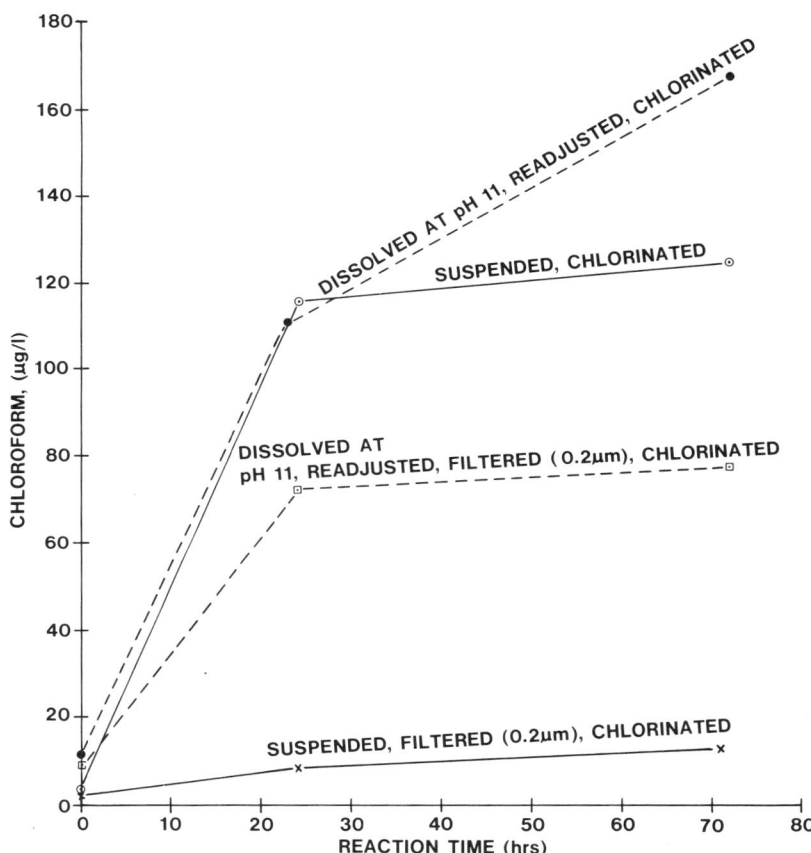

Figure 5. Chloroform production from filtered and unfiltered humic acid mixtures, 5 mg/l. Conditions: chlorine dose, 10 mg/l; pH 7.

Effect of pH on Reaction Rate and Precursor Identity

General

Because the rate-determining step of the classical haloform reaction is enolization of a ketone, the rate of trihalomethane formation is pH-dependent. For example, the reaction of acetone with hypochlorite to form chloroform proceeds at a

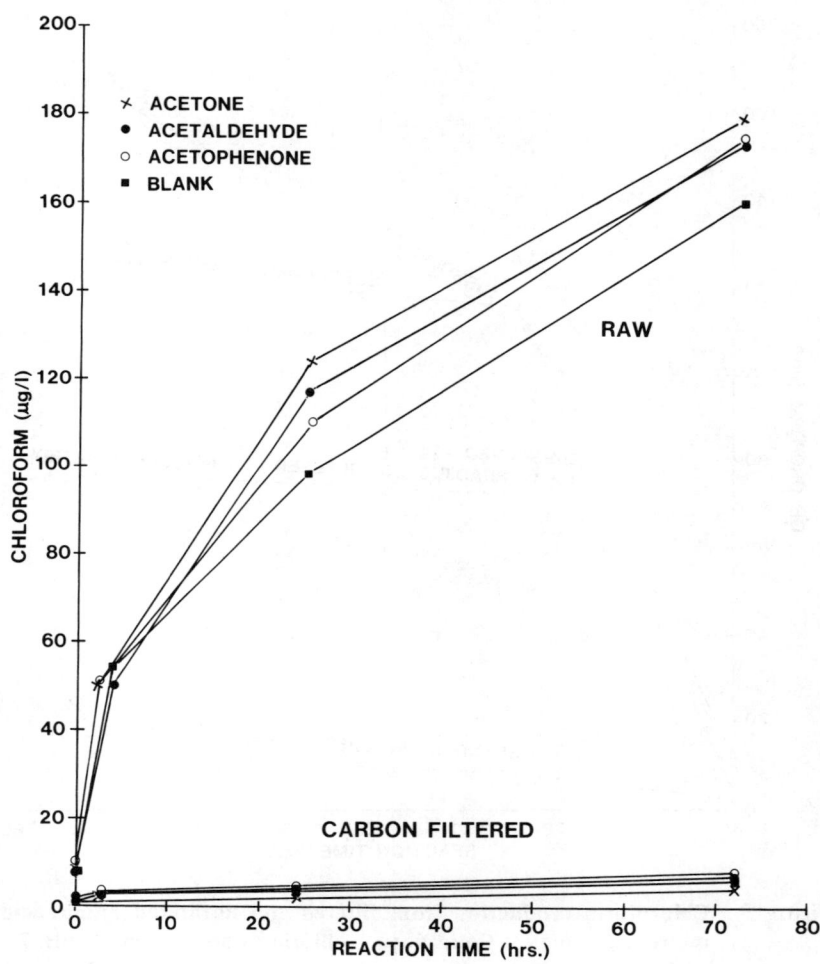

Figure 6. Chloroform production from raw and carbon-filtered water spiked at 5 μM with low-molecular-weight acetyl compounds. Conditions: chlorine dose, 10 mg/l; pH 7.

faster rate at pH 11.5 than at pH 6.5. Experimentally, a sample of settled water was buffered at pH 6.5 and another at pH 11.5 and both were chlorinated at an initial concentration of 10 mg/l. The results (Figure 7) illustrate that the rate of

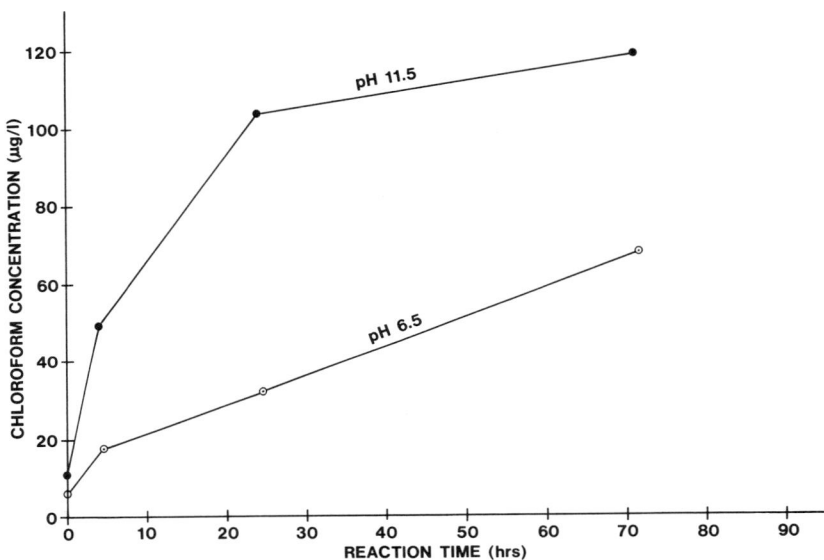

Figure 7. Effect of pH on chloroform production from settled water. Conditions: chlorine dose, 10 mg/l; 25°C.

formation of chloroform increases with an increase in pH. This could be explained simply by an increase in the humic acid reaction rate, as would be expected by the classical mechanism. Another possibility, however, is that other compounds such as acetone in the source water (settled) that do not react readily at pH 6.5 become significant contributors to the overall reaction rate (chloroform formation) at pH 11.5. An indication of the latter possibility was previously noted in the work of Fairless et al. (1975) in which acetone was shown to react at a significant rate at pH 9.5, but not at a pH near 7. Because chlorination is carried out at high pH in some water supplies, especially where lime softening or excess lime softening is practiced, further investigation of the effect of pH was necessary.

Humic Acid

Figure 8 illustrates the reaction rate curves for formation of total trihalomethanes (TTHM) from three concentrations of

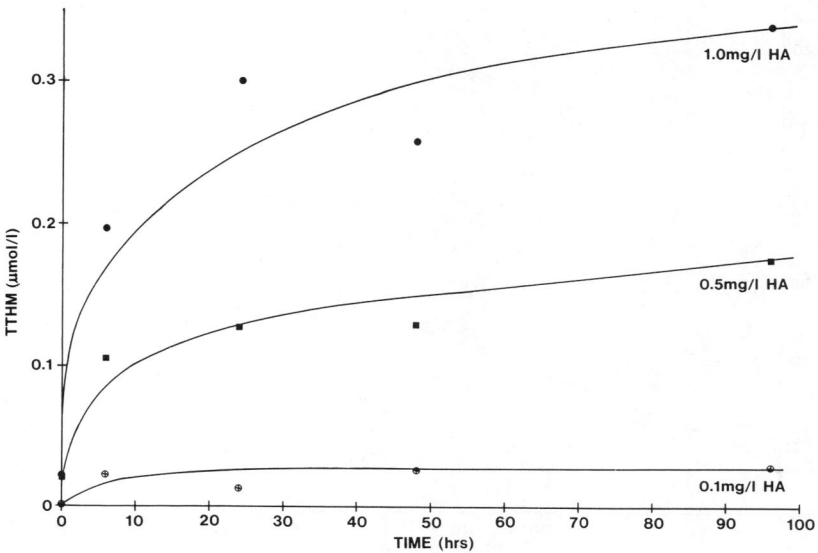

Figure 8. Effect of humic acid concentration on trihalomethane production. Conditions: chlorine dose, 10 mg/l; pH 6.7; 25°C.

humic acid (0.1, 0.5 and 1.0 mg/l) spiked in GAC-filtered water in the presence of excess chlorine (10 mg/l with less than 10% change during the course of the experiment). An apparent first-order rate-dependence on initial humic acid concentration is graphically demonstrated; that is, at any given time between any two curves, the ratios of concentration of TTHM produced are equal to the respective ratios of initial humic acid concentrations. The change in rates with apparent exhaustion of reaction sites can also be seen as nearly constant TTHM concentrations are approached.

In Figure 9 the pH-dependency of reaction rate at one of these concentrations (1 mg/l) is illustrated. The same curve characteristics were observed at all pH values. As noted above, one can assume from the shape of the curves that the reaction was nearly complete at pH 6.7 or was proceeding very slowly relative to the initial rate. Because the reaction is essentially complete at pH 6.7 at the end of the experiment, the nearly

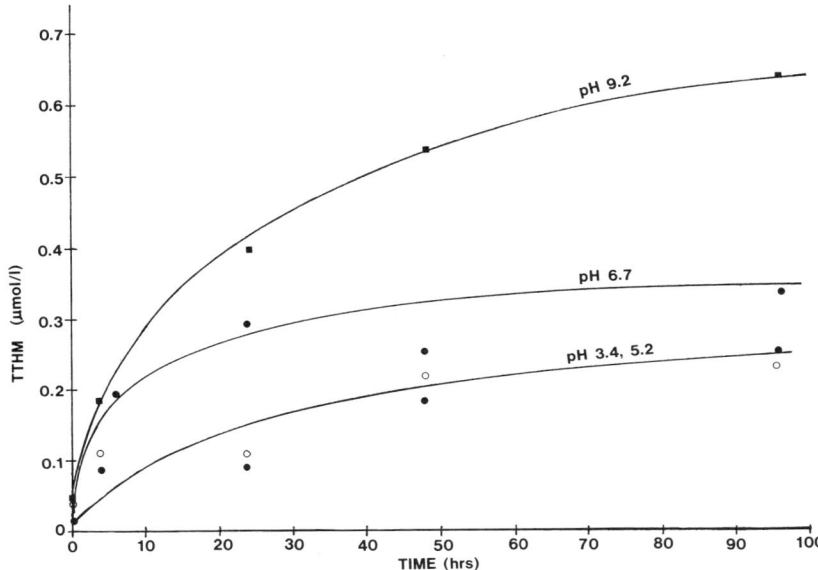

Figure 9. Effect of pH on trihalomethane production from humic acid, 1 mg/l. Conditions: chlorine dose, 10 mg/l; 25°C.

twofold increase in final product concentration at pH 9.2 can only be explained by the presence of certain reactive sites on the complex humic acid molecule that react at insignificant rates at the lower pH but are reactive at higher pH. The concentration of significant reactive sites in the reaction mixture, when expressed as equivalents per liter, is therefore at least twice as high at the higher pH. Based on this analysis, and considering humic acid to be 60% carbon, 0.7% and 1.4% of the carbon present reacts ultimately to become trihalomethane at the low and high pH values, respectively.

Acetone

Reactions of acetone with chlorine can be compared quantitatively with those of humic acid in an evaluation of the potential role of acetone as a precursor because the similarity of the humic acid reaction to that of the natural material in the source water has already been demonstrated (see Figure 4).

Figure 10 shows the pH-dependency of the rate of reaction of 1 mg/l acetone. At pH 6.7 the TTHM concentration from acetone after 96 hr is about one-third of that observed from 1 mg/l humic acid in the same 96-hr period (see middle curve, Figure 9). These numbers might seem to indicate that acetone could be a significant precursor at pH 6.7. Because the rate of trihalomethane production from acetone through the classical haloform reaction mechanism is known to be proportional to acetone concentration, however, 3 mg/l of acetone would be required to give the same TTHM concentration at 96 hr as that from 1 mg/l of humic acid. Therefore, approximately 15 mg/l of acetone would be required to give the concentration of chloroform observed for the raw water (Figure 4). Thus, if acetone were the important precursor at pH 6.7, sufficient acetone would be required in solution to account for over 9 mg/l of NVTOC, which far exceeds the 2 to 3 mg/l NVTOC

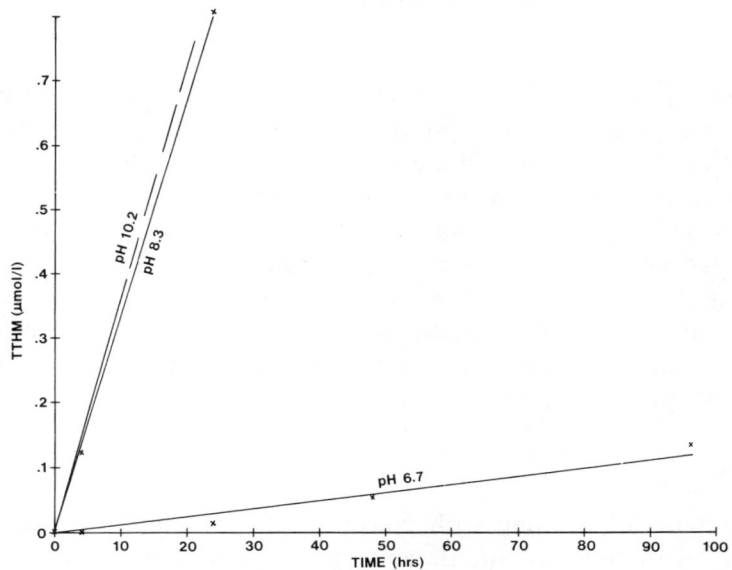

Figure 10. Effect of pH on trihalomethane production from acetone, 1 mg/l. Conditions: chlorine dose, 10 mg/l; 25°C.

usually found in the source water (acetone is not easily lost in the CO_2 stripping during NVTOC sample preparation).

Furthermore, the reaction rate curve for acetone at pH 6.7 is nearly a straight line, which indicates no change in rate during the experiment. By again using the assumption that acetone reacts by the classical haloform reaction mechanism and from the final trihalomethane concentration observed, less than 1% of the acetone initially present was calculated to have reacted. Because this change of acetone concentration was insignificant, its effect on reaction rate was not observed in this experiment. An insignificant change was expected, based on calculations using a reported rate expression for acetone in the haloform reaction (Manufacturing Chemists Association, 1972). Therefore, if acetone was the most important precursor and if its concentration was high enough to account for the observed rate of trihalomethane production from the source water, the characteristic rate curve would be linear as plotted. For these two reasons acetone is not likely to be a significant precursor at pH 6.7.

At pH values much higher than 6.7, however, the situation could be different. Figure 10 has been plotted on the same numerical scale as Figure 9, so that a direct comparison of reaction rates between acetone and humic acid at the various pH values is possible. A comparison of the curves on these figures, representing the trihalomethane formation rates at the higher pH values, reveals a much larger increase in reaction rate of acetone with changing pH than that observed with the same concentration of humic acid. The 30-fold observed increase (graphically measured) in acetone reaction rate was also expected from calculations based on the reported rate expression (Manufacturing Chemists Association, 1972). A rate increase of this magnitude could allow as little as 500 μg/l (*i.e.,* 15 mg/l divided by 30) of acetone to account for the trihalomethanes formed on raw water chlorination at pH 10.2. Therefore, low-molecular-weight compounds containing the acetyl moiety that have haloform reaction rates similar to that of acetone can become significant contributors to total trihalomethane production where chlorination is carried out at high pH. Thus, both possible explanations for the effect of pH on reaction rate noted in the discussion of Figure 7 are valid.

The question of precursor identity is, therefore, complicated because the "precursor" is actually a mixture of compounds with differing reactivities at varying pH values, solubilities and other physical and chemical characteristics. The relative contributions of the various constituents of a given water depend somewhat on the treatment practiced as well as on the source of the water. The probable diverse nature of precursor also may hamper efforts to find a single general organic parameter for unit process control that will predict effective removal of precursor.

Temperature

The effect of temperature on the rate of reaction of precursors present in Ohio River water was investigated to assess the potential effect of wide seasonal temperature variations in raw and treated waters. The winter-to-summer water temperature variation in the raw and finished water at Cincinnati, Ohio, is approximately 26°C (from less than 2°C to greater than 28°C). The results presented in Figure 11 show that this temperature differential could easily account for most of the winter-to-summer variation in chloroform concentration (less than 30 μg/l to greater than 200 μg/l) observed in Cincinnati tap water over the past year when raw water chlorination with a 3- to 4-day chlorine contact time was practiced. Some other factors, such as seasonal variation in precursor concentration, certainly have some additional effect, however.

Disinfectant

Work is progressing with measurement of the effects on trihalomethane production on the use of oxidants other than chlorine as disinfectants (*e.g.*, O_3, ClO_2). When completed, the results of these experiments will be the subjects of future reports. The work reported herein was confined to a study of the effect of chlorination practice, given the presently recognized need for maintenance of a chlorine residual in the distribution system. Chlorination in the presence of added ammonia is practiced in some locations in an attempt to maintain residuals

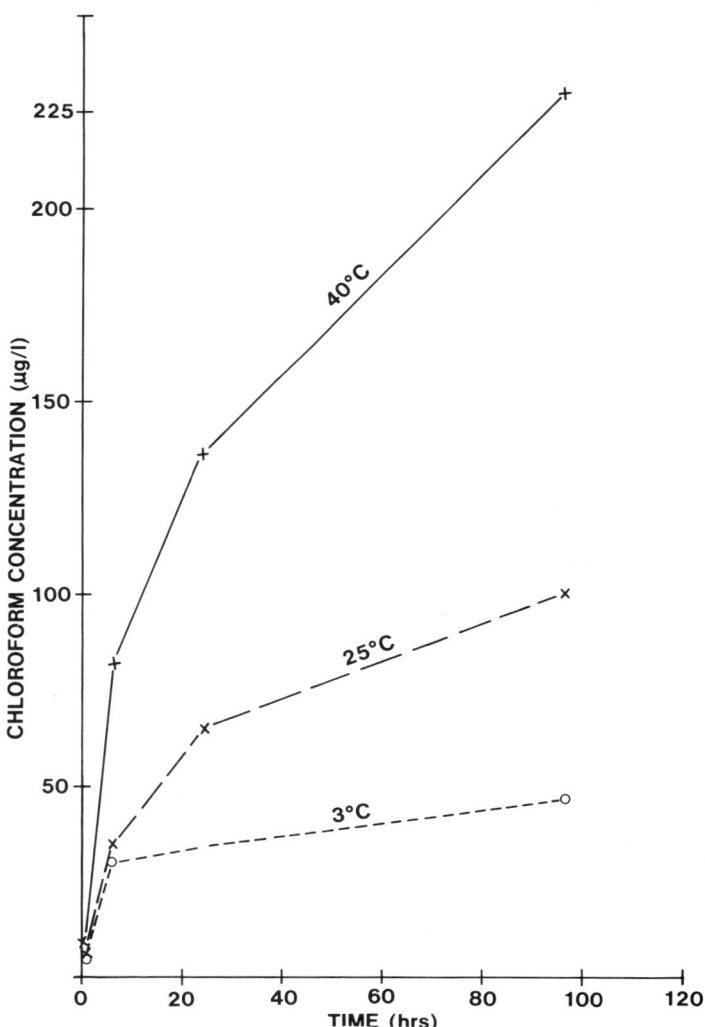

Figure 11. Effect of temperature on chloroform production from raw water. Conditions: chlorine dose, 10 mg/l; pH 7.

(as chloramine) for extended periods of time. Figure 12 illustrates the result of an attempt to form trihalomethanes with chlorine added in the presence of added ammonia. Chlorine was added at 5.5 mg/l (measured) to raw water and to raw

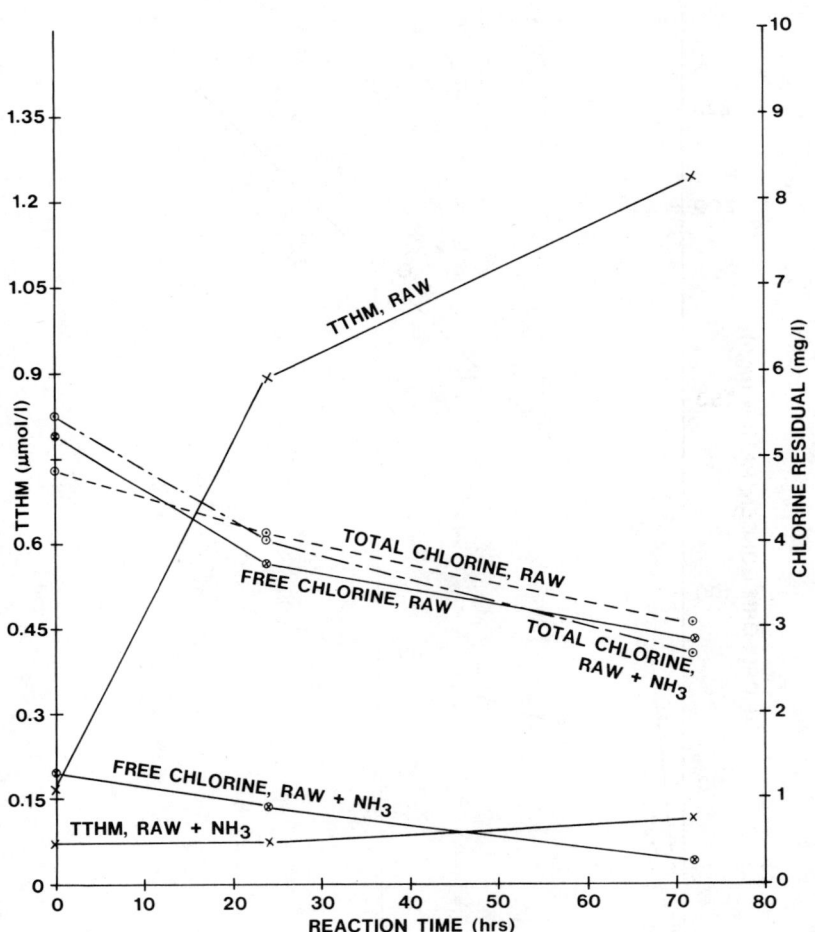

Figure 12. Effect of free vs combined chlorine on TTHM production at pH 7.

water spiked with 20 mg/l NH_4Cl (ammonia-nitrogen, 5.2 mg/l). The results of the measurements for trihalomethane production and free- and combined- (mostly NH_2Cl) chlorine residuals in Figure 12 show that when combined chlorination was practiced, trihalomethane production was minimized. Therefore, during chlorination of water where ammonia breakpoint is not achieved, trihalomethane production may not be a problem. At this time, however, ammoniation is not recommended as a technique to avoid trihalomethane formation because of the relatively poor disinfecting power of chloramines when compared with that of free chlorine.

Full-Scale Plant Operation

The reduction of ultimate trihalomethane concentration in finished drinking water is the primary goal of ongoing field research at a number of water treatment plants in the United States. Preliminary results of this work indicate that the conclusions drawn above with regard to the role of coagulation and settling in reducing precursor concentration are valid, although analyses of data are not yet completed. The details of this field work will be the subject of future papers.

SUMMARY AND CONCLUSIONS

The precursor to trihalomethane production during the chlorination process in drinking water treatment is probably a complex mixture of humic substances and simple low-molecular-weight compounds containing the acetyl moiety. The relative importance and contribution to trihalomethane production of each of the specific precursor compounds are pH-dependent. Where chlorination following clarification is carried out at pH values near 7, effective coagulation and sedimentation may be sufficient to reduce the precursor concentration to levels where ultimate trihalomethane concentrations are below the yet undefined adverse health effect levels. Where chlorination is carried out at high pH (as in a lime- or excess lime softening plant), treatment for precursor removal is more complicated. In these cases, removal of relatively water-soluble low-molecular-weight compounds (concentrations of which would not be expected to be significantly affected by coagulation and settling processes) is also necessary before chlorination. Thus, the point of chlorination in the treatment process, being a significant factor in trihalomethane production, probably represents the most important variable to be considered for change in attempts to reduce ultimate trihalomethane concentrations in finished drinking water.

To date, GAC has been used with only limited success to remove precursor compounds. Because its effectiveness is limited to only a few weeks after being placed in filters, its use would require frequent activation or replacement cycles.

Work is continuing in an effort to determine ways to reduce the extent of trihalomethane reaction through precursor removal or control of reaction rates. The final evaluation of the success of this work must, however, await more precise health effect information regarding the significance of the presence of trihalomethanes in drinking water.

ACKNOWLEDGMENTS

We acknowledge the assistance of the Research Sanitary Engineers, O. T. Love and J. K. Carswell and accompanying staff, who were responsible for pilot plant aspects of this work; B. L. Smith, Physical Science Technician, for NVTOC analyses and some chlorine residual measurements; J. M. Symons and J. K. Carswell for review of the manuscript, and Mrs. M. Lilly for its preparation.

REFERENCES

American Public Health Association (APHA). 1971. *Standard Methods for the Examination of Water and Wastewater*, 13th ed., New York.

Bellar, T. A. and J. J. Lichtenberg. 1974. "The Determination of Volatile Organic Compounds at the μg/l Level in Water by Gas Chromatography." EPA-670/4-74-009. U. S. Environmental Protection Agency, National Environmental Research Center, Cincinnati, Ohio.

Bellar, T. A. and J. J. Lichtenberg. 1974. "Determining Volatile Organics at the μg/l Level in Water by Gas Chromatography." *J. Am. Water Works Assoc.* 66:739.

Bellar, T. A., J. J. Lichtenberg and R. C. Kroner. 1974. "The Occurrence of Organohalides in Chlorinated Drinking Water." *J. Am. Water Works Assoc.* 66:703.

Bunn, W. W., B. B. Haas, E. R. Deane and R. D. Kleopfer. 1975. "Formation of Trihalomethanes by Chlorination of Surface Water." *Environ. Lett.* 10:205.

Fairless, B. 1975. U.S. Environmental Protection Agency, Region V, Central Regional Laboratory, Chicago, Illinois. Personal Communication.

Glaze, W. H. 1975. North Texas State University, Denton, Texas. Personal Communication.

Love, O. T., Jr., J. K. Carswell, A. A. Stevens and J. M. Symons. 1975. "Treatment of Drinking Water for Prevention and Removal of Halogenated Organic Compounds (An EPA Progress Report)." Presented at the 95th Annual Conference of the American Water Works Association, June 8-13, Minneapolis, Minnesota.

Manufacturing Chemists Association. 1972. "The Effect of Chlorination on Selected Organic Chemicals." Project 12020 EXG 03/72, U.S. Environmental Protection Agency, Washington, D. C.

Rook, J. J. 1974. "Formation of Haloforms During Chlorination of Natural Waters." *Water Treat. Exam.* 23(2):234.

Rook, J. J. 1975. "Formation and Occurrence of Chlorinated Organics in Drinking Water." Presented at the 95th Annual Conference of the American Water Works Association, June 8-13, Minneapolis, Minnesota.

Schnitzer, M. and S. U. Khan. 1972. *Humic Substances in the Environment.* Marcel Dekker, Inc., New York.

Stevens, A. A. and J. M. Symons. 1975. "Analytical Considerations for Halogenated Organic Removal Studies," p. XXVI-1. In *Proc. Am. Water Works Assoc.* Water Quality Technology Conference, December 2-3, Dallas, Texas.

Symons, J. M., T. A. Bellar, J. K. Carswell, J. DeMarco, K. L. Kropp, G. G. Robeck, D. R. Seeger, C. J. Slocum, B. L. Smith and A. A. Stevens. 1975. National Organics Reconnaissance Survey for Halogenated Organics in Drinking Water. Water Supply Research Laboratory and Methods Development and Quality Assurance Laboratory, National Environmental Research Center, U.S. Environmental Protection Agency, Cincinnati, Ohio. *J. Am. Water Works Assoc.* 67:634.

U.S. Environmental Protection Agency (EPA). 1974. Lower Mississippi River Facility, New Orleans Area Water Supply Study (Draft Analytical Report), Slidell, Louisiana.

DISCUSSION

John R. J. Sorenson, Quad Corporation. I am still concerned with the purity of chlorine used in water purification. What grade of chlorine did you use in your studies? Isn't that a purer grade of chlorine than used in waste purification? Finally, would you comment on the possibility that the concentration of chlorinated organic compounds in drinking water might be reduced by using purer chlorine gas, without substantial amounts of CH_2Cl_2, $CHCl_3$, CCl_4, C_2Cl_6, chlorinated aromatic hydrocarbons, etc.?

Stevens. We are using a high-purity grade and, of course, we still get the trihalomethanes. They are produced on reaction of the chlorine with the precursor compounds. This is clear. It was one time suspected that in some commercial chlorine, carbon tetrachloride was a contaminant. If this were the case you should see not an increase with time of carbon tetrachloride, but as soon as you chlorinate, you should see an immediate increase that stays constant. That is the same with any other contaminant compound. We've never observed this; I'm not sure we've ever really carefully looked for it. In the Nationwide Organics Reconnaissance Survey, we did look for it with the two compounds that weren't formed in chlorination that I mentioned, carbon tetrachloride being one. We don't feel we have strong evidence or any significant results that would show that any finished waters contain higher carbon tetrachloride levels than raw waters. We see variations in the data that might look, if you don't think about them in the context in which the numbers were taken, like carbon tetrachloride was higher in some cities. But those numbers are all very small values and usually approaching or even below the detection limit on the day the raw water was run. So we have to be careful about interpretation of that. In any case, we saw no significant increases of carbon tetrachloride. Now, our study was for 80 different cities. I'm not saying what you say is impossible. We just haven't seen it.

Joseph J. Delfino, University of Wisconsin. After the EPA survey last fall, there was a lot of talk about what these 300 mg/l meant and then a lot of discussion came up about cough syrup containing 1% chloroform. What is the significance of 1% $CHCl_3$ in cough syrup compared with 100 ppb in drinking water in terms of health? Does the EPA have any toxicological data as to the significance of, say, somebody drinking a certain amount of chloroform that way in terms of relating it to drinking water consumption?

Stevens. Well, as I said in my last statement, we don't really know the health significance of the chloroform levels we have observed in the water. We don't know what the health ramifications are of drinking that water. You can obviously calculate what you take from cough syrup, and what you get in drinking water at so many hundred micrograms per liter. The chronic versus the acute problem hasn't been evaluated. We don't have those answers. We're waiting for them from those who are working on them.

Keith Lawson Murphy, McMaster University. Were you working with a straight humic material compound or did it contain fulvic acid as well?

Stevens. We were working with what Aldrich Chemical Company markets as humic acid. It is relatively insoluble in water. I think that if it contained fulvic acids, as you call them, you would have seen a soluble portion and an insoluble portion. I think it's humic acid according to the classical definition, which is pretty much insoluble at pH 7. As I said, we went to alkaline solutions, say around 0.01 N sodium hydroxide, to dissolve it and then diluted and readjusted the pH. We got essentially the same curves for the reaction of that mixture as we did when we just suspended it by ultrasonification. However, filtration has a different effect. So, going to the base first rendered the molecule more soluble. Whether it was the rate of going into the solution or some kind of alkaline hydrolysis, we can't really say.

David Friedman, Food and Drug Administration. Has anyone compared the ratio of chlorinated haloforms to brominated haloforms to the chloride-bromide ratio in the water?

Stevens. I'm not surprised the question has come up. The problem is getting the data on bromide content of the water. The extremes we saw were extremely low concentrations of bromoforms, that is, none found in well over 100 mg/l of chloroform, versus the other extreme for the finished water at Brownsville, Texas, which had 12 mg/l chloroform, 116 of bromodichloro, 106 dibromochloro and 92 bromoform. A lot of the southwestern cities and South Texas cities, if you look in the report which will be available in the November issue of the *Journal of the American Waterworks Association*—a lot of those cities had high bromo-compound concentrations. We suspect their ground water sources are probably high in bromide. We suspect that's probably the reason. I think Bill Glaze is going to talk about some of this. We have also done work which gives presumptive evidence such as that at our Kansas City EPA Laboratory in which Missouri River water, which normally gives the usual thing that I was describing, was spiked with KI, KBr, KCl and KF and all ten possible haloforms were observed, except there were no fluorocompounds. But all 10 possible combinations of chlorine, bromine and iodine were observed. It is clear to us that is how it happens. We just don't have a good, fast, easy method for bromide determination in the presence of all that chloride.

Max Eisenberg, Maryland State Department of Health. In your comparison of the chloroform concentrations of the chlorinated raw waters versus the filtered waters, have you carried your chlorinated raw water through filtration (activated carbon) to determine the removal of chloroform?

Stevens. We have used granulated activated carbon filtration and it does work. We can run Cincinnati tap water through—I keep saying Cincinnati tap water only because that is where we are, O.K.? We remove the chloroform rather easily with fresh carbon for a few weeks. Then it will start to come through fairly quickly. So carbon used in that mode would have to be reactivated or replaced rather frequently. As for following chlorinated water through a plant, we can do that anyway and we have done it in many plants where they do chlorinate their raw water. Of course, you get a continual increase with reaction time.

6

CHLORINATION OF ORGANICS IN COOLING WATERS AND PROCESS EFFLUENTS

Robert L. Jolley, Guy Jones, W. Wilson Pitt and James E. Thompson

    Chemical Technology Division
    Oak Ridge National Laboratory
    Oak Ridge, Tennessee 37830

ABSTRACT

Many water-soluble chlorine-containing organic compounds of low volatility were found to be present in samples of chlorinated cooling waters from electric power-generating plants and chlorinated effluents from domestic sanitary sewage treatment plants. Both types of samples had been chlorinated to milligram-per-liter chlorine concentrations in the laboratory under conditions similar to those used for treatment of cooling waters and disinfection of sewage effluents. The chlorinated constituents were separated from concentrates of the water samples by high-pressure liquid chromatography. Chlorination yields (as Cl) of the chloro-organic compounds, determined by using the radioactive tracer ^{36}Cl, ranged from 0.5 to 3.1% of the chlorine dosage. The formation of chloro-organics and the reaction yields correlated with the chemical compositions of the water samples. Several chloro-organics were identified in the typical domestic sewage effluents and were quantified at the microgram-per-liter level. Comparison of the chromatograms of the chlorinated constituents in the cooling water samples with those of the sewage process effluent samples revealed a high degree of correspondence with respect to the elution positions of the separated constituents. A compilation of relevant data concerning organic constitutents in natural waters and sewage process effluent is presented. The chemical species subject to chlorination during the cooling water and sewage treatment are discussed.

INTRODUCTION

For several decades, chlorine has been the principal means for controlling water-borne diseases via disinfection of drinking waters and sewage effluents (Morris, 1971; White, 1972). More recently, with the rapid expansion of the electric power industry, chlorine has achieved major importance as an antifoulant for the cooling systems of electric power-generating plants (Draley, 1972; White, 1972). Concomitant with the increased understanding of the ecological problems associated with the ubiquitous chlorinated pesticides and polychlorinated biphenyls (Van Middelem, 1966; Risebrough et al., 1968; Veith and Lee, 1970) has been the increasing concern that chloro-organics may be formed in the environment through the chlorination of various waters and process effluents (Ingols and Jacobs, 1957; Dugan, 1972; Weber, 1972; Jolley, 1973). Evidence to confirm that chloro-organics are produced during the chlorination of sewage treatment plant effluents (Glaze et al., 1973; Jolley, 1973, 1974, 1975), cooling waters (Jolley, Gehrs and Pitt, 1975) and potable waters (Rook, 1974; Bellar, Lichtenberg and Kroner, 1974) has been presented only recently, because of the need to develop new methodologies for carrying out the required analyses.

In this paper we will review the available information concerning organic constituents in natural waters, including effluents from sewage treatment plants, and discuss selected aspects of aqueous chlorination. We will also summarize previously detailed chlorination studies with sewage effluents (Jolley, 1973, 1974, 1975) and cooling water (Jolley, Gehrs and Pitt, 1975) and present results from a recent study with a sample of cooling water. Finally, several general conclusions will be drawn concerning the total environmental impact of water chlorination and possible ecological ramifications.

SOLUBLE ORGANIC CONSTITUENTS

Assessment and prediction of chlorination effects on organic compounds present in waters of environmental interest require a knowledge of the identity, concentration and nature of such compounds. Several major efforts have been made to collect the known and published data relative to the soluble organic constituents in a variety of waters. Vallentyne (1957) comprehensively summarized the information available at that time concerning organic matter in natural waters, sewage and soil. Little (1970) prepared a report which listed the organic compounds present in fresh waters principally as a result of pollution. Hood (1970) and Faust and Hunter (1971) edited proceedings of conferences which made significant contributions to the understanding of the complex nature and activity of organic substances in the aquatic environment. The EUROCOP report (1973) compiled qualitative and quantitative data on naturally occurring organic compounds and industrial pollutants in process effluents, natural waters and tap water. Pitt, Jolley and Katz (1974) and, later, Pitt, Jolley and Scott (1975) presented quantitative information about the soluble organic constituents of low volatility in effluents from both primary and secondary stages of municipal sewage treatment plants. The report by the WHO International Reference Centre for Community Water Supply (1975) tabulated qualitative and quantitative data for 289 organic constituents that have been identified in wastewater, river water and drinking water. The U.S. Environmental Protection Agency (1975) has prepared an exhaustive compilation of the known organic constituents in water supplies. Symons *et al.* (1975) recently presented quantitative data about the volatile halogenated organic compounds present in the drinking water of 80 metropolitan areas. In a recent compilation, Junk and Stanley (1975) summarized data collected from the literature published through April 30, 1975, for potable and river waters but did not include organics in sewage effluents.

Of these source of information, only the conference proceedings (Hood, 1970; Faust and Hunter, 1971) deal in any significant way with humic substances. These materials are reported to comprise 50% of the soluble organic matter in sewage effluents (Rebhun and Manka, 1971; Manka *et al.*, 1974) and 90% of the soluble organic matter in surface waters (Junk and Stanley, 1975). Humic materials are a generic type of organic substances classified according to solubility. These complex polymers, which range in molecular weight from several hundred to many thousand, are composed of a variety of subunits such as aromatic and alicyclic moieties containing alcoholic, carbonyl, carboxylic and phenolic functional groups (Steelink, 1963; Christman and Minear, 1971; Schnitzer and Khan, 1972; Stevenson and Goh, 1972; Wershaw and Goldberg, 1972).

Voluminous tables of data are available in the cited references and will not be reproduced in this paper. Only relevant data for selected chemical species and groups will be considered during the discussion of aqueous organic chlorination reactions.

AQUEOUS CHLORINATION REACTIONS

The aqueous chemistry of chlorine and possible organic reactions in aqueous systems have been already treated in considerable depth by previous speakers at this conference. Therefore, we will only briefly summarize the chemistry germane for understanding chlorination effects on organic constituents in cooling waters and sewage effluents.

The reactive chlorine-containing species formed by the addition of chlorine to natural waters or process effluents that are at near-neutral pH values and also contain ammonia and organic nitrogen compounds are generally considered to be the following: hypochlorous acid (HOCl), hypochlorite ion (OCl$^-$), monochloramine (NH_2Cl), dichloramine ($NHCl_2$), organic monochloramine (RNHCl or R_2NCl, in which R represents any organic moiety) and organic dichloramines ($RNCl_2$). At the concentrations of chlorine normally used for the chlorination of cooling water, molecular chlorine (Cl_2) and the chloramine nitrogen trichloride (NCl_3) are not significant except at pH values less than 4. Equilibrium conditions are established very rapidly for HOCl, OCl$^-$ and NH_2Cl, but somewhat more slowly for $NHCl_2$. The equilibrium concentrations of these species are dependent on pH, temperature and initial chlorine and ammonia concentrations (Morris, 1967; Jolley, 1973). The nature and concentrations of the reactive chlorine-containing species are of critical importance in determining the formation and yield of chloro-organic compounds. HOCl is known to be an effective chlorinating agent for aromatic organic compounds in aqueous solution (Burttschell *et al.*, 1959; Lee and Morris, 1962). Morris (1967) estimated HOCl to be four orders of magnitude more effective as a chlorinating agent than NH_2Cl. Several authorities indicate that NH_2Cl is principally an aminating agent in aqueous solutions (Colton and Jones, 1955; Drago, 1957; Theilacker and Wegner, 1964; Kovacic, Lowery and Field, 1970) but no information concerning the effectiveness of $NHCl_2$ as a chlorinating agent in aqueous solution is available. The presence of organic chloramines is, of course, dependent on the presence of organic nitrogen compounds, such as amines and amides, in the water. Although amides form the corresponding chloramine slowly, amines are known to react rapidly; in turn, the chloramine product is a very effective chlorinating agent for phenolic compounds (Morris, 1967). Hence the presence of organic amines in some waters may be significant with respect to the formation of chloro-organic compounds. If HOCl is the principal chlorinating agent, then it follows that the formation and yield of chlorine-containing organics in the cooling waters and process effluents should be proportional to the HOCl concentration. Conversely, all other things being equal, a smaller quantity of chlorinated organics would be anticipated in solutions with high ammonia concentrations (Jolley, Gehrs and Pitt, 1975) because of the extremely rapid reaction of HOCl with ammonia to form NH_2Cl (Weil and Morris, 1949).

Assuming that HOCl is the major chlorinating agent, the possible chemical reactions with organic constituents in aqueous solution may be grouped into several general types according to Jolley (1973), namely:

       1. Oxidation
       2. Substitution
          a. Formation of N-chlorinated compounds
          b. Formation of C-chlorinated compounds
       3. Addition

Oxidation may be the predominant type of reaction occurring between HOCl and organic compounds in natural and process waters (Jolley, 1973), although some conflicting evidence has been reported (Zaloum and Murphy, 1974; Murphy, Zaloum and Fulford, 1975). Many organic compounds are subject to oxidative reactions (Holst, 1954). The carbohydrates and carbohydrate-related compounds are examples of organic constituents in natural waters and process effluents which are probably subject only to oxidative reactions in aqueous chlorine solutions (and not to substitution and addition reactions). These constituents are present in parts-per-billion (ppb) concentrations (Vallentyne, 1957; Pitt, Jolley and Scott, 1975). Table I lists the identities and concentrations of carbohydrate-related constituents determined in a typical effluent from the primary stage of a municipal sewage treatment plant (Jolley, Pitt et al., 1975). Although these types of compounds probably contribute to the chlorine demand of cooling waters and process effluents, they would not result in the formation of environmentally significant chloro-organic compounds.

The chemistry of the formation of N-chlorinated compounds has been detailed by Morris (1967). Various organic nitrogen-containing compounds have been found to be present in natural waters (Vallentyne, 1957) and sewage effluents (Jolley, Katz et al., 1975). Table II summarizes the identities and concentrations of nitrogen-containing organic compounds determined in either primary or secondary sewage effluents (Pitt, Jolley and Scott, 1975). Many of these compounds would be subject to the generic N-chlorination reactions shown in Figure 1 (Morris, 1967). Furthermore, proteinaceous material such as that in bacterial cell walls, for example, would probably be subject to chlorine substitution reactions, which lead to the formation of N-chlorinated proteinaceous material. However, the kinetics for the formation of N-chloro compounds from amides is considerably slower than the formation of such compounds from amines (Morris, 1967).

The substitution of chlorine into organic compounds and the resulting formation of C-chlorinated compounds have been summarized by Jolley (1973), Carlson et al. (1975) and Morris (1975). The two principal reaction types of interest are substitution into aromatic or heterocyclic compounds and the

Table I. Carbohydrate and Carbohydrate-Related Organic Constituents in Primary Domestic Sewage Effluent

Constituent	Identification Method[a]	Concentration[b] (ppb)
Carbohydrates		
Galactose	AC, GC	-
Glucose	AC, GC	-
Maltose	AC, GC	0.5
Polyols		
Erythritol	AC, GC, MS	5
Ethylene glycol	AC, GC, MS	3
Galactitol	AC, GC, MS	2
Glycerin	AC, GC, MS	15-19
Aliphatic organic acids		
3-Deoxyarabinohexonic acid	MS	7
3-Deoxyerythropentonic acid	MS	4
2-Deoxyglyceric acid	MS	7
2,5-Dideoxypentonic acid	MS	6
3,4-Dideoxypentonic acid	MS	13
2-Deoxytetronic acid	MS	6
4-Deoxytetronic acid	MS	6
Glyceric acid	MS	5
4-Hydroxybutyric acid	GC, MS	6
2-Hydroxyisobutyric acid	GC, MS	4
Oxalic acid	AC, GC, MS	2
Quinic acid	MS	50
Ribonic acid	MS	4
Succinic acid	AC, GC, MS	24

[a]AC—anion exchange chromatography; GC—gas chromatography on two columns; MS—mass spectrometry.
[b]Based on flame ionization detector response during gas chromatography.

haloform reaction. As previously mentioned, humic substances comprise the major portion of the soluble organic matter in cooling waters and sewage effluents. These complex molecular substances contain aromatic moieties as indicated by such degradation products as the following: benzoic acid, catechol, 3,4-dihydroxybenzoic acid, 4-hydroxybenzoic acid, 2-methylphenol, 4-methylphenol, resorcinol, syringic acid and vanillin (Christman and Minear, 1971). Aromatic compounds such as phenols and aromatic acids are readily chlorinated in aqueous

Table II. Organic Nitrogen-Containing Constituents in Sewage Effluents

Constituent	Identification Methods[a]	Concentration[b] (ppb)
Amides		
Urea	AC, GC, MS	16-43
Amino acids		
Phenylalanine	AC, GC, MS	50-90[c]
Tyrosine	AC, GC, MS	34
Indoles		
3-Hydroxyindole	MS	2
Indican	AC, GC, F	1, 2[c]
Indole-3-acetic acid	MS	13
Pyridine derivatives		
N^1-Methyl-2-pyriodone-5-carboxamide	AC, CC, UV, GC	20[c], 25
N^1-Methyl-4-pyriodone-3-carboxamide	AC, UV, GC	10[c], 14
Purine derivatives		
Adenosine	AC, CC, UV, GC, MS	13
Caffeine	AC, CC, GC, MS	10[c], 29-46
1,7-Dimethylxanthine	AC, CC	-
Guanosine	AC, CC, UV, GC, MS	4-28, 50[c]
Hypoxanthine	AC, GC, MS	12-42, 25[c]
Inosine	AC, CC, UV, GC, MS	11-23, 50[c]
1-Methylinosine	AC, CC, UV	80[c]
1-Methylxanthine	AC, CC, UV	70[c]
3-Methylxanthine	AC, CC	-
7-Methylaxanthine	AC, CC, GC	2, 90[c]
Theobromine	AC, CC	-
Uric acid	AC, GC, MS	20[c]
Xanthine	AC, CC, UV, GC, MS	2-7, 70[c]
Pyrimidine derivatives		
5-Acetylamino-6-amino-3-methyluracil	AC, CC, UV, GC	140[c]
Orotic acid	AC, UV, GC, MS	2, 5[c]
Thymine	AC, CC, GC, MS	7[c], 9-28
Uracil	AC, CC, UV, GC, MS	16-58, 40[c]

[a]AC—anion exchange chromatography; CC—cation exchange chromatography; UV—ultraviolet spectroscopy; GC—gas chromatography on two columns; MS—mass spectrometry; F—fluorometry.

[b]Based on flame ionization detector response during gas chromatography, unless otherwise designated.

[c]Based on UV absorbance of anion exchange chromatographic peak.

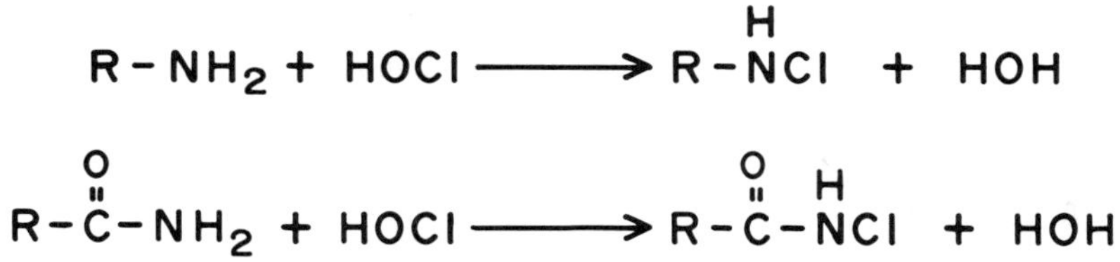

$$R-NH_2 + HOCl \longrightarrow R-NCl\overset{H}{} + HOH$$

$$R-\overset{O}{\underset{\|}{C}}-NH_2 + HOCl \longrightarrow R-\overset{O}{\underset{\|}{C}}-NCl\overset{H}{} + HOH$$

Figure 1. Chlorine substitution reactions with organic nitrogen-containing compounds.

solution by HOCl or OCl⁻. Thus, if humic substances do contain such aromatic moieties, they chould chlorinate readily at activated sites. Furthermore, free phenols and aromatic acids have been detected in natural waters and sewage effluents (Vallentyne, 1957; Pitt, Jolley and Scott, 1975). Data concerning the identities and concentrations of aromatic organic compounds found to be present in sewage effluents are presented in Table III. Chlorination of such compounds with HOCl should readily produce their chlorinated analogs. For example, the reactions of phenol and resorcinol are presented in Figure 2 (Chandelon, 1883; Lee and Morris, 1962), and the substitution reactions of benzoic, salicylic and phthalic acids are given in Figure 3 (Hopkins and Chisholm, 1946; Goodrich, 1949).

Some organic nitrogenous compounds may also undergo chlorine substitution reactions (Jolley, 1973; Morris, 1975). For example, the pyrimidines cytosine and uracil (Table II) react with aqueous HOCl to form the 5-chloro analog (Patton *et al.*, 1972; Jolley, Pitt and Thompson, 1975) as shown in Figure 4, in addition to forming more complex degradation products (Ramage and Landquist, 1959). Purines (Table II) may react with aqueous HOCl to form the chlorinated analog, as shown for xanthine and theobromine in Figure 5. Actually, the formation of the chloro-compound in the case of these two purines has not been corroborated in dilute HOCl solutions; however, the reactions have been studied in nonpolar solvents and in acetic acid (Howard, 1969). The chlorination of purine and pyrimidine bases in nucleic acid material of bacteria, plankton and decomposing plant and animal matter resulting in

Table III. Aromatic Organic Acids and Phenolic Compounds in Sewage Effluents

Constituent	Identification Method[a]	Concentration[b] (ppb)
Aromatic organic acids		
Benzoic acid	AC, GC, MS	3
2-Hydroxybenzoic acid	AC, GC, MS	2, 7[c]
3-Hydroxybenzoic acid	AC, GC, MS	7. 40[c]
4-Hydroxybenzoic acid	AC, GC	1
4-Hydroxyphenylacetic acid	AC, UV, GC, MS	16-52, 190[c]
3-Hydroxyphenylhydracrylic acid	AC, UV, GC, MS	10-22
3-Hydroxyphenylpropionic acid	AC, GC, MS	6-20[c]
Phenylacetic acid	AC, GC	10[c]
o-Phthalic acid	AC, UV, MS	200[c]
Phenolic compounds		
Catechol	MS	1
p-Cresol	AC, GC, MS	20[c], 29
Phenol	AC, GC, MS	6[c], 12

[a]AC—anion exchange chromatography; GC—gas chromatography; UV—ultraviolet spectroscopy; MS—mass spectrometry.
[b]Based on Flame Ionization Dector response during gas chromatography, unless otherwise designated.
[c]Based on UV absorbance of anion exchange chromatographic peak.

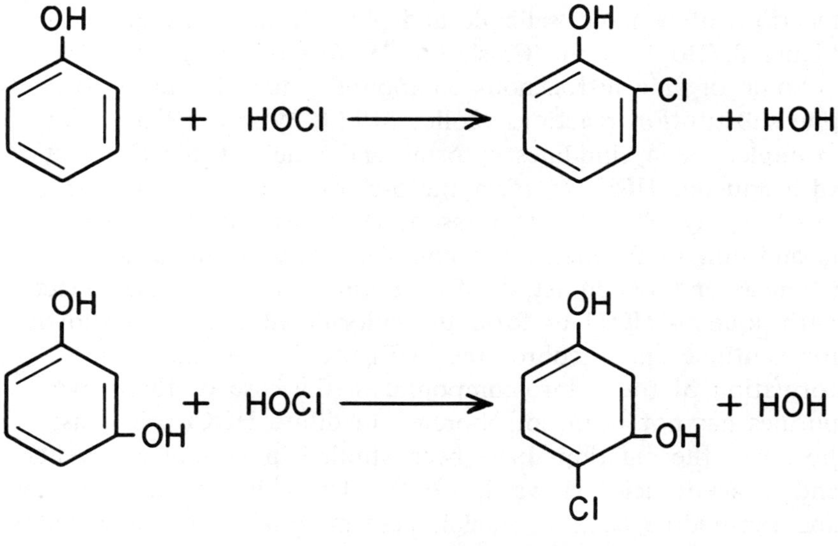

Figure 2. Chlorine substitution reactions with phenolic compounds.

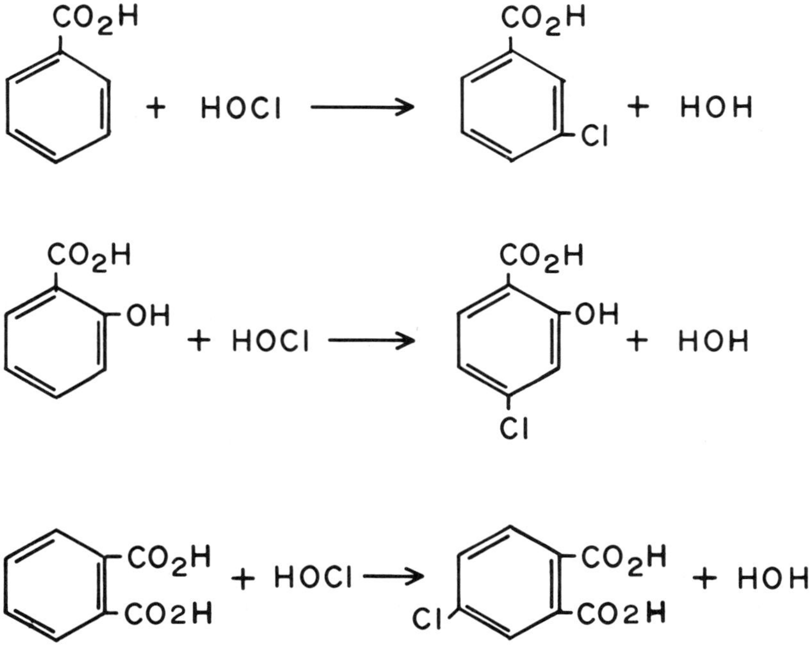

Figure 3. Chlorine substitution reactions with aromatic organic acids.

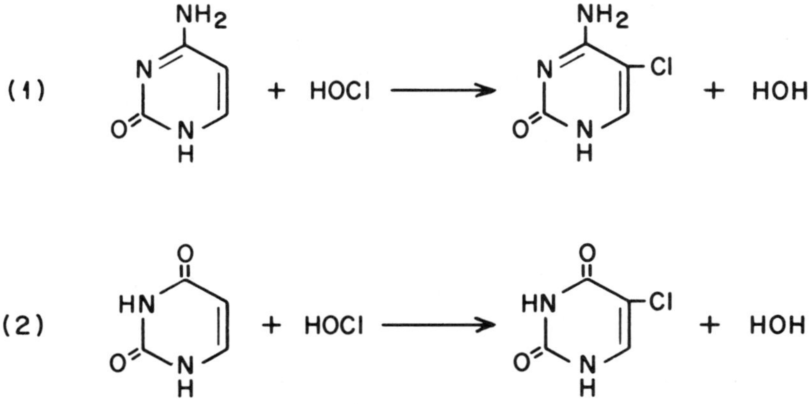

Figure 4. Chlorine substitution reactions with pyrimidines.

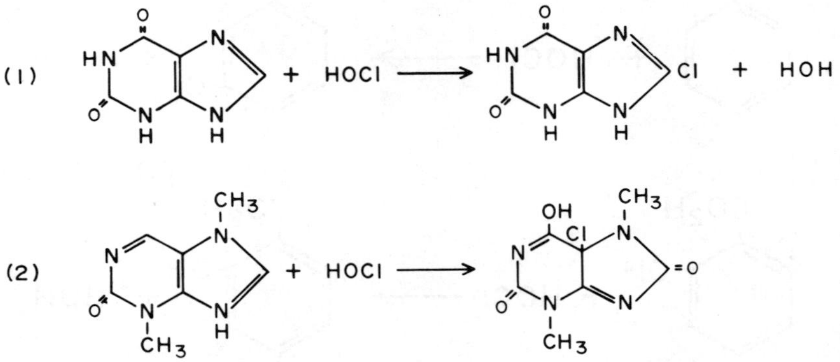

116 WATER CHLORINATION

Figure 5. Chlorine substitution reactions with purines.

chlorinated nucleic acid fragments or polymers may also occur. Prat, Nofre and Cier (1965) identified 5-chlorouracil and 5-chlorocytosine in acid hydrolysates of nucleic acids separated from bacteria after disinfection with parts-per-million (ppm) chlorine concentrations.

A major pathway for chlorine substitution may be the haloform reaction with humic materials, as postulated by Rook (1974). The haloform reaction (see Figure 6) has been discussed in detail by Morris (1975). Thus, a principal product of the chlorination of sewage effluents and cooling waters may be the volatile chloro-organics such as chloroform.

The last generic type of organic reaction, *i.e.*, addition, may occur with organic compounds containing reactive double bonds, for example, unsaturated fatty acids in sewage effluents and natural waters (Vallentyne, 1957; Pitt, Jolley and Scott, 1975). The reaction products are chlorohydrin and dichloro compounds (Houben and Weyl, 1962).

CHLORINATION OF SEWAGE EFFLUENTS AND COOLING WATERS

The detection and examination of chlorination effects on organic constituents that are present in natural waters and effluents of sewage treatment plants at ppb concentrations

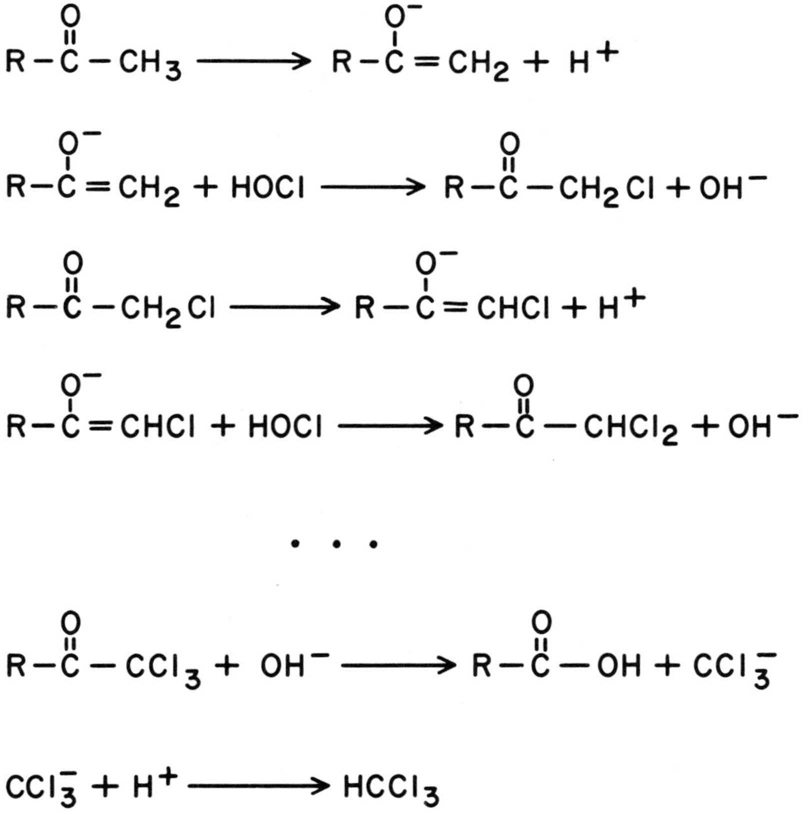

$$R-\overset{O}{\underset{\|}{C}}-CH_3 \longrightarrow R-\overset{O^-}{\underset{|}{C}}=CH_2 + H^+$$

$$R-\overset{O^-}{\underset{|}{C}}=CH_2 + HOCl \longrightarrow R-\overset{O}{\underset{\|}{C}}-CH_2Cl + OH^-$$

$$R-\overset{O}{\underset{\|}{C}}-CH_2Cl \longrightarrow R-\overset{O^-}{\underset{|}{C}}=CHCl + H^+$$

$$R-\overset{O^-}{\underset{|}{C}}=CHCl + HOCl \longrightarrow R-\overset{O}{\underset{\|}{C}}-CHCl_2 + OH^-$$

· · ·

$$R-\overset{O}{\underset{\|}{C}}-CCl_3 + OH^- \longrightarrow R-\overset{O}{\underset{\|}{C}}-OH + CCl_3^-$$

$$CCl_3^- + H^+ \longrightarrow HCCl_3$$

Figure 6. Haloform reaction.

represent a challenging and formidable task. Definitive information concerning these effects (Glaze et al., 1973; Jolley, 1973) has been obtained only recently as a result of the adaptation and development of new methodologies. In our laboratory we developed the following stepwise procedure (Jolley, 1973, 1974) to provide realistic, reproducible results for samples chlorinated under conditions simulating those used for disinfection of sewage effluents and/or antifoulant treatment of cooling waters of electric power-generating plants:

1. Chlorination of the water sample in the laboratory with ^{36}Cl-tagged chlorine gas or hypochlorite solution.
2. Concentration of the radioactive chlorinated solution via low-temperature vacuum distillation.
3. High-pressure, high-resolution anion exchange chromatographic separation (high-pressure liquid chromatography, HPLC) of the chlorination reaction products in the concentrate of the reaction mixture.
4. Detection and quantitative measurement of the ^{36}Cl-tagged chlorinated constituents using sensitive liquid scintillation counting.

Using this analytical technique we examined the chlorination effects on the organic constituents in both primary and secondary effluents from a municipal sewage treatment plant and two cooling water samples. The experimental results from the analyses of sewage treatment plant effluents and one cooling water sample have been previously presented (Jolley, 1973, 1975; Jolley, Gehrs and Pitt, 1975). The major aspects and conclusions from these studies will be summarized below. In addition, previously unpublished results from the recent analysis of a second sample of cooling water will be discussed.

Sewage Treatment Plant Effluents

Over 50 chloro-organic constituents were separated during each HPLC analysis of 500- to 1000-fold concentrations of primary and secondary sewage effluents which had been chlorinated in the laboratory with ppm concentrations of chlorine. The effluent samples, which were grab samples collected at the Oak Ridge Waste Treatment Plants, contained essentially no industrial pollutants. Each sample was chlorinated in the laboratory, using either chlorine gas or hypochlorite solution, to chlorine dosages of 2.5 and 6 mg/l for secondary and primary effluents, respectively; the corresponding chlorine residuals (orthotolidine) were 1 and 2 mg/l. The chlorination contact times were approximately equivalent to those used for disinfection at the sewage treatment plants. The chlorination reagents contained 0.03 to 0.12 millicuries (mCi) of ^{36}Cl (Jolley, 1973, 1974, 1975).

A typical HPLC chromatogram of the ^{36}Cl-tagged chlorine-containing constituents separated from chlorinated secondary sewage effluents is shown in Figure 7. Seventeen chloro-organics, principally chlorinated purines, pyrimidines, phenols and aromatic acids, were tentatively identified and are indicated at their respective elution positions in the chromatogram. The identifications were made by: (1) comparing anion exchange elution positions of the chromatographic peaks with chloro-organic reference standards, and (2) establishing the presence of an unchlorinated analog or progenitor organic compound in the unchlorinated sewage effluent (Jolley, 1973, 1975). We have also obtained corroborative evidence for the presence of 5-chlorouracil by comparison of cation exchange elution positions. More than 99% of the radioactivity was associated with the chloride ion peak, which was eluted at 18 hr. This high chloride concentration is supporting evidence that a major reaction mechanism of chlorine with organics in sewage effluents is oxidation. However, this has not been proved because chloride may also result from the decomposition reactions of monochloramine. About 1% of the radioactivity was associated with the other chromatographic peaks and some residual activity not removed from the resin by the chromatographic eluent. These peaks and the residual activity represent stable chloro-organic compounds as deduced from an extensive study of chlorinated secondary effluents (Jolley, 1973, 1975). In that study it was shown that less than 1% of the activity associated with the chromatographic peaks other than chloride was contributed by chloride-metal or inorganic chloramine-metal complexes, or by isotopic exchange with chlorine-containing constituents present in the effluents prior to chlorination. Apparently, these constituents are quite stable since their separation and detection are accomplished only after a rigorous sample preparation procedure and subsequent chromatographic separation. Assuming complete isotopic dilution of the ^{36}Cl-tagged chlorinating agent with the inert chloride in the effluent samples, the reaction yields of chloro-organics (as Cl) during the chlorination of both primary and secondary sewage effluents were calculated to be about 1% of the chlorine dosage for the reaction times customarily used for disinfection (Jolley, 1973, 1974, 1975).

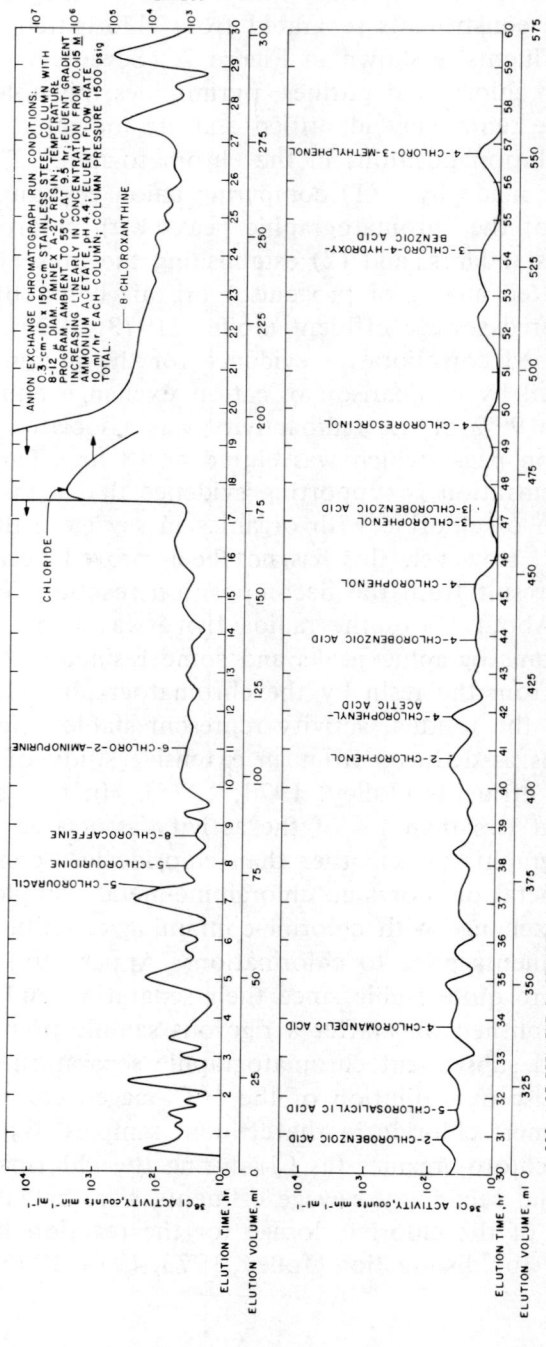

Figure 7. Chromatogram showing the ^{36}Cl-tagged chlorine-containing constituents in a sample of chlorinated secondary effluent from a domestic sanitary sewage treatment plant.

The following major conclusions were deduced from the analytical data obtained during these studies of chlorinated sewage treatment plant effluents (Jolley, 1973, 1975):

1. Stable chloro-organic compounds are formed during the chlorination of sewage effluents at ppm chlorine concentration.
2. The chlorination yield of chloro-organic compounds (as Cl) is about 1% of the chlorine dosage when disinfection reaction conditions are used.
3. The types of organic products formed included chlorinated phenols, purines, pyrimidines and aromatic acids at the ppb concentration level.

Watts Bar Lake: Cooling Water for Kingston Steam Plant

Over 50 chloro-organic constituents were separated by HPLC from a 1500-fold concentrate of Watts Bar Lake water which had been chlorinated in the laboratory with ^{36}Cl-tagged chlorine gas (Figure 8). The water sample was collected at the cooling water inlet of the Kingston Steam Plant, which is a coal-fired electric power-generating plant operated by the Tennessee Valley Authority and located on Watts Bar Lake at Kingston, Tennessee. A 1.5-liter aliquot of the grab sample of cooling water was chlorinated, at a Cl dosage of 2.1 mg/l, to a chlorine residual (orthotolidine) of 1 mg/l. The chlorinating reagent was tagged with 0.01 mCi of ^{36}Cl per mg of Cl. After a 75-min reaction time, the chlorine residual was destroyed by using thiosulfate (Jolley, Gehrs and Pitt, 1975).

As in the chlorination of sewage effluents, more than 99% of the radioactivity was associated with the chloride peak which eluted at 19 hr. About 0.5% of the ^{36}Cl activity was associated with the other 52 chromatographic peaks, which represent stable chloro-organic constituents as deduced by analogy with the extensive study of chlorinated sewage effluents. Examination of the chromatogram (Figure 8) indicates considerable similarity in chromatographic profile with many of the separated chloro-organic constituents corresponding in elution position to those of compounds separated from sewage treatment plant effluents (Figure 7). The chlorination reaction yield (as Cl) of chloro-organics in this experiment was 0.78% of the chlorine

122 WATER CHLORINATION

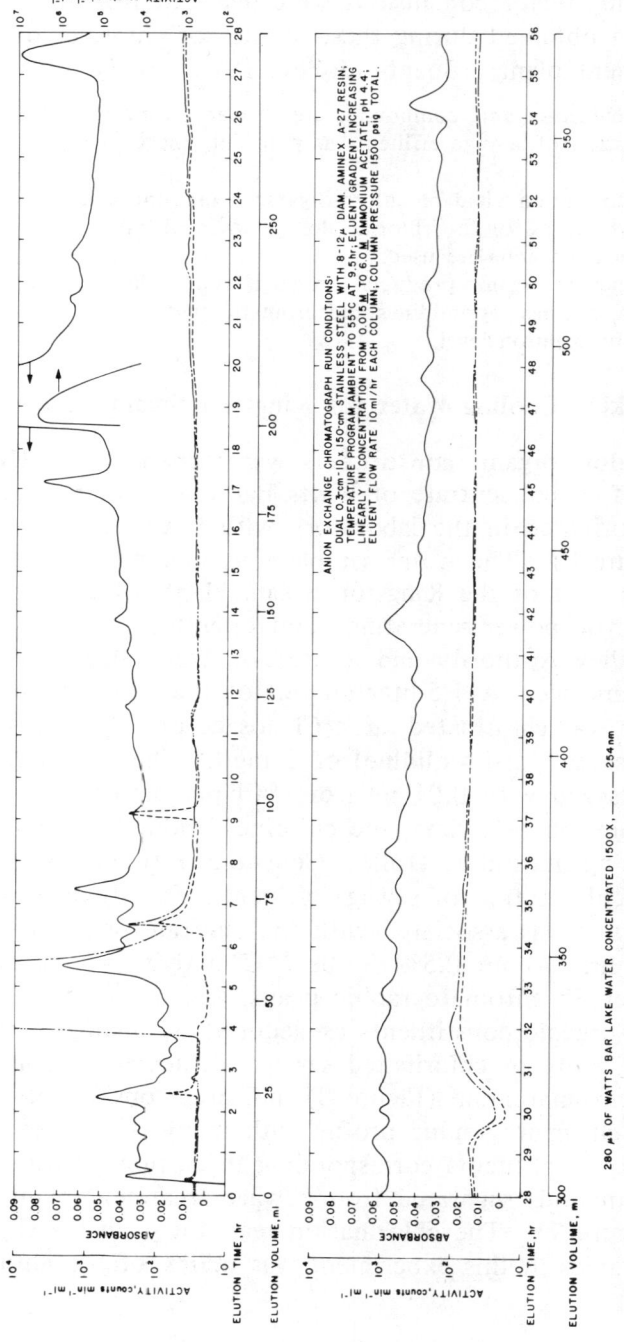

Figure 8. Chromatogram showing both ^{36}Cl-tagged chlorine-containing and UV-absorbing constituents in a sample of Watts Bar Lake water collected from the Kingston Steam Plant cooling water inlet and chlorinated in the laboratory. A chromatogram of the UV-absorbing constituents in the unchlorinated cooling water is included for comparison. Both samples were chromatographed simultaneously on the dual-column UV-Analyzer (Jolley, Gehrs and Pitt, 1975).

dosage. The reaction yield under actual plant antifoulant operating conditions was estimated to be about 0.5% (Jolley, Gehrs and Pitt, 1975).

The following major conclusions were deduced from the analytical data obtained in this study (Jolley, Gehrs, and Pitt 1975):

1. Stable chloro-organic compounds are formed during the chlorination of cooling waters at ppm chlorine concentrations.
2. The chlorination yield of chloro-organic constituents (as Cl) is about 0.5% of the chlorine dosage under reaction conditions simulating those used for antifoulant treatment of the cooling system of an electric power-generating plant.
3. The HPLC chromatographic profile and peak elution positions were similar to those obtained for chlorinated sewage effluents.

Mississippi River: Cooling Water for Allen Steam Plant

As with the previous cooling water sample, over 50 stable chloro-organic constituents were separated by HPLC from a 1470-fold concentrate of Mississippi River water which had been chlorinated in the laboratory (Figure 9). This concentrate had been prepared from a sample collected from the cooling water inlet (prior to chlorination) of the Allen Steam Plant, which is a coal-fired electric power-generating plant operated by the Tennessee Valley Authority and located on the Mississippi River near Memphis, Tennessee. The total chlorination yield of chloro-organics (as Cl) was 3.1% of the chlorine dosage after 15-min contact time. This yield was calculated assuming that isotopic dilution of the ^{36}Cl-tagged chlorinating agent occurs rapidly and essentially completely with the inert chloride in the cooling water sample.

After collection, the water sample was stored at -60°C and thawed just prior to use. Analytical data on the sample are presented in the Discussion section and compared with the results obtained for sewage effluents and the previous cooling water sample. The chlorine demand of the sample was determined using *Standard Methods* (American Public Health Association, 1971). The linear equation expressing the relationship between the chlorine dosage and the chlorine residual of the chlorinated cooling water after 5-min contact time is:

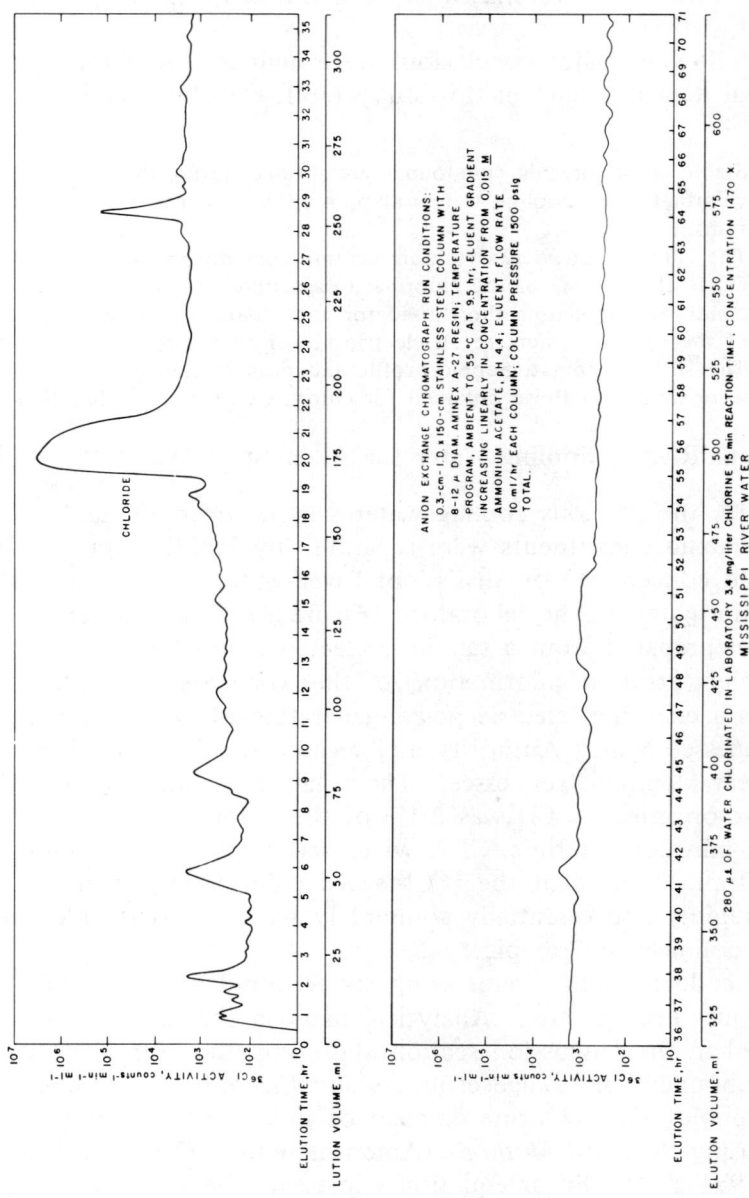

Figure 9. Chromatogram showing the ^{36}Cl-tagged chlorine-containing constituents in a sample of chlorinated Mississippi River water collected from the Allen Stream Plant cooling water inlet and chlorinated in the laboratory.

$$Y = 0.38X - 0.09$$

in which Y is the chlorine residual in mg/l and X is the chlorine dose in mg/l. A 1.0-liter aliquot of the cooling water was chlorinated with 3.4 mg of chlorine gas containing 0.033 mCi of ^{36}Cl radioactive tracer. After a chlorination contact time of 15 min, the chlorine residual of 1 mg/l was destroyed with a slight stoichiometric excess of thiosulfate solution. The sample was then concentrated 1470-fold by vacuum distillation at temperatures ranging from ambient to 35°C (Jolley, 1974). The chromatogram shown in Figure 9 was obtained by HPLC analysis of 0.28 ml of the radioactive concentrate.

Comparison of this chromatogram (Figure 9) with the chromatograms obtained for sewage effluent (Figure 7) and the Watts Bar Lake cooling water (Figure 8) reveals a considerable similarity in profile. The chromatographic peak elution positions show much correspondence; however, some significant differences exist. For example, the peak eluting at 29 hr is much larger in the chromatogram of the Mississippi River water sample (Figure 9) than in either of the other chromatograms. As in the chlorination of sewage effluents and the sample of Watts Bar Lake water, about 99% of the ^{36}Cl activity was associated with the chloride peak. The remaining 1.3% was associated with the other chromatographic peaks and the strongly sorbed material on the resin. By analogy with the previous studies, this activity represents stable ^{36}Cl-tagged chloro-organic products.

The following major conclusions were deduced from the analytical data in this study:

1. Stable chloro-organic compounds are formed during the chlorination of cooling waters at ppm chlorine concentrations.
2. The chlorination yield of chloro-organic products (as Cl) is about 3% of the chlorine dosage under reaction conditions simulating those used for antifoulant treatment of the cooling system of an electric power-generating plant.
3. The HPLC chromatographic profile and the peak elution positions were similar to those obtained for chlorinated sewage effluents and Watts Bar Lake water.

DISCUSSION

Similarities in the HPLC chromatographic profiles of the chlorinated sewage effluents and cooling waters suggest that many of the same chloro-organic products are formed in each medium during chlorination. A possible explanation is that natural waters must contain many of the progenitor organic compounds that occur in sewage effluents. This could be reasonably expected since the natural waters used for cooling systems contain a vast complex of microflora and microfauna, animal excreta, plant and animal metabolites, and decomposing plant and animal matter. In many ways, the composition of such waters resembles that of dilute sewage effluents.

Based on a comparison of chromatographic profiles and peak elution positions, the peaks for the samples of Watts Bar Lake and Mississippi River cooling water corresponding to the chloro-organic constituents identified in the sewage effluents were quantified. The quantitative data for the chlorinated secondary sewage effluent and cooling water samples are presented in Table IV. Two compounds, 5-chlorouracil and 4-chlororesorcinol, were chosen for initial assessment of toxicity and possible environmental effects. The chloropyrimidine was selected because of possible incorporation into genetic material; the chlorophenol was selected because of probable toxicity. Results of these studies have been previously reported (Gehrs et al., 1974; Gehrs and Jolley, 1975) and will be discussed later in this conference. It is interesting that the estimated concentrations of chlorinated purines and pyrimidines are comparable for chlorinated secondary sewage effluent and chlorinated cooling waters. The concentrations of chlorinated phenols and aromatic acids are highest for the sample of Mississippi River water. They may be the result of a higher concentration of humic material in that water.

Because of the large number of unidentified chloro-organic constituents and the variety of those that have been tentatively identified, it appears that a number of complex chlorination reactions take place during the disinfection of sewage effluents and the antifoulant treatment of cooling waters. The yields of chloro-organic products are determined by the concentrations

Table IV. Chloro-Organic Constituents in Chlorinated Secondary Sewage Effluent and Cooling Waters

Identifications	Concentration[a] (μg/liter)		
	Secondary Sewage Effluent[b]	Watts Bar Lake Sample	Mississippi River Sample
Nucleoside			
5-Chlorouridine	1.7	0.6	7
Purine			
8-Chlorocaffeine	1.7	1.1	6
6-Chloro-2-aminopurine	0.9	1.0	3
8-Chloroxanthine	1.5	3	-
Pyrimidine			
5-Chlorouracil	4	0.6	3
Aromatic acid			
2-Chlorobenzoic acid	0.3	1.1	10
3-Chlorobenzoic acid	0.6	0.2	8
4-Chlorobenzoic acid	1.1	0.3	8
3-Chloro-4-hydroxybenzoic acid	1.3	0.8	3
4-Chloromandelic acid	1.1	1.8	6
4-Chlorophenylacetic acid	0.4	3	20
5-Chlorosalicylic acid	0.2	3	18
Phenol			
4-Chloro-3-methylphenol	1.5	0.2	0.7
2-Chlorophenol	1.7	0.2	4
3-Chlorophenol	0.5	0.2	6
4-Chlorophenol	0.7	0.2	2
4-Chlororescorcinol	1.2	0.5	7

[a] Calculations based on assumption of complete isotopic exchange of the ^{36}Cl-tracer in the chlorinating agent with the nonradioactive chloride in the water samples.
[b] Jolley 1973, 1975.

of the available chlorinating agent and organic reactant, the thermodynamics and kinetics of the individual reactions, and reaction parameters such as time, temperature, light and catalysts. Most of the chloro-organic compounds separated by the coupled ^{36}Cl tracer–HPLC analytical technique used in these studies are considered to be nonvolatile or, at the least, to have a relatively low volatility. Their molecular weights are probably less than several thousand. Volatile chloro-organics, such as chloroform, which might be anticipated as reaction products from the chlorination of natural waters and which have been identified by Glaze and Henderson (1975) in chlorinated sewage effluents, would not have been detected or measured by our analytical technique. Our analytical methodology would also have been inadequate for detecting or quantifying the chlorination effect on large polymers such as nucleic acids and humic acids. As previously discussed, chlorine substitution in the organic bases uracil and cytosine might be anticipated in nucleic acid polymers. In addition, humic substances might be expected to chlorinate at active sites on the aromatic rings.

Several general conclusions may be drawn concerning reaction yields of chloro-organic products from the disinfection of sewage effluents and the antifoulant treatment of cooling waters. Table V presents selected analytical data for the water samples used in these studies and gives the chlorination results in terms of chlorine dosage, chlorine residuals and reaction yields. The chlorination yield (*i.e.,* that fraction of the chlorine dosage associated with the chloro-organic products after termination of the reaction) is relatively much lower for the sewage effluent, which has an ammonia (as N) concentration of 11 mg/l, than for the Mississippi River cooling water, which contains 0.15 mg/l. This supports an earlier observation that, when all other factors are equal, smaller quantities of chloro-organics would be anticipated in waters containing higher concentrations of ammonia. Two factors—the extremely rapid formation of monochloramine and the resulting low equilibrium concentration of HOCl—are involved. The significant effect of ammonia on equilibrium concentrations of the reactive chlorine-containing species is shown in Table VI. If HOCl is the effective chlorinating agent for organic compounds in aqueous solutions, as we suspect, then a relatively greater reaction yield should be obtained for Watts Bar Lake and Mississippi River water samples than for the secondary sewage effluent. The data obtained for Mississippi River water samples in our study bear out this expected relationship. On the other hand, the yields we obtained for the secondary sewage effluent were higher than those for the Watts Bar Lake water; however, we believe that this apparent discrepancy is due to the much higher concentrations of soluble organic constituents in the sewage effluent (*e.g.,* see the organic carbon values, Table V).

Table V. Comparison of Selected Analytical Data[a] and Chlorination Results for Secondary Sewage Effluent and Cooling Water Samples

	Secondary Sewage Effluent[b]	Watts Bar Lake Sample[c]	Mississippi River Sample
pH	7.4	7.5	7.3
Chloride	22.0	0.5	8.0
Organic carbon	12.0	2.7	9.0
Organic nitrogen	5.8	2.7	<0.05
Ammonia (as N)	11.0	0.5	0.15
Chlorine dosage	3.2	2.5	3.4
Chlorine residual (OT)	1.0	1.0	1.2
Chlorination yield of Cl as chloro-organics	1.0%/45 min	0.8%/75 min (0.5%/15 min)[d]	3.1%/15 min

[a]Concentrations are given in mg/l.
[b]Jolley, 1973, 1975.
[c]Jolley, Gehrs and Pitt, 1975.
[d]Estimated value after a chlorination contact time of 15 min.

Table VI. Equilibrium Concentrations[a] of Selected Constituents and Reaction Yields of Chloro-Organic Products for Chlorinated Secondary Sewage Effluent and Cooling Water Samples

	Secondary Sewage Effluent[b]	Watts Bar Lake Sample	Mississippi River Sample
Chlorine residual (OT)	1.0	1.0	1.2
NH_3 (as N)	11.0	0.5	0.15
$HOCl$[c]	< 0.0001	0.0016	0.2886
OCl^-[c]	< 0.0001	0.0016	0.1843
NH_2Cl[c]	0.9979	0.8907	0.0321
$NHCl_2$[c]	0.0020	0.1061	0.6950
Reaction yield after 15 min	0.9	~0.5	3.1

[a]Concentrations are given in mg/l.
[b]Jolley, 1973, 1975.
[c]Equilibrium values were calculated by using a computer program developed by J. E. Draley, Argonne National Laboratory, and modified by R. H. Rainey, Oak Ridge National Laboratory.

The progressive decrease of NH_2Cl and concomitant increase of $NHCl_2$ as the ammonia content of each water sample decreased may be significant. Unfortunately, little is known concerning the aqueous chemistry of dichloramine. If $NHCl_2$ is a chlorinating agent for organics in aqueous solutions, then the relatively high equilibrium concentration of dichloramine in the Mississippi River water sample may have contributed to the high reaction yield of chloro-organics. This aspect of the chemistry of $NHCl_2$ should be studied.

ENVIRONMENTAL SIGNIFICANCE

If we assume that 100,000 to 200,000 tons of chlorine are used annually in the United States for disinfection and antifoulant purposes, and that the reaction yields determined in our studies are representative, we can reasonably estimate that several thousand tons of chloro-organics are produced each year and released to aquatic ecosystems. Although these compounds may be present individually at only ppb concentrations, their environmental effect on a collective basis may be very significant. A large gap exists in our knowledge concerning the nature and concentrations of the compounds produced by these water treatment processes. Acute toxicity, chronic or low-level toxicity and mutagenicity studies are currently being conducted in several laboratories. The results obtained in some of these studies will be presented in this conference.

Another area of concern is the creation of noise in or interference with chemical communications and pheromone systems in aquatic ecosystems exposed to chlorination. It is possible that this may represent a major biological effect accompanying the chlorination of cooling waters. For example, anadromous fish apparently use chemical sensing to find breeding areas (Sutterlin and Gray, 1973 Sutterlin, 1974). If this homing instinct is dependent on sensing organic constituents, the chlorination of cooling waters could cause confusion because of the large variety of possible chlorination reactions. Biological changes such as species shifts have been documented below chlorinated sewage outfalls (Tsai, 1968, 1970). Although these effects are usually attributed to the toxicity of the chlorine residual and to the decreased oxygen concentration associated with an increased concentration of organic matter, a contributing cause might be interference with pheromone systems or other chemical communication systems. Phthalates have been found to be either crowding factor pheromones or mimics of natural crowding factors (Pfuderer, Williams and Francis, 1974; Pfuderer and Francis, 1975; Pfuderer, Janzen and Rainey, 1975). Phthalates are also subject to chlorine substitution reactions in aqueous solutions (Goodrich, 1949).

One of the goals of this conference is to define environmental problems associated with water chlorination. To help establish the significance of the problem, we recommend that the efforts to identify and quantify the chloro-organics (and possibly the bromo-organics) that are formed during water chlorination be intensified, that the ecological consequences of possible pheromone or communications effects be investigated, and that analytical methodology for detection, isolation and identification of organics down to the parts-per-quadrillion level be developed to facilitate these efforts. Obviously, if the environmental aspects of water chlorination are found to be unfavorable, we must either live with the most suitable compromise or develop alternative biocides and/or disinfection techniques.

CONCLUSIONS

The major conclusions of the experimental studies summarized in this paper may be stated as follows:

1. The chlorination reaction yields of chloro-organic products (as Cl) in chlorinated cooling waters and sewage effluents range from 0.5 to 3.1%.
2. Annually, the environmental impact of water chlorination on the aquatic ecosystems of the United States is estimated to include the introduction of several thousand tons of chloro-organic compounds.
3. Any or all of a large number of possible aqueous chlorination reactions may occur during water chlorination, depending on the presence of organic constituents, reaction kinetics and thermodynamics, and other reaction parameters.
4. Complex mixtures of chloro-organic compounds are produced during chlorination, each at ppb concentration or less.

5. The nature of the chloro-organic products formed during the chlorination of sewage effluents and cooling waters suggests a variety of possible effects relative to (a) genetics; (b) toxicity; and (c) population, through altered chemical communications in aquatic ecosystems.

ACKNOWLEDGMENTS

The authors wish to thank C. D. Scott and S. Katz for their encouragement and J. E. Attrill, C. W. Hancher, S. Katz and M. G. Stewart for critical analysis of this paper.

The original research reported here was sponsored by the Energy Research and Development Administration, U.S. Environmental Protection Agency, and National Science Foundation—RANN. The work was carried out at Oak Ridge National Laboratory, which is operated for the Department of Energy under contract with the Union Carbide Corporation.

REFERENCES

American Public Health Association. 1971. *Standard Methods for the Examination of Water and Wastewater*, 13th ed. Washington, D.C., 874 p.

Bellar, T. A., J. J. Lichtenberg and R. C. Kroner. 1974. "The Occurrence of Organohalides in Chlorinated Drinking Water." *J. Am. Water Works Assoc.* 66:703-706.

Burttschell, R. H., A. A. Rosen, F. M. Middleton and M. B. Ettinger. 1959. "Chlorine Derivatives of Phenol Causing Taste and Odor." *J. Am. Water Works Assoc.* 51:205-213.

Carlson, R. M., R. E. Carlson, H. L. Kopperman and R. Caple. 1975. "Facile Incorporation of Chlorine into Aromatic Systems During Aqueous Chlorination Processes." *Environ. Sci. Technol.* 9:674-675.

Chandelon, T. 1883. "Ueber die Durch Einwirkung Alkalischer Hypochlorite auf Phenol Gebildeten Chlorphenole." *Chem. Ber.* 16:1749.

Christman, R. F. and R. A. Minear. 1971. "Organics in Lakes," p. 119-143. In S. J. Faust and J. V. Hunter (Ed.) *Organic Compounds in Aquatic Environments*. Marcel Dekker, Inc., New York.

Colton, E. and M. M. Jones. 1955. "Monochloramine." *J. Chem. Educ.* 32:485-487.

Drago, R. S. 1957. "Chloramine." *J. Chem. Educ.* 34:541-545.

Draley, J. E. 1972. "The Treatment of Cooling Water with Chlorine." ANL/ES-12. Argonne National Laboratory, Argonne, Illinois. 11 pp.

Dugan, P. R. 1972. *Biochemical Ecology of Water Pollution.* Plenum Press, New York. 159 pp.

EUROCOP. 1973. "List of Substances Which Have Been Identified in Various Waters." EUROCOP-COST Project 64B "Micro-pollutants." Appendix to first annual report from International Co-ordinating Laboratory, Hertfordshire, England. September.

Faust, S. D. and J. V. Hunter. 1971. *Organic Compounds in Aquatic Environments.* Marcel Dekker, Inc., New York. 638 pp.

Gehrs, G. W., L. D. Eyman, R. L. Jolley and J. E. Thompson. 1974. "Effects of Stable Chlorine-Containing Organics on Aquatic Environments." *Nature* 249:675-676.

Gehrs, G. W. and R. L. Jolley. 1975. "Chlorine-Containing Stable Organics: New Compounds of Environmental Concern." *Verh. Internat. Verein. Limnol.* 19:2185-2188.

Glaze, W. H. and J. E. Henderson. 1975. "Formation of Organochlorine Compounds from the Chlorination of a Municipal Secondary Effluent." *J. Water Poll. Control Fed.* 47:2511-2515.

Glaze, W. H., J. E. Henderson, J. E. Bell and V. A. Wheeler. 1973. "Analysis of Organic Materials in Wastewater Effluents after Chlorination." *J. Chromatogr. Sci.* 11:580-584.

Goodrich, B. F. Co. 1949. "Nuclear Chlorination of Aromatic Carboxy Acids." Brit. Pat. 628, 401. August 29. (*Chem. Abstr.* 44:2561 g).

Holst, G. 1954. "The Chemistry of Bleaching and Oxidizing Agents." *Chem. Rev.* 54:169-194.

Hood, D. V. 1970. *Organic Matter in Natural Waters.* Institute of Marine Science, University of Alaska. 625 pp.

Hopkins, C. Y. and M. J. Chisholm. 1946. "Chlorination by Aqueous Hypochlorite." *Can. J. Res.*, Sect. B 24:208-210.

Houben, J. and T. Weyl. 1962. "Halogen-Verbindungen," p. 760-811. In J. Houben and T. Weyl, *Methoden der Organischen Chemie* (Houben-Weyl) *Vierte Auflage*, Band 5, Teil 3. George Thieme Verlag, Stuttgart.

Howard, G. 1960. "Purines and Related Ring Systems," p. 1635-1759. In E. H. Rodd (Ed.) *Chemistry of Carbon Compounds*, Vol. IV, Part C. Heterocyclic compounds. Elsevier Publishing Co., New York.

Ingols, R. S. and G. M. Jacobs. 1957. "BOD Reduction by Chlorination of Phenol and Amino Acids." *Sew. Ind. Wastes* 29:258-262.

Jolley, R. L. 1973. *Chlorination Effects on Organic Constituents in Effluents from Domestic Sanitary Sewage Treatment Plants.* Ph.D. Diss., University of Tennessee, Knoxville. (ORNL/TM-4290. Oak Ridge National Laboratory, Oak Ridge, Tennessee.) 340 pp.

Jolley, R. L. 1974. "Determination of Chlorine-Containing Organics in Chlorinated Sewage Effluents by Coupled ^{36}Cl Tracer–High Resolution Chromatography." *Environ. Lett.* 7:321-340.

Jolley, R. L. 1975. "Chlorine-Containing Organic Constituents in Sewage Effluents." *J. Water Poll. Control Fed.* 47:601-618.

Jolley, R. L., C. W. Gehrs and W. W. Pitt. 1975. "Chlorination of Cooling Water: a Source of Chlorine-Containing Organic Compounds with Possible Environmental Significance," p. 21-28. In C. E. Cushing (Ed.) *Radioecology and Energy Resources.* Dowden, Hutchinson & Ross, Inc., Stroudsburg, Pennsylvania.

Jolley, R. L., S. Katz, J. E. Mrochek, W. W. Pitt, Jr. and W. T. Rainey. 1975. "Analyzing Organics in Dilute Aqueous Solutions." *Chem. Technol.* 1975:312-318.

Jolley, R. L., W. W. Pitts, Jr., C. D. Scott, G. Jones, Jr. and J. E. Thompson. 1975. "Analysis of Soluble Organic Constituents in Natural and Process Waters by High-Pressure Liquid Chromatography," p. 247-253. In D. D. Hemphill (Ed.) *Trace Substances in Environmental Health*, Volume IX. University of Missouri, Columbia, Missouri.

Jolley, R. L., W. W. Pitt and J. E. Thompson. 1975. "Synthesis of ^{36}Cl-Tagged 5-Chlorouracil," p. 162-165. In C. D. Scott (Ed.) *Experimental Engineering Section Semiannual Progress Report*, March 1, 1974 to August 31, 1974. ORNL/TM-4777. Oak Ridge National Laboratory, Oak Ridge, Tennessee.

Junk, G. A. and S. E. Stanley. 1975. *Organics in Drinking Water*, Part 1. Listing of identified chemicals. IS-3671. Ames Laboratory, Ames, Iowa. July. 84 pp.

Kovacic, P., M. Lowery and K. W. Field. 1970. "Chemistry of N-Bromamines and N-Chloramines." *Chem. Rev.* 70:639-665.

Lee, G. F. and J. C. Morris. 1962. "Kinetics of Chlorination of Phenol-Chlorphenolic Tastes and Odors." *Int. J. Air Water Poll.* 6:419-431.

Little, Arthur D. Inc. 1970. *Organic Pollution of Freshwater.* Water Quality Criteria Data Book, Vol. 1. U.S. Environmental Protection Agency. Washington, D. C.

Manka, J., M. Rebhun, A. Mandelbaum and A. Bortinger. 1974. "Characterization of Organics in Secondary Effluents." *Environ. Sci. Technol.* 8:1017-1020.

Morris, J. C. 1967. "Kinetics of Reactions Between Aqueous Chlorine and Nitrogen Compounds," p. 22-53. In S. D. Faust and J. V. Hunter (Ed.) *Principles and Applications of Water Chemistry.* John Wiley and Sons Inc., New York.

Morris, J. C. 1971. "Chlorination and Disinfection–State of the Art." *J. Am. Water Works Assoc.* 63:769-774.

Morris, J. C. 1975. *Formation of Halogenated Organics by Chlorination of Water Supplies.* EPA-600/1-75-002. U.S. Environmental Protection Agency, Washington, D.C. 54pp.

Murphy, K. L., R. Zaloum and D. Fulford. 1975. "Effect of Chlorination Practice on Soluble Organics," *Water Res.* 9:389-396.

Patton, W., V. Bacon, A. M. Duffield, B. Halpern, Y. Hoyano, W. Pereira and J. Lederberg. 1972. "Chlorination Studies. I. The Reaction of Aqueous Hypochlorous Acid and Cytosine." *Biochem. Biophys. Res. Commun.* 48:880-884.

Pfuderer, P. and A. A. Francis. 1975. "Phthalate Esters: Heartrate Depressors in the Goldfish." *Bull. Environ. Contam. Toxicol.* 13:275-278.

Pfuderer, P., S. Janzen and W. T. Rainey, Jr. 1975. "The Identification of Phthalic Acid Esters in the Tissues of Cyprinodont Fish and Their Activity as Heartrate Depressors." *Environ. Res.* 9:215-223.

Pfuderer, P., P. Williams and A. A. Francis. 1974. "Partial Purification of the Crowding Factor from *Carassius auratus* and *Cyprinus carpio*." *J. Exp. Zool.* 187:375-382.

Pitt, W. W., R. L. Jolley and S. Katz. 1974. *Automated Analysis of Individual Refractory Organics in Polluted Water.* EPA-660/2-74-076. U.S. Environmental Protection Agency, Washington, D. C. 98 pp.

Pitt, W. W., R. L. Jolley and C. D. Scott. 1975. "Determination of Trace Organics in Municipal Sewage Effluents and Natural Waters by High-Resolution Ion-Exchange Chromatography." *Environ. Sci. Technol.* 9:1068-1073.

Prat, R., C. Nofre and A. Cier. 1965. "Effet de L'Hypochlorite de Sodium les Constituants Pyrimidiques des Bacteries." *Comp. Rend. Acad. Sci.* Paris 260(May):4859-4861.

Ramage, G. R. and J. K. Landquist. 1959. "The Pyrimidine Group," p. 1257-1298. In E. H. Rodd (Ed.) *Chemistry of Carbon Compounds*, Vol. IV, Part C. Heterocyclic compounds. Elsevier Publishing Co., New York.

Rebhun, M. and J. Manka. 1971. "Classification of Organics in Secondary Effluents." *Environ. Sci. Technol.* 5:606-609.

Risebrough, R. W., P. Rieche, D. B. Peakall, S. G. Herman and M. N. Kirven. 1968. "Polychlorinated Biphenyls in the Global Ecosystem." *Nature* 220:1098-1102.

Rook, J. J. 1974. "Formation of Haloforms During Chlorination of Natural Waters." *Water Treat. Exam.* 23(2):234-243.

Schnitzer, M. and S. U. Khan. 1972. *Humic Substances in the Environment.* Marcel Dekker, New York. 327 pp.

Steelink, C. 1963. "What is Humic Acid?" *J. Chem. Educ.* 40:379-384.

Stevenson, F. J. and K. M. Goh. 1972. "Infrared Spectra of Humic and Fulvic Acids and Their Methylated Derivatives: Evidence for Nonspecificity of Analytical Methods for Oxygen-Containing Functional Groups." *Soil Sci.* 113:334-345.

Sutterlin, A. M. 1974. "Pollutants and the Chemical Senses of Aquatic Animals—Perspective and Review." *Chem. Senses Flavor* 1:167-178.

Sutterlin, A. M. and R. Gray. 1973. "Chemical Basis for Homing of Atlantic Salmon (*Salmo salar*) to a Hatchery." *J. Fish Res. Bd. Can.* 30:985-998.

Symons, J. M., T. A. Bellar, J. K. Carswell, J. Demarco, K. L. Kropp, G. G. Robeck, D. R. Seeger, C. J. Slocum, B. L. Smith and A. A. Stevens. 1975. "National Organics Reconnaissance Survey for Halogenated Organics in Drinking Water." *J. Am. Water Works Assoc.* 67:634-647.

Theilacker, W. and E. Wegner. 1964. "Organic Syntheses Using Chloramine," p. 303-317. In W. Foerst (Ed.) *Newer Methods of Preparative Organic Chemistry.* (Transl. by H. Birnbaum.) Academic Press, Inc., New York.

Tsai, C. F. 1968. "Effects of Chlorinated Sewage Effluents on Fishes in Upper Patuxent River, Maryland." *Chesapeake Sci.* 9(2):83-93.

Tsai, C. F. 1970. "Changes in Fish Population and Migrations in Relation to Increased Sewage Pollution in Little Patuxent River, Maryland." *Chesapeake Sci.* 11(1):34-41.

U.S. Environmental Protection Agency. 1975. *Suspect Carcinogens in Water Supplies.* Office of Research and Development. Interim report with appendices. April 1975.

Vallentyne, J. R. 1957. "The Molecular Nature of Organic Matter in Lakes and Oceans with Lesser Reference to Sewage and Terrestrial Soils." *J. Fish. Res. Bd. Can.* 14:33-82.

Van Middelem, C. H. 1966. "Fate and Persistence of Organic Pesticides in the Environment," p. 228-249. In *Organic Pesticides in the Environment,* Advances in Chemistry Series No. 60. American Chemical Society, Washington, D. C.

Veith, G. D. and G. F. Lee. 1970. "A Review of Chlorinated Biphenyl Contamination in Natural Waters." *Water Res.* 4:265-269.

Weber, W. J., Jr. 1972. *Physiochemical Processes for Water Quality Control.* Wiley Interscience, New York. 640 pp.

Weil, I. and J. C. Morris. 1949. "Kinetic Studies on the Chloramines. I. The Rates of Formation of Monochloramine, N-Chloromethylamine, and N-Chlorodimethylamine." *J. Am. Chem. Soc.* 71:1664-1674.

Wershaw, R. L. and M. C. Goldberg. 1972. "Interaction of Organic Pesticides with Natural Organic Polyelectrolytes," p. 149-158. In

S. D. Faust (Symp. Chairman) *Fate of Organic Pesticides in the Environment.* Advances in Chemistry Series, Number 111. American Chemical Society, Washington, D. C.

White, G. C. 1972. *Handbook of Chlorination.* Van Nostrand Reinhold Company, New York. 744 pp.

WHO International Reference Centre for Community Water Supply. 1974. "Current Knowledge of Concentrations of Organic Contaminants in Waste Water, River Water, and Drinking Water." Working Document No. 4. *Working Meeting on Health Effects Relating to the Re-Use of Waste Water for Human Consumption.* Amsterdam. January 13-16, 1975.

Zaloum, R. and K. L. Murphy. 1974. "Reduction of Oxygen Demand of Treated Wastewater by Chlorination." *J. Water Poll. Control Fed.* 46:2770-2777.

DISCUSSION

Alan A. Stevens, U.S. Environmental Protection Agency. I would like to point out that Bob's estimates of chlorination yields are probably low because the HPLC technique does not detect the volatile compounds, for example, the trihalomethanes. In our work, roughly 1-3% of the chlorine used, when Ohio River water is chlorinated to meet the 96-hr demand, becomes trihalomethanes.

Jolley. I think that is very significant. Although I did not show any material balances for our chlorination experiments, the number 3% is about the yield of volatile chlorinated compounds that would have been estimated from the Watts Bar water experiment. Essentially, we had 97% material balance and so, indeed, maybe this did represent the formation of chloroform and other haloform compounds.

Robert S. Ingols, Georgia Institute of Technology. May I ask whether you believe that you have any proteins or any organized organic matter, not the monomers as it were, involved in any of your analytical procedures? Have you hydrolyzed or modified any of the large molecules such as proteins in which tyrosine or tryptophan are present? I assume you meant this was present as free tyrosine.

Jolley. That is correct.

Robert S. Ingols. And the tryptophan would also be readily chlorinated, if present as the free amino acid. I have good reason to believe, on the basis of some of the work done in the early fifties, that chlorine goes on to the tyrosine moiety within the protein molecule because it will not develop color in the presence of chlorine dioxide.

Jolley. Thank you, Dr. Ingols. No. We have not looked at proteinaceous materials. We have separated some rather large molecules by HPLC but we are dealing essentially with the compounds that are less than 1000 in molecular weight. We have not analyzed many free amino acids in these waters. I included in my tables only those confirmed by mass spectral data.

J. J. Nelson, Energy Research and Development Administration. I would like to first comment for the record that, if our instrumentation for detection of chloro-organics approached the order of magnitude of sensitivity available in radioactivity detection equipment, then our questions on impact would be much more well defined. My question is, what was the basis for your suggestion that organo-chlorides may have an impact on the pheromone communication among animals?

Jolley. I am speculating. However, phthalic acid esters are known to be crowding factor pheromones, or at least to mimic them, in several aquatic species, and they are known to be susceptible to chlorination. This leads me to think that such organic compounds and others which may be pheromones may be chlorinated during cooling water treatment. Thus, one effect of chlorinating cooling waters may be that of disrupting the natural chemical communications system.

J. Carrell Morris, Harvard University. It would be interesting to note while we're on the question of tonnage production here, that the use of chlorine in the bleaching of pulp and paper is about 10 times the use in water and sewage treatment. And probably there is, resulting from this, a production of the order of 10^6 tons of halogenated methanes per year. This is of the order of magnitude, but possibly only about a third, of the total global production of chloroform from all sources. So it is still moot as to whether marine bacteria or man is producing the most chloroform.

7

ANALYSIS OF NEW CHLORINATED ORGANIC COMPOUNDS FORMED BY CHLORINATION OF MUNICIPAL WASTEWATER

William H. Glaze, James E. Henderson, IV and Garmon Smith
Institute of Applied Sciences and Department of Chemistry
North Texas State University
Denton, Texas 76203

ABSTRACT

The effect of chlorination on secondary municipal wastewater effluents has been investigated using two analytical techniques. Total organic chlorine (TOCl) is measured before and after chlorination by a microcoulometric procedure. Concentrated extracts of the effluent before and after chlorination at various chlorine dose levels are pyrolyzed and titrated in the Dohrmann halide analyzer. The TOCl results show a significant increase in the level of organic chlorine after chlorination, particularly using large doses of chlorine (2000 to 4000 ppm). More explicit information regarding the nature of the new organic chlorine-containing compounds is obtained by gas chromatography-mass spectrometry studies of the concentrates obtained by XAD-2 resin extractions of the effluents. The GCMS results confirm that chlorination causes the formation of many new chlorinated organics, the structures of over 50 of which have been identified. Whereas the majority of the compounds are aromatic halides, many are not derivatives of "activated" aromatics such as phenol but are simple derivatives such as chlorobenzenes, -toluenes and -alkylbenzenes. Nonaromatic chlorides have also been identified. Particular attention has been focused in this work on the effect of heavy doses of chlorine in the range of 2000 to 4000 ppm. The TOCl and GCMS results indicate that treatment of this type causes a very large increase in organic chlorine content. Also reported are GCMS data on XAD extracts of Denton, Texas, drinking water.

Of particular interest are the occurrence of three iodine-containing compounds, *viz.* dichloroiodomethane, dibromoiodomethane and bromochloroiodomethane in the finished water.

INTRODUCTION

The use of aqueous chlorine as a water and wastewater treatment agent has come under scrutiny lately, primarily due to the discovery that chlorination of residual organics in such waters causes the formation of new chlorinated organics of unknown toxicities (Glaze *et al.*, 1973; Jolley, 1973; Bellar and Lichtenberg, 1974; Rook, 1974). Of course, the notion that organic compounds of certain types will react with aqueous chlorine is not new; workers as early as 1883 recognized this fact. Morris (1975), one of the contemporary pioneers in this area, has recently reviewed the chemistry of aqueous chlorine, including typical reactions with organic functionalities. The more recent attention of environmental chemists to the phenomena of chlorination, therefore, may be viewed as a result of increased federal participation in the water and wastewater treatment arena, as well as the wider availability of sensitive instrumentation with which to study materials and processes at the trace level.

Under the auspices of an Environmental Protection Agency grant, the Trace Analysis Laboratory at North Texas State University has been involved for the past two years in a study of the effects of chlorination of municipal wastewaters. Our early work under the sponsorship of the NTSU Faculty Research Fund was published in 1973.

In this paper we describe more recent data on the formation of new chlorinated organics, particularly with the use of large doses of chlorine (1000 to 4000 mg/l). We also report on the development of a method for monitoring total organic halogen (TOCl) in chlorinated wastewaters, and show the application of this method to the analysis of TOCl in superchlorinated municipal wastewaters.

EXPERIMENTAL METHODS

Total Organic Chlorine Determinations*

Figure 1 is a scheme for the concentration of organics from water samples. The scheme consists of a concentration step involving Rohm and Haas XAD macroreticular resins, and subsequent analysis of the organic eluate from the resin. A 2-liter wastewater sample is obtained from the treatment plant before chlorination, usually after final clarification. The sample is filtered through coarse paper and divided into two equal portions. One portion serves as the system control (BEFORE); the other portion (AFTER) is chlorinated to whatever level desired. Chlorine contact time is usually 1 hr. Both BEFORE and AFTER samples are treated with sodium sulfite to quench active chlorine, and then allowed to flow through an XAD-2 resin bed at approximately 30 ml/min. The scrupulously cleaned** resin (ca. 2 g) is contained in an 8-cm x 1-cm-i.d. glass column fitted with a 24/40 ℑ outer joint at top and a Teflon stopcock at the bottom. The water sample is introduced into the column from a 1-liter separatory funnel, after which the separatory funnel is rinsed with a few milliliters of purified diethylether.*** These washings and more ether, totaling approximately 20 to 30 ml, are used to elute the organic materials trapped on the XAD-2 column. The ether eluate is collected in a flask of the design recommended by Junk *et al.*(1974) and concentrated to final volume (1 to 5 ml) using a three-ball Snyder column.

Microcoulometric analysis of the organic halides (excluding fluorides) in the eluate is accomplished using a Dohrmann-Envirotech C-300 microcoulometer with S-300 pyrolysis furnace.

---

*We shall refer to the TOCl method as if only chlorinated organics were being measured; in fact, the method measures organic bromides and iodides as well, and should be termed a TOX (total organic halogen) method.

**The resin is cleaned and regenerated using the methods described by Junk *et al.* (1974).

***Ether (analytical reagent-grade) is distilled in a 1.5-m Oldershaw column and the center cut taken.

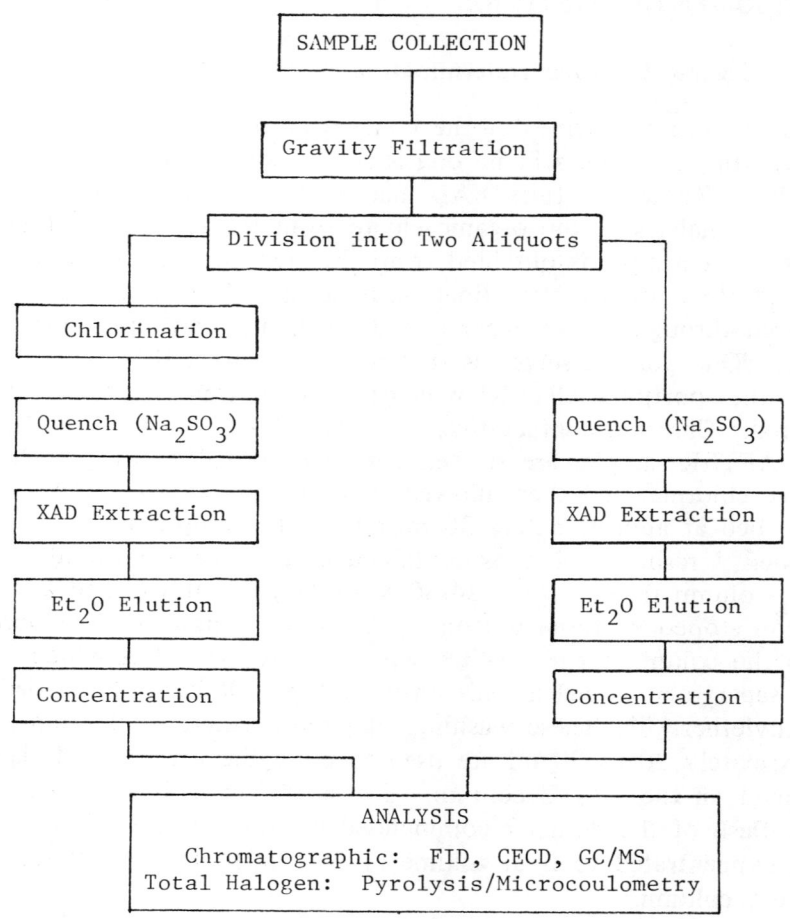

Figure 1. Scheme for the extraction and analysis of chlorinated organic compounds from water and wastewater samples.

Aliquots (1 to 25 μl) of the ether eluate are injected directly into the furnace through a septum or using a platinum or quartz boat sample injection system. In each case, the sample is entrained in a stream of argon into an inlet furnace operated at 800°C (lower temperatures are used in some cases), where the gas stream is mixed with oxygen. Oxidative pyrolysis is completed as the sample components flow through the middle and outlet areas of the furnace, which are at 800°C. The pyrolyzed sample plus carrier gas then flows into the coulometric cell where halides other than fluoride are titrated. Readout of the system is digital and is given in terms of nanograms of chlorine. A readout of the signal vs time is also displayed on a strip-chart recorder. Minimum detectable limit of the system is quoted as 2 ng chlorine (S/N = 10).

Injections of samples to be titrated may also be made directly into the coulometer cell, both for calibration of the system and for monitoring of inorganic halide levels in aqueous samples.

Mass Spectrometry-Gas Chromatography

Wastewater samples are also treated according to the scheme shown in Figure 1. After elution from the resin with ether and subsequent concentration to 1 ml, the organic extract is surveyed by gas chromatography using both flame ionization and Coulson electrolytic conductivity detectors. The Coulson detector is coupled to a Hewlett-Packard Model 5710 chromatograph; the FI detector is a part of the Finnigan Model 3200 Mass Spectrometer/Model 8500 Gas Chromatograph System with Model 600 Digital Interactive Control and Graphic Output System. The chromatograph column is usually 6-ft x 3-mm-i.d. glass packed with 3% Dexsil 300 GC coated on 100/120 mesh Supelcoport (Supelco Inc.). The helium flow rate is 30 ml/min. The temperature program conditions are as follows: (1) isothermal at room temperature (ca. 27°C) for 4 min after injection; (2) program ballistically to 50°C; (3) 50° to 300°C at 6°/min; (4) isothermal at 300°C until completion of run.

The injector temperature was 255°C; the detector temperature was 300°C. The Coulson block temperature was 300°C, and the furnace temperature was 830°C. The detector was operated in the reductive mode with 80 ml/min of hydrogen added to the GC column effluent prior to pyrolysis. The bridge current was 30 V. The FID detector was run at a range of 10^{-11} amps/mV and an attenuation of X8 to X32.

Extraction of Drinking Water Samples

The procedure followed that used for wastewater samples except that no laboratory chlorination was carried out. Tap water (1 to 50 liters) was passed directly into the column containing XAD resin; workup was as described for wastewater samples.

RESULTS

Evaluation of the TOCl Method

The applicability of the TOCl procedure depends on several factors, *viz.*: (1) the efficiency of adsorption of organics on the XAD-2 resins; (2) the recovery of the organics from the resin by ether elution; (3) the efficiency of the pyrolysis/ coulometric titration procedure; and (4) the magnitude of interferences in these processes. Junk *et al.* (1974) have addressed themselves to items (1) and (2) and have shown that overall recovery of a large number of organics, including organochlorides, is quite good (average recovery of 80 to 90% from 10- to 100-ppb aqueous samples). It should be noted that no data are available in their paper on volatile organics such as chloroform, nor on substances of high molecular weight such as polymeric materials. In both of these cases, one may assume that the recovery efficiencies would be less than those obtained on the test compounds used by Junk *et al.* (1974).

With respect to item (3), the efficiency of the pyrolysis/ microcoulometric titration procedure, there are limited data published although apparatus similar to ours have been utilized by other workers for TOCl determinations of liquid-liquid extracts (Greve and Haring, 1972).* The manufacturer (Dohrmann, private communication) reportedly has found good efficiency for a wide variety of organic halides, but it is clear that the system should be checked with a variety of compounds before the TOCl method is widely applied. Further information on the efficiency of the Dohrmann pyrolyzer/microcoulometer

---

*Kuhn and Sontheimer (1973, 1975) have described TOCl methods based on adsorption of organic halides by activated carbon. A sample of the carbon is then pyrolyzed in a steam/air atmosphere and the resulting hydrogen halides measured either by microcoulometry or by the halide-specific ion electrode.

system will be published by us in a forthcoming paper. However, we may report on the basis of these results that it would appear that one should limit injection volumes to 5 to 25 μl and chloride amounts to something less than 1000 ng. Both conclusions are consistent with the recommendations of the manufacturer. Our data also show that 1 to 2% precision is possible (3 injections); however, it should be noted that the injection process requires care and there is considerable variation from one operation to another.

Figures 2 and 3 show recorder traces of the coulometric integral for a chlorobenzene/ether standard and a wastewater extract in ether. It is of particular interest to note the shapes of these curves, both of which were obtained using the same platinum boat injection procedure.* The shape of the chlorobenzene integral is typical of volatile organic halides which apparently are swept into the furnace as a narrow plug. The wastewater extract curves in Figure 3 show a much different shape, which apparently reflects the presence of higher-molecular-weight organohalides in wastewater extracts. Thus, the XAD resin extraction procedure does not appear to be limited to low-molecular-weight materials, although the data here do not give any evidence of the overall efficiency of the extraction process.

Because of the apparent presence of high-molecular-weight materials in the XAD extracts, it was important to determine the time required for complete combustion of the sample using the platinum boat procedure. Figure 4 is a plot of integrator response (ng Cl) vs time for the injection of a 1-μl sample of two different "superchlorinated" wastewater extracts. Inlet furnace temperatures were 800° and 200°C. As indicated in the figure, the 200° curve has not reached a limiting value after 960 sec (the present integration limit), although it appears that the signal is leveling off. For the data obtained at 800°C, however, only 420 sec are required for the integral to reach a value of 95% of the 960-sec value.

Table I shows the results of a limited number of runs using two inorganic halides. In these runs aqueous solutions of NH_4Cl and NaCl were injected into platinum boats and analyzed

---
*The boat is pushed into the edge of the pyrolysis furnace until the recorder signal "peaks," then the boat is pushed into the furnace to its mechanical limit.

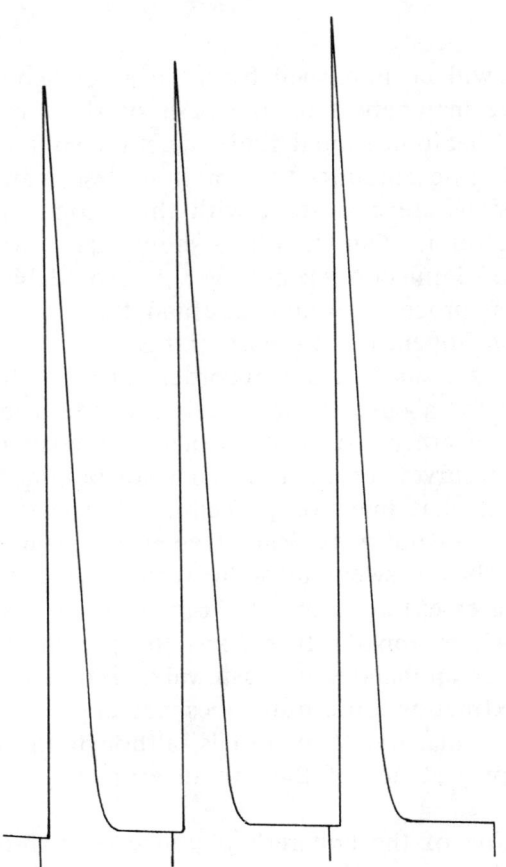

Figure 2. Response of microcoulometric halide analyzer: three repetitive samples of chlorobenzene in isooctane (10 µl of 100 ng/µl). Integral values: 965.4, 955.7 and 966.7 ng Cl. Inlet furnace temperature, 200°C.

as usual. The results in Table I clearly show the interference of inorganic chloride in the pyrolysis/microcoulometry system. Of particular interest are the results with ammonium chloride where quantitative detection is observed even at an inlet furnace temperature of only 200°C.

On the basis of the results shown in Table I, it became imperative to test the recovery of ammonium chloride using the XAD procedure described earlier. Table II shows the results of such experiments. One-liter samples of ammonium chloride in water (100- and 500-mg/l concentrations) were processed

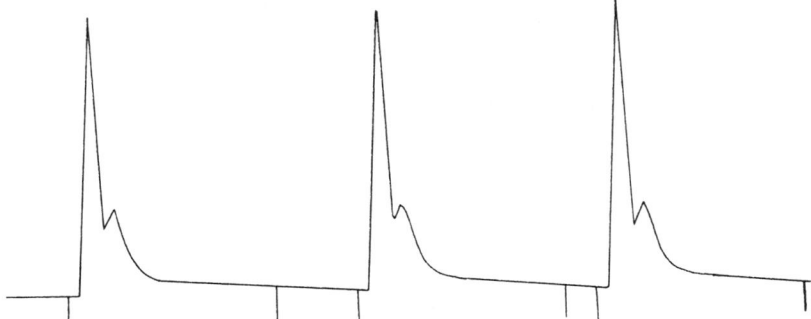

Figure 3. Response of microcoulometric halide analyzer: three repetitive samples of Denton, Texas, wastewater extract after chlorination. 25 ppm (20-μl injections). Integral values: 363.6, 356.5 and 361.5 ng Cl. Inlet furnace temperature, 200°C.

using the scheme shown in Figure 1. Also processed was a distilled water blank. The total quantities of ammonium chloride passed through the resin were 1×10^8 and 5×10^8 ng respectively for the 100 and 500 ppm solutions. As shown in the figure, the ether extract contained only traces of chloride which could not be distinguished from blank levels. Thus, it appears that while inorganic halide would represent a significant interference if one attempted direct aqueous pyrolysis, the use of the XAD/TOCl procedure would seem to eliminate this interference.

Finally, it should be noted that the Dohrmann pyrolysis/ microcoulometer system is sensitive to other elements, particularly sulfur and nitrogen, which may be in wastewater extracts. However, the relative sensitivities for halogens, sulfur and nitrogen are reported to be $1:0.01:10^{-4}$ respectively. Thus, it would appear that the XAD/TOCl method is applicable for the determination of TOCl values in both wastewater and potable waters, since interferences of these magnitudes are not expected to occur in XAD extracts.

Table III shows the results of the application of the method for the analysis of municipal wastewaters after chlorinations. Figure 5 is a recorder trace for a typical wastewater extract

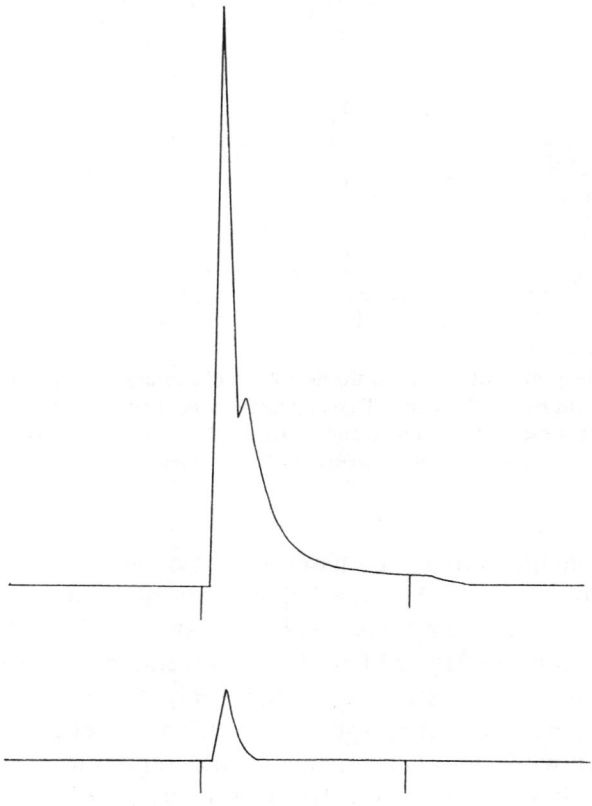

Figure 4. Response of microcoulometric halide analyzer. Upper trace: Denton, Texas, wastewater extract after chlorination (2000 ppm). Lower trace: system blank.

Table I. Microcoulometer Response: Direct Aqueous Injections into Platinum Boat

NH_4Cl/H_2O (200°C Inlet Furnace)		$NaCl/H_2O$ (800°C Inlet Furnace)	
Quantity Injected (ng)	Quantity Detected (ng)	Quantity Injected (ng)	Quantity Detected (ng)
1000	860	100	119
100	104	100	92
100	107	100	84
		100	93
		100	87
		100	85

Table II. Retention of NH$_4$Cl by XAD-2 Resin

Nanograms NH$_4$Cl (1 liter H$_2$O)	Nanograms Recovered in Et$_2$O Eluate (5-ml volume)	% Recovery
Distilled H$_2$O	4.9 ± 1.0 x 10^2	
1 x 10^8 (100 ppm)	5.5 ± 0.5 x 10^2	5.5 x 10^{-4}
5 x 10^8 (500 ppm)	1.2 ± 0.05 x 10^3	2.4 x 10^{-4}

Table III. Total Organic Chlorine: Denton, Texas, Wastewater Effluents

Pyrolysis Temp. (°C)	Chlorine Dose	TOCl (µg/l) Before Chlorination	TOCl (µg/l) After Chlorination	Blank
800	2000	10.96 ± 0.2	906.30 ± 73.6	4.4 ± 0.2
800	1200	22.40 ± 0.3	608.95 ± 5.3	
800	3500	35.40 ± 0.4	163.80 ± 1.9	21.8 ± 0.2
800	1200	36.90 ± 0.8	337.70 ± 20.2	
800	1400	23.00 ± 0.2	402.80 ± 1.0	
200	2000	13.00 ± 0.1	479.70 ± 6.8	1.7 ± 0.1
200	25	16.20 ± 0.5	18.80 ± 0.1	

and a system blank. As indicated in Table III, system blanks (using distilled water) are quite variable and reflect the impurity levels of distilled water as well as the ether. Nonetheless, it is clear from these data that chlorination of wastewaters, particularly using "superchlorination" conditions, results in a significant quantity of organic halogen as compared to nonchlorinated samples. Further work in this area will attempt to extend these measurements to survey various wastewaters and potable waters, and to improve the overall efficiency of the XAD/TOCl procedure.

GCMS Studies on Chlorinated Waters

More explicit information concerning the new chlorinated organic compounds in "superchlorinated" wastewaters has been obtained by combined gas chromatography-mass spectrometry. This method should be viewed as complementary to the high-pressure ion exchange chromatography method used by Jolley

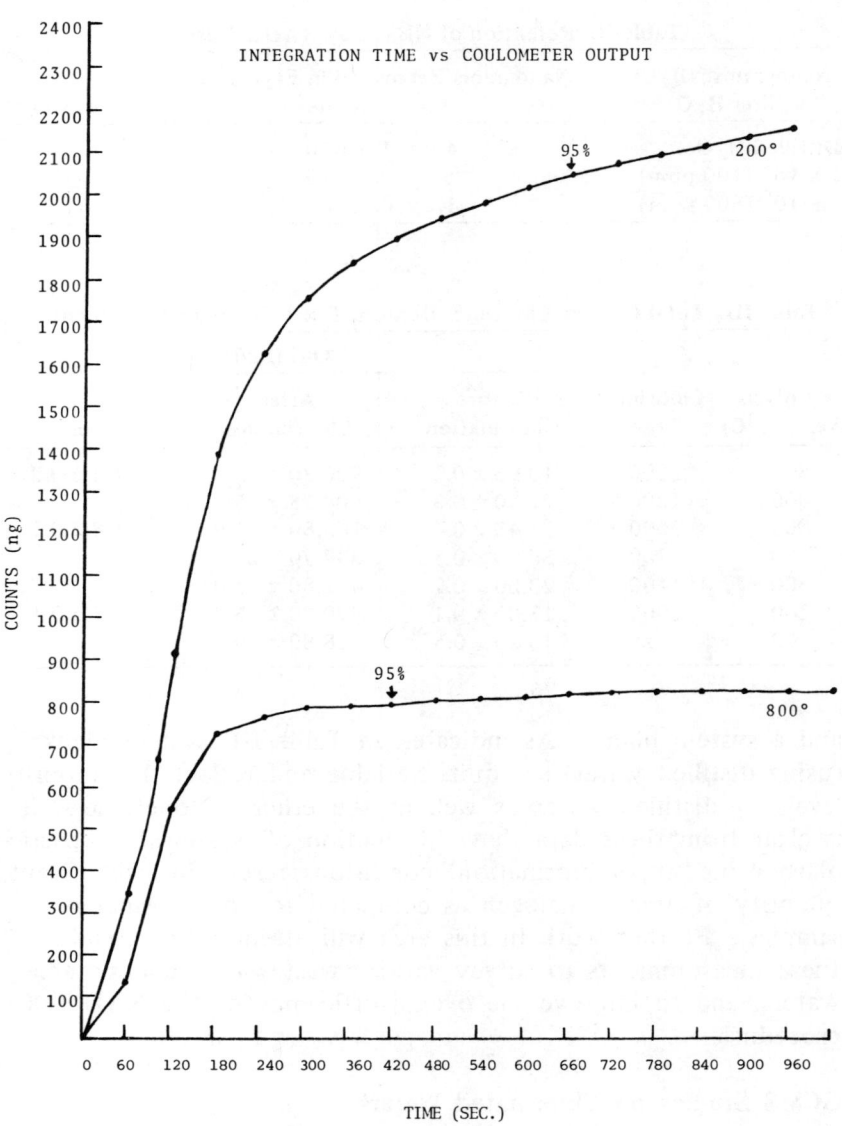

Figure 5. Response of pyrolysis/microcoulometry system. Denton, Texas, wastewater extract. Upper curve, Sample I: 200°C inlet furnace temperature. Lower curve, Sample II: 800°C inlet furnace temperature.

(1973, 1974, 1975), in that XAD-2 is more efficient at trapping nonpolar compounds. Thus, Jolley characterized several new chlorinated aromatics including carboxylic acids and phenols. Of particular interest was the finding of chloroderivatives of uracil, uridine, caffeine, guanine and xanthine, which may possess significant physiological activity.

The results of GCMS investigations of "superchlorinated" wastewaters have been published by us recently and will be reviewed here (Glaze and Henderson, 1975).* As shown in Figure 6, the FID chromatogram of a typical wastewater XAD extract contains innumerable species, only some of which are chlorinated. To distinguish chlorinated species, the Coulson electrolytic conductivity detector was employed, the result of which is shown in Figure 7. The very large number of chlorinated species in the upper spectrum is consistent with the TOCl results discussed in the earlier portion of this paper.

Figure 8 is a total ion monitor of the gas chromatograms of another superchlorinated extract and control. In this case, the time axis is replaced by a spectrum number axis, indicating that a total of 450 mass spectra have been taken and stored in computer memory during the chromatographic run. Analysis of these mass spectra is then accomplished by manual interpretation and comparison with standard spectra (Registry of Mass Spectral Data, 1974 and EPA Mass Spectral Search System). Some of the compounds identified by this procedure are listed in Table IV.

It is clear from an inspection of Table IV that most of the compounds identified thus far are aromatic derivatives. However, the compounds are by no means derivatives of "activated" aromatics in every case. The chloroderivatives of benzene, toluene and benzyl alcohol are evidence that superchlorination may lead to substitution of unactivated aromatic moieties. Moreover, we are particularly interested in the formation of several nonaromatic derivatives, such as chlorocyclohexane, a chloroalkyl acetate, and perhaps most significant, three chlorinated acetone derivatives. The latter may be precursors of chloroform which we have shown in previous work to be formed

*Figures 6 to 8 published by permission of the Journal of the Water Pollution Control Federation.

152 WATER CHLORINATION

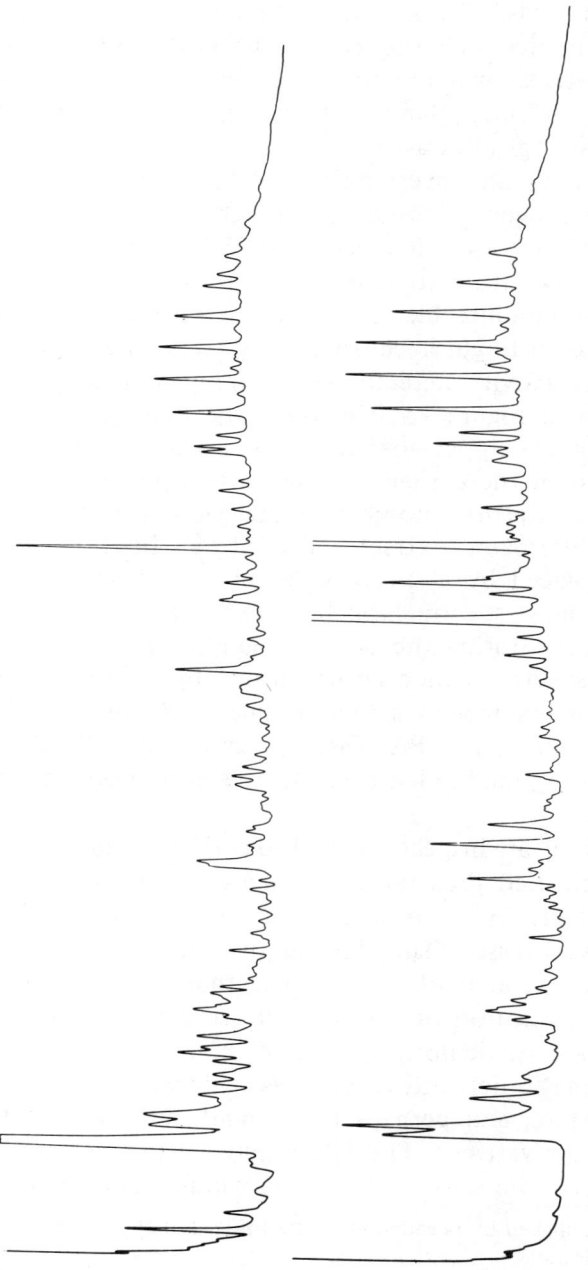

Figure 6. Gas chromatograms (FID detector, $10^{-11} \times 16$ amps/fs) of Denton, Texas, wastewater extract (11-12-74). Bottom, before chlorination: top, after 2000 mg/l chlorine dose for 1-hr contact period. Analytical conditions described in text.

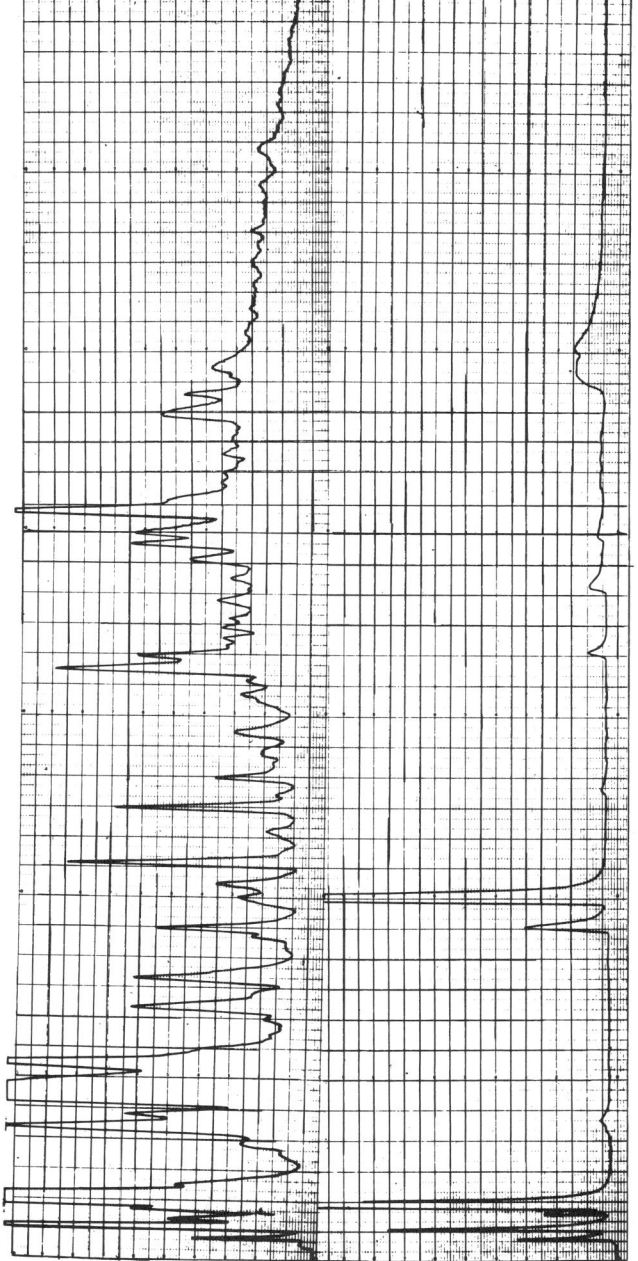

Figure 7. Gas chromatograms (Coulson electrolytic conductivity detector, halogen mode, x 4) of Denton, Texas, wastewater extract (11-22-74). Bottom, before chlorination; top, after 2000 mg/l chlorine dose for 1-hr contact period. Analytical conditions described in text.

154 WATER CHLORINATION

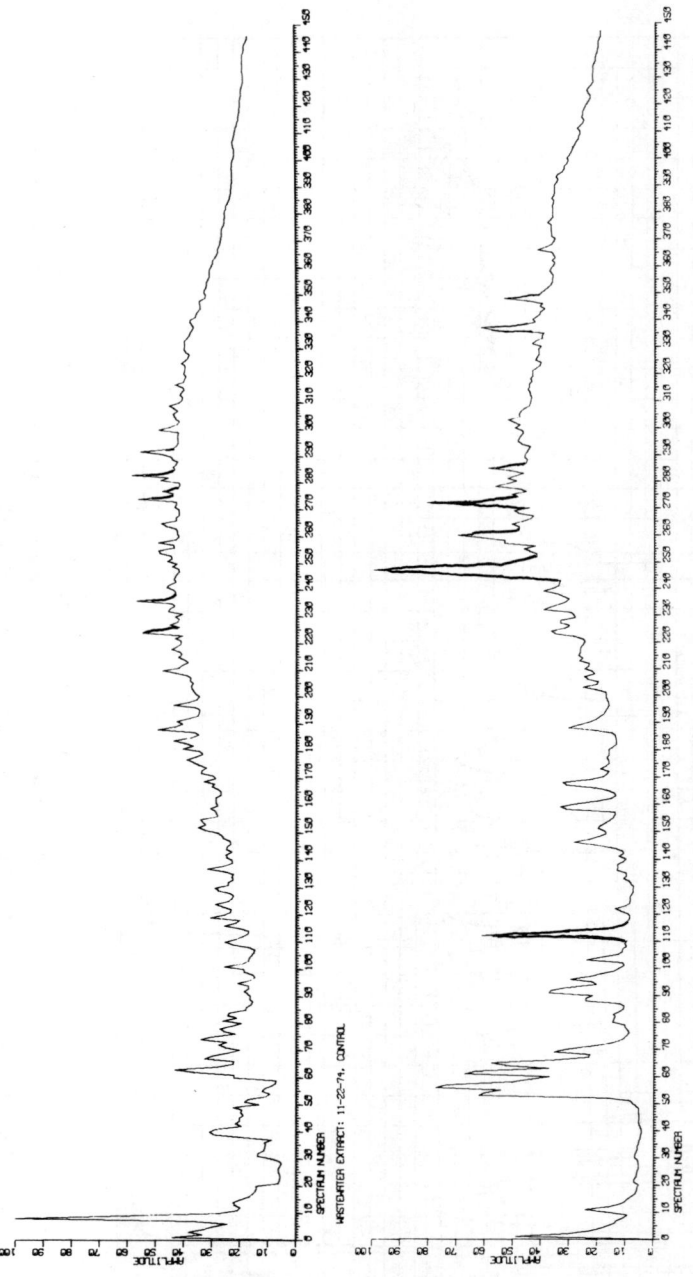

Figure 8. Gas chromatograms (mass spectrometer total ion monitor) of Denton, Texas, wastewater extract (11-22-74). Peaks normalized by computer to make largest peak full scale.

Table IV. Compounds Identified in Superchlorinated Municipal Wastewaters (Glaze and Henderson, 1975)

Nonaromatics

Chloroform	Chlorocyclohexane
Dibromochloromethane	Tetrachloroacetone
Dichlorobutane	Pentachloroacetone
3-Chloro-2-methylbut-1-ene	Hexachloroacetone

Aromatics

o-Dichlorobenzene	Trichloroethylbenzene[a]
p-Dichlorobenzene	Trichlorocumene[a]
Chloroethylbenzene[a]	Dichlorotoluene[a]
Dichloroethylbenzene[a]	Chlorocumene[a]
N-Methyl-trichloroaniline[a]	Trichlorophenol[a]
Trichlorodimethoxybenzene[a]	Tetrachlorophenol[a]
Tetrachloroethylstyrene[a]	Tetrachlorodimethoxybenzene[a]
Trichloromethylstyrene[a]	Trichlorophthalate[a]
Chloro-α-methylbenzyl alcohol[a]	Tetrachlorophthalate[a]
Dichloro-α-methylbenzyl alcohol[a]	

[a]Isomer not specified.

in wastewater superchlorinations and which has been shown by other workers to result from the chlorination of organics in drinking water.

Finally, we note that the concentrations of the compounds listed in Table IV are in the $\mu g/l$ range (ppb) in agreement with earlier works in this area.

Mass Spectra of Volatile Organohalides in Drinking Water

Figure 9 is a gas chromatogram of an ether extract of drinking water obtained from a tap in the NTSU Chemistry Building. In addition to those compounds found in earlier work, it is interesting to note the presence of three iodine-containing compounds, dichloroiodomethane, bromochloroiodomethane, and dibromoiodomethane. The source of these compounds in Denton, Texas, drinking water is unknown; presumably they are formed by chlorine treatment. Further work in this area is currently in progress.

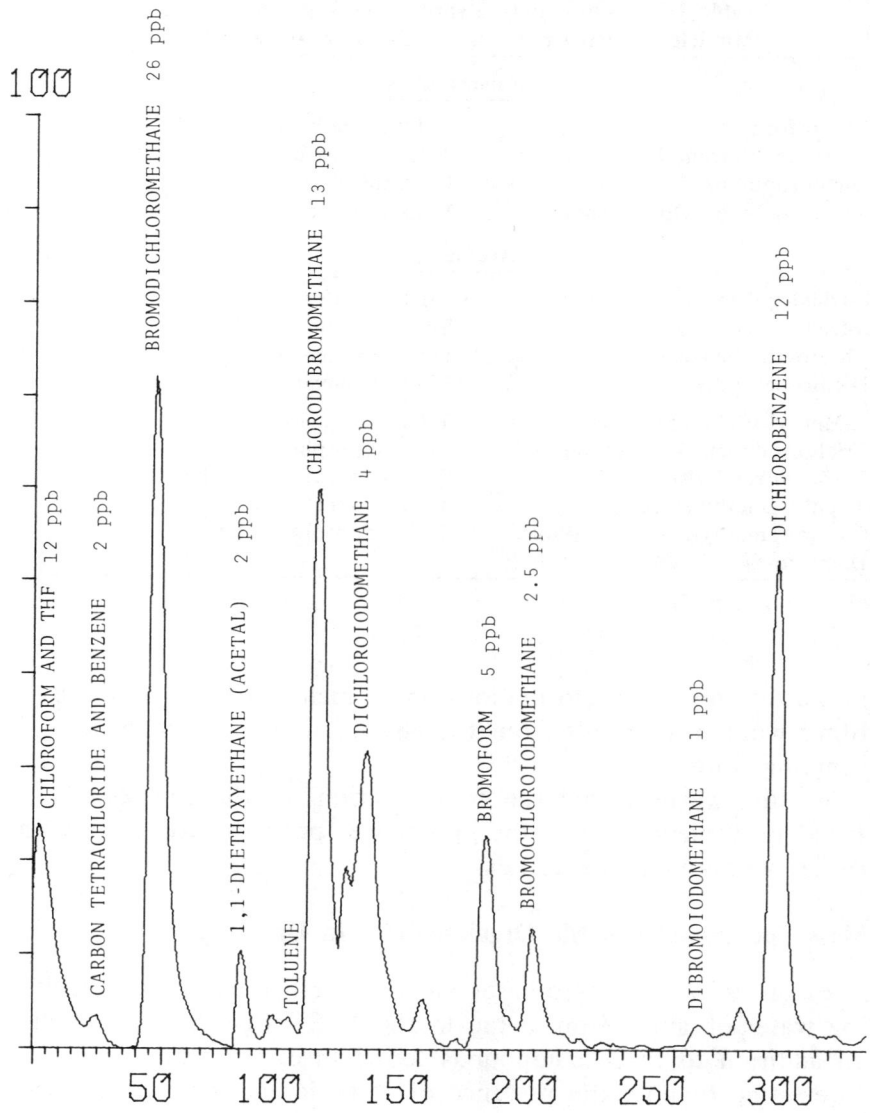

Figure 9. Gas chromatogram of tap water taken at NTSU Chemistry Building (8-13-75); Finnigan Model 3200 MS/GC System; 3% Dexsil 300 GC column on 100/120 mesh Supelcoport.

ACKNOWLEDGMENTS

This research was supported in part by Grant No. R803007 from the Environmental Protection Agency and in part from a grant from the NTSU Faculty Research Committee. We are grateful to Dr. Craig Shew of the Robert Kerr Water Research Laboratory in Ada, Oklahoma, for assistance in the GCMS work.

REFERENCES

Bellar, T. A. and J. J. Lichtenberg. 1974. "The Determination of Volatile Organic Compounds at the µg/l Level in Water by Gas Chromatography." *J. Am. Water Works Assoc.* 66:739-744.

E.P.A. Mass Spectral Search File, Cyphernetic Corp., Ann Arbor, Michigan.

Glaze, W. H., J. E. Henderson IV, J. E. Bell and V. A Wheeler. 1973. "Analysis of Organic Materials in Wastewaters After Chlorination." *J. Chromatogr. Sci.* 11:580-584.

Glaze, W. H. and J. E. Henderson IV. 1975. "Formation of Organochlorine Compounds from the Chlorination of a Municipal Secondary Effluent." *J. Water Poll. Control Fed.* 47:2511-2515.

Greve, P. A. and B. J. A. Haring. 1972. "Die Mikrocoulometrische Bestimmung von organische gebunden Halogen in Oberflachen–und anderen Gewassern." *Schriftenr. Ver. Wasser. Boden Lufthyg. Berlin-Dahlem* 37:59-64.

Kuhn, W. and H. Sontheimer. 1973. "Einige Untersuchungen Zur Bestimmung von organische Chlorverbindungen auf Aktivkohlen." *Vom Wasser* 41:65-79.

Kuhn, W. and H. Sontheimer. 1975. "Zur analytischen Erfassung organischer Chlorverbindungen mit der temperaturprogrammierten Pyrohydrolyse." *Vom Wasser* 43:327-341.

Jolley, R. L. 1973. *Chlorination Effects on Organic Constituents in Effluents from Domestic Sanitary Sewage Treatment Plants.* ORNL/TM-4290. Oak Ridge National Laboratory, Oak Ridge, Tennessee.

Jolley, R. L. 1974. "Determination of Chlorine-Containing Organics in Chlorinated Sewage by Coupled ^{36}Cl tracer–High Resolution Chromatography." *Environ. Lett.* 7:321-340.

Jolley, R. L. 1975. "Chlorine-Containing Organic Constituents in Chlorinated Effluents." *J. Water Poll. Control Fed.* 47:601-618.

Junk, G. A., J. J. Richard, M. D. Grieser, D. Witiak, J. L. Witiak, M. D. Arguello, R. Vick, H. J. Svec, J. S. Fritz and G. V. Calder. 1974. "Use of Macroreticular Resins in the Analysis of Water for Trace Organic Contaminants." *J. Chromatogr.* 99:745-762.

Morris, J. C. 1975. United States Environmental Protection Agency Report. EPA-600/1-75-002.

Rook, J. J. 1974. "Formation of Haloforms During Chlorination of Natural Waters." *Water Treat. Examin.* 23(2):234-243.

Stenhagen, E., F. Abrahamsson and F. W. McLafferty. 1974. *Registry of Mass Spectral Data.* Wiley-Interscience, New York.

DISCUSSION

George Clifford White, Consulting Engineer. This has a very significant practical application in wastewater disinfection. Normally chlorinators operate their injectors with effluent. To put it in perspective, an 8000-lb/day chlorinator that would dose at the rate of 8 to 10 mg/l chlorine uses 200 gal/min backwater that would be producing the organics that you showed there.

Glaze. You mean the make-up water?

White. The water going through the injector. In other words, for an 8000-lb/day chlorinator the ratio of chlorinated effluent injector water to sewage effluent would be about 200 gal/min effluent containing 3500 ppm chlorine solution to about 80,000,000 gal/day sewage. So you could work out what the organic pickup was on account of the use of effluent for injector water.

Robert B. Dean, U.S. Environmental Protection Agency. I am glad you are going after the amino acids. It is very important to look at chlorinated proteins after hydrolysis to amino acids. It is easy to graft chlorine onto a protein and produce a product that is resistant to bacterial attack. This production is based on the extreme ease of producing iodinated tyrosine and thyroxine hormone by adding iodine to dry proteins. We have been looking for compounds only where we have a convenient analytical technique and have not looked at the rest of the compounds that are not volatile or elutable from XAD resin.

George Helz, University of Maryland. I have several questions about the XAD resin. First of all, what happens to the chloramine? Second, do you know if a very large chlorine concentration generates chloro-organics on the resin?

Glaze. Well, we destroy the chlorine with sodium sulfite before we use the resin. So the answer to the last question is presumably no. The more general question of what happens to chloramines on XAD, again they are presumably destroyed by sulfite. So we don't see them. Whether if they were not destroyed they would be trapped, eluted and preserved in this method, I don't know. We have not looked into that.

Max Deinzer, Oregon State University. What do you think is the mechanism for formation of symmetrically substituted chloroacetones in water supplies?

Glaze. I suppose we are really under very acid conditions. I don't have a mechanism. Perhaps some other expert in the audience has a mechanism.

Sidney Katz, Oak Ridge National Laboratory. Since recovery of unknown compounds from XAD resins is an unknown, how do you quantify your results?

Glaze. Any kind of concentration, whether it would be carbon or XAD, suffers from that disadvantage. I don't even want to apologize for it. That is a fact of life. What one can do, and what Fritz's group has done, is to look at a wide variety of model compounds and look at efficiencies. In the *Journal of Chromatography* in 1974, p. 745, there is a very very fine paper by that group which lists efficiencies for many compounds. The efficiencies are all high. They didn't go to high-molecular-weight ranges. They didn't look at the volatiles. So there are some holes in that work, obviously. They just didn't have that much time, I guess. So you do what you can do and, fortunately for us, they have done it at Iowa State, and it looks like efficiencies for most of the compounds that we're identifying are really not bad, not bad at all. So we have quantified these, by the way, and if you want the numbers I'll give them to you. But I agree with you that they are probably only good to within a factor of two or three or so.

David Friedman, Food and Drug Administration. Just a comment on XAD resins. We have tried the resins with highly chlorinated materials. They are very good at taking material out of the water, but you can't get them back off the resin.

Glaze. What kind of material are you talking about?

Friedman. Chlorinated paraffins.

Glaze. Chlorinated paraffins, yes, because they presumably penetrate the polymer. That's a valid criticism. I think we cannot say that we get all of the material out. It's one technique, like carbon, which has some merit.

8

CHEMISTRY OF HALOGENS IN SEA WATER

James H. Carpenter and Donald L. Macalady
   Rosenstiel School of Marine and
     Atmospheric Science
   University of Miami
   Miami, Florida 33149

ABSTRACT

There has not been sufficient research to provide a satisfactory understanding of the reactions that occur when +1 oxidation state chlorine is added to sea water. However, present information suggests that the bromide ion is oxidized and, perhaps, disproportionates to several oxidation states. Formation of brominated or mixed brominated-chlorinated organic compounds can be expected, but the extent and speciation of such reactions remain to be determined.

Our experiments show that present analytical procedures do not measure all of the inorganic "residuals" present in chlorinated sea water.

INTRODUCTION

Other presentations at this conference review the chemistry of chlorine in fresh waters and wastewaters. This paper focuses on the limited information that is available for reactions that occur when chlorine is added to sea water. The material that is reviewed here is intended to supplement the fresh water and wastewater data and those facts will not be repeated here. The principal reason for differences between the sets of reactions that may occur in sea water in contrast to fresh waters and wastewaters is, of course, the particular composition of sea water.

Since this presentation is addressed to individuals of diverse backgrounds, a brief and superficial description of the nature of sea water and oceanography may be helpful. The large mass of liquid that fills the ocean basins has been produced by processes acting over long periods of time and it is, therefore, rather uniform. However, the composition of this natural solution is rather complex. Elemental analysis has shown 75 of the chemical elements to be present in concentrations ranging from grams per liter to picograms per liter. The ionic and molecular speciation is only partly known for the inorganics and the several hundred identified organic compounds only account for part of the total organic content. Furthermore, sea water contains suspended and motile particules ranging from colloidal inorganic minerals through single-celled organisms to large mammals. In short, sea water studies would probably not be attractive to thoughtful scientists and the discipline of oceanography has been developed to provide a haven for the foolhardy.

Much of the current knowledge in chemical oceanography has been summarized in the treatises edited by Riley and Skirrow (1975) and this reference leads to the following generalizations with regard to sea water as a reaction medium:

1. The ionic strength is 0.7 molal.
2. The abundant (mg/l or greater) cations are sodium, magnesium, calcium, potassium and strontium ions and the abundant anions are chloride, sulfate, bromide, bicarbonate, borate, silicate and fluoride ions.
3. The pH is close to 8 if the sea water is in equilibrium with the atmosphere.
4. The alkalinity is 2 milliequivalents (meq).
5. Dissolved oxygen provides an "apparent" redox potential of approximately 0.5 V.
6. The total organic carbon content is approximately 1 mg/l or less.
7. Transition metals (Fe, Mn, Cu, Ni, Zn, Mo) are present at μg/l concentrations; *i.e.*, at potentially catalytic concentrations but substantial fractions of the metals are present as colloidal hydroxides and organic complexes.
8. The sun shines on the oceans periodically.

OXIDATION-STATE CHLORINE REACTIONS

Turning to the question of what reactions occur when +1 oxidation state chlorine is added to sea water, the first alternative is the addition of chlorine gas or solutions of salts of hypochlorous acid. If chlorine gas is added, the familiar hydrolysis reaction would be expected,

$$Cl_2 + H_2O \rightleftharpoons HOCl + H^+ + Cl^-$$

with a concurrent adjustment of the primary proton equilibrium in sea water involving bicarbonate ion,

$$2HCO_3^- \rightleftharpoons CO_3^= + H_2CO_3$$
$$\downarrow \uparrow$$
$$H_2O + CO_2 \rightleftharpoons CO_2 \text{ (atm)}$$

In order to see the characteristics of these reactions, consider the addition of chlorine to make the solution 3.5 ppm added chlorine. The presence of 0.53 M chloride tends to repress the hydrolysis. The assumed added chlorine to 50 μM is computed to cause a resultant concentration of dissolved molecular chlorine of 6 pM and a transient pH change of less than 0.1 unit until equilibrium with the atmosphere occurs.

The addition of a similar quantity of hypochlorous acid or hypochlorite salts would produce approximately the same final equilibrium. Disequilibrium during the transient initial injection of chlorine where local higher concentrations would be expected would permit lower pH and higher chlorine concentrations but cannot be generalized.

The high ionic strength and pH of sea water would encourage the dissociation of the hypochlorous acid and, assuming an activity coefficient of 0.6 for hypochlorite ions, the equilibrium would be 90% hypochlorite ion and 10% hypochlorous acid at 20°C, using pK of 7.54 (Morris, 1966). Sugam and Helz (1975) have examined the apparent ionization of hypochlorous acid in "sea water" (no bromide ion present) and found that some ion-pairing of hypochlorite ions with the cations was present to the extent of approximately 20%.

Thus, it seems clear that the most pertinent chemistry is that of the halogen oxyacids. The statements by A. C. Downs and C. J. Adams in Part 26 of the series, *Comprehensive Inorganic Chemistry* (Pergamon Press) on page 1347 provide an excellent introduction to this subject.

"Both thermodynamic and kinetic factors are important in the chemistry of the halogen oxyacids, and particularly in the interrelationships of the various oxidation states of the halogens in the condensed phases; moreover, both the thermodynamic and kinetic parameters which control the reactions of the acids and anions in solution are critically sensitive to pH." We may reasonably add the idea that other aspects of the reaction medium as outlined above will need consideration. Grounds for intimidation are apparent if reproducible and interpretable results are sought.

Three broad categories of further possible reactions are:

1. Decomposition
2. Reaction with organics
3. Reaction with inorganics
   a. ammonium ions
   b. other halogens

The decomposition of hypochlorite solutions can occur through different competing reactions:

$$2ClO^- \rightarrow 2Cl^- + O_2 \qquad (1)$$

$$2ClO^- \rightarrow ClO_2^- + Cl^- \qquad (2)$$

$$ClO^- + ClO_2^- \rightarrow ClO_3^- + Cl^- \qquad (3)$$

Reaction 1 does not occur readily for the hypochlorite ion, as shown by the stability of commercial bleaching solution. The analogous reaction for the free acid occurs more readily and, if it were predominant, the environmental effects of our use of chlorine would be greatly reduced.

Reaction 2 is reported to be slow and solutions of hypochlorite ions can be expected to contain chlorite ion concentrations that increase with the age of the solution. The use of aged commercial bleaching solutions or aged reagent-grade sodium hypochlorite solutions for chlorination in bioassay tests or biofouling control usually involves a determination of the strength of the solutions by the *Standard Methods* procedure (1971, sec. 114A). As pointed out by Kolthoff (1922), rapid titration of solutions buffered with acetic acid-acetate ion misses the chlorite ion content, since it reacts slowly with iodide ion. Chlorite ion may be toxic, and differences between experiments and experimenters may arise from the failure of the present "strength" standardization procedure to measure the chlorite ion content of the solutions.

For sea water, the high chloride ion concentration may influence the extent of both Reactions 2 and 3. The addition of chlorine may be expected to yield a mixture of hypochlorite, chlorite and chlorate ions with the relative abundances of the three species depending on reaction conditions and time. Photolysis may be especially important.

With regard to reaction with organics, there have not been any extensive systematic studies using sea water. Dr. Peter Williams of the Scripps Institution of Oceanography has observed the transformation of fluorescein to the brominated product, eosin, and the formation of brominated phenols in chlorinated sea water.

With regard to reaction with inorganics, the formation of chloramines from ammonium ion would be expected and this subject has been thoroughly reviewed by other participants. It should be noted that the ammonium ion concentrations in sea water are variable and lower than in domestic wastewaters. For example, Chesapeake Bay waters vary seasonally from 50 to 5 μM. Florida Current waters contain less than 1 μM ammonium ion.

Reactions with other halogens would not involve iodide, since the abundant form of iodine in sea water is iodate and the concentration is 5 μM. Oxidation of fluoride ion would not be expected on thermodynamic grounds. Reaction with bromide would be the principal one to be expected, as pointed out by Johannesson (1955) and Lewis (1966). The resulting bromine would be 99.9% hydrolyzed to hypobromous acid, or tenfold less than the same reaction for chlorine under sea water reaction conditions. The hypobromous acid would be 40% dissociated in sea water. The relative reactivity of bromine in picomolar concentrations and hypobromous acid and hypobromite ion in micromolar concentrations with organics will not be considered here, but poses some interesting questions.

The problem of further reaction of these bromine species is analogous to those discussed previously for chlorine, except the literature is more sparse. Bromamines may be formed. Direct decomposition will compete with disproportionation reactions. The reaction reported by Weszelszky (1900) and modified by van der Meulen (1931) is interesting:

$$Br^- + 3\ ClO^- \rightarrow BrO_3^- + 3\ Cl^- \qquad (4)$$

The requirement that the solution must contain a large excess of sodium chloride for complete reaction from left to right is, to use the vernacular of the American teenagers, somewhat "mind-expanding." The pH is critical and various authors have recommended buffers such as bicarbonate, borate, dihydrogen phosphate, etc., with pH in the range 5.5 to 8.5. Szabo and Csanyi (1952) report that the reaction takes place in two main steps:

$$2\ Br^- + Cl_2 \rightarrow Br_2 + 2\ Cl^-$$
$$Br_2 + H_2O \rightarrow BrOH + H^+ + Br^- \qquad (5)$$
$$BrOH + Cl_2 \rightarrow BrCl + HOCl$$

and

$$3\ BrOH \rightarrow BrO_3^- + 3\ H^+ + 2\ Br^-$$
$$BrCl + 2Cl_2 \rightarrow BrCl_5 \qquad (6)$$
$$BrCl_5 + 3\ H_2O \rightarrow BrO_3^- + 6\ H^+ + 5\ Cl^-$$

The first steps are favored by pH 6.5 to 7.5 and the second steps are favored by somewhat higher pH. The details of the proposed reactions are probably not in close agreement with current reaction mechanism theory but clearly illustrate the multistep system with which we must deal.

Recent unpublished work at the University of Miami has sought the identification of bromate as a product of the chlorination of sea water. With 30-min reaction time, polarographic observations after destruction of hypochlorite ion with thiosulfate show that bromate is *not* present at concentrations as large as 10 μM.

Another possible product is bromite, but there is no evidence cited in the literature for bromite formation in chlorinated sea water.

In addition to the oxyanions, there is the possibility that interhalogen complexes are formed in sea water. Pungor et al. (1959) have presented spectroscopic evidence for the existence of chloride ion complexes of bromine chloride in aqueous solution. Identical spectra were found in 0.5 N NH$_4$Cl and 0.5 N NaCl. They suggest the formula BrCl$_6^{5-}$ for the complex, but Gutmann et al. (1968) express the view that other chloride complexes with less chloride per bromine may also account for the spectra. The bromine chloride complexes show lower redox potentials than "free bromine chloride" by 0.2 V and, if such species are formed when chlorine is added to sea water, possible additional thermodynamic and kinetic phenomena will be present. Also, bromine chloride is reported to react more rapidly with aromatic and olefinic compounds.

The impression that we have gleaned from the literature is that chlorination of sea water may be expected to produce several inorganic species containing bromine (these compounds may be directly toxic to aquatic organisms) and, in view of the reactivity with organics of the expectable bromine species, a variety of brominated or mixed chlorobromo compounds is probably being produced with present chlorination of power plant cooling waters and chlorinated wastewater discharges.

WHAT DO PRESENT ANALYTICAL METHODS MEASURE?

Colorimetric methods are popular because of the apparent simplicity. The most widely used reagent is orthotolidine and some state agencies require chlorination of wastewaters to residual levels as determined with this reagent. As reported by Bellanca and Bailey (1975), the use of orthotolidine produced a serious underestimation of residual chlorine and resulted in extensive fish kills in the James River estuary. Eppley et al. (1975) found that chlorinated sea water samples, judged to be chlorine-free with the orthotolidine method, were still inhibitory to phytoplankton photosynthesis. Carpenter et al. (1972) observed that chlorination of sea water used for power plant cooling, at levels that produced no detectable residuals at the outfall, decreased phytoplankton productivity by 79%. It seems clear that the orthotolidine procedures are badly misrepresenting the "residuals" in wastewaters and sea water.

Various methods have been compared as reported in *Standard Methods* (1971). The iodometric method is considered the standard against which other methods are judged. Standards were distributed to 32 laboratories and the results show a relative standard deviation of ca. 27% for the iodometric method —an appalling situation. The amperometric titration method is "rated among the most accurate for the determination of free or combined available chlorine." The iodometric method includes the use of starch for visual endpoint estimation in the presence of 0.012 M potassium iodide and the amperometric method includes the use of a galvanic cell in a deadstop endpoint procedure in the presence of 0.0015 M potassium iodide. The amperometric endpoint detection is more sensitive than visual starch indication, as pointed out by Bradbury and Hambly (1952). However, the electroactive species appears to be free diatomic iodine, rather than triiodide ion that is formed in the presence of high concentrations of iodide ion. The lower concentration of iodide used in the amperometric method reflects this characteristic. However, the volatility of free iodine is well known and severe errors may result from the violent agitation used in commerical amperometric instruments. The magnitudes of the errors depend on the particular ways in which various people carry out the manipulations and, with practice, precise but erroneous data can be generated by any one individual, as demonstrated by Carritt and Carpenter (1966). Present methods do not appear to be satisfactory if close compliance with regulatory statutes is to be demonstrated by analysis of samples from wastewater treatment plants and electric generating stations.

We have carried out some tests of methods for estimating "residual chlorine" in sea water and other solutions in the following manner.

EXPERIMENTAL PROCEDURE

In each of our experiments, sufficient chlorine reagent was added to a measured volume (usually about 250 ml) of the test solutions to achieve a chlorine concentration in the range 2 to 12 ppm. In some of the early trials, Fisher Chlorine Test

Solution, U.S.P., was used. For later trials, a chlorine reagent was made by diluting a saturated (15°C) solution of Cl_2 gas in distilled water. The dilute chlorine reagent was prepared fresh daily and buffered with 0.1 g Na_2CO_3 per 100 ml. (Results were observed not to be related to the origin of the Cl_2 reagent.)

The five different test solutions were prepared as follows:

1. Ion-exchanged and distilled water, with 2.0×10^{-3} M $NaHCO_3$ added.
2. 1.0×10^{-3} M KBr in 2.0×10^{-3} M $NaHCO_3$.
3. 35°/oo NaCl in $2.0 \times 10^{-3} M$ $NaHCO_3$. Due to impurities in the reagent-grade NaCl, this solution was also about 10^{-5} M in Br⁻.
4. 35°/oo NaCl in 2.0×10^{-3} M $NaHCO_3$ with KBr added to make the solution 1.0×10^{-3} M in Br⁻.
5. Florida Current sea water, filtered through a 0.22-μ Teflon Millipore filter.

During a typical experiment, identical quantities of the chlorine reagent were added simultaneously to each of the five test solutions. The chlorinated test solutions were then allowed to react with gentle stirring at room temperature for 30 min. Analysis for "total residual chlorine" was then made by an iodometric procedure using standardized sodium thiosulfate solution as the reducing titrant and 5% starch solution as an indicator.

Three different chemical conditions for generation of iodine were created for each set of analyses.

In some tests, analytical conditions were established to duplicate those prescribed in *Standard Methods*, Section 114B, the amperometric titration method. Here 1.0 ml of a 50-g/l solution of KI and 1 ml of acetic acid/sodium acetate buffer (pH = 4) were added sequentially to the 250 ml of test solution immediately prior to titration.

In other test sets, analytical conditions similar to *Standard Methods*, Section 114A, were achieved by adding 2 ml of a 50-g/l KI solution followed by 2 ml of the pH 4 buffer immediately prior to titration.

In a third group of experiments, iodine was generated by adding 2 ml of 50 g/l KI followed by sufficient 10 M H_2SO_4

to achieve a pH of 2.0. Titration was again begun immediately following the addition of the KI and acid.

After titration to an initial endpoint (disappearance of I_3^--starch blue color), the titration flasks were stoppered and allowed to stand unstirred in the laboratory. Any additional blue color which appeared after 30 min was then titrated with additional thiosulfate. In many of the tests, the flasks were then allowed to stand for longer periods of time, with subsequent titration of reappearing blue color.

Periodic checks for air oxidation of KI solutions were made by adding starch directly to buffered KI solutions and noting any appearance of blue color. If a color developed, fresh solutions were prepared.

Standardization of the thiosulfate was carried out according to the biiodate method outlined in *Standard Methods*, Section 114A.

RESULTS

The results of one experiment, in which 1.8 ppm of chlorine was added to the five different solutions, are shown in Figure 1. The differences observed with titrations of distilled water solutions under different pH and potassium iodide concentrations are larger than we have found in subsequent experiments. The data in Figure 1 were collected using a piston microburette that permits very rapid titration. With slower titration (1 to 2 min), permanent endpoints are found and the three titration conditions show differences of less than 8% with the solution containing the lower concentration of potassium iodide requiring the least titrant.

The sea water solution showed a "residual" of 0.7 ppm chlorine or an apparent demand of 1.1 ppm, with the recommended potassium iodide concentration for amperometric titration. The Florida Current sea water that was used would be expected to contain only 0.1 ppm of total organic carbon and the "chlorine demand" is seen to be related to the titration conditions; *i.e.*, the pH 2, higher potassium iodide concentration results indicate a demand of only 0.2 ppm and the apparent large demand is an artifact of the particular titration conditions. This viewpoint is strengthened by our observation that when the same sea water

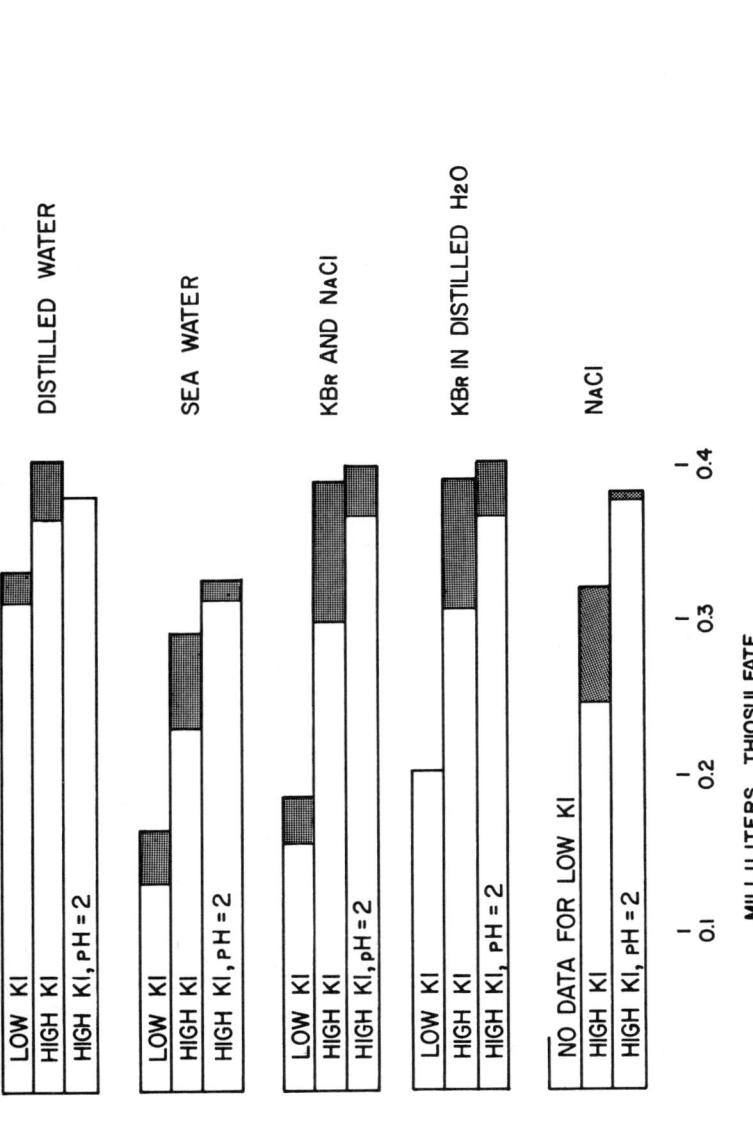

Figure 1. Titration of solutions that had 1.8 ppm chlorine added. Rapid titration shown as the clear bar and additional titration after 30 min shown as cross-hatched.

was chlorinated to an added level of 11.4 ppm, the "apparent demand" was 4.9 ppm, which cannot be accounted for in terms of reducing substances in such clean sea water. When the titrated solution was stored in the dark for 48 hr, the starch triiodide blue color reappeared and titration to the "new" endpoint showed an apparent demand of 2.6 ppm.

The same disappearance of the added chlorine was found in potassium bromide solutions and in potassium bromide-sodium chloride solutions when they were titrated at pH 4 with the lower concentration of potassium iodide. With higher concentration of potassium iodide or lower pH, these solutions showed no significant "loss of chlorine" during the 30-min reaction time.

The results for the sodium chloride solution are somewhat perplexing. In subsequent experiments, the lower concentration of potassium iodide produced some apparent chlorine demand for additions of 1 to 3 ppm chlorine, but this effect was smaller when additions of 10 to 26 ppm chlorine were made. With slower titration of several minutes, results for the sodium chloride solution were similar to those for distilled water. If the solutions were stored for 24 to 48 hr, there was no significant chlorine demand. The sodium chloride was not free of sodium bromide and some of the kinetic effects could be due to the bromide rather than the chloride content of the solutions.

The slow reaction of the test solutions with potassium iodide can occur if chlorite is present. As pointed out by Kolthoff *et al.* (1957), if the solutions are acidified before adding the potassium iodide, low results are obtained owing to the formation of chlorine dioxide. We examined our chlorination solutions for evidence of the presence of chlorite ions by comparing the results with acidification before and after the addition of potassium iodide. As shown in Figure 2, there was no evidence of significant chlorite concentrations. Also, with 30 min of reaction time after chlorination, no evidence for chlorite was found.

A possible reason for the smaller titrant volumes required for chlorinated sea water compared to chlorinated distilled water could be the formation of substances that oxidize iodide ion slowly but oxidize thiosulfate ion to sulfate ion, rather than to tetrathionate ion. Phenylarsine oxide does not have the multiple oxidation state characteristics that thiosulfate does.

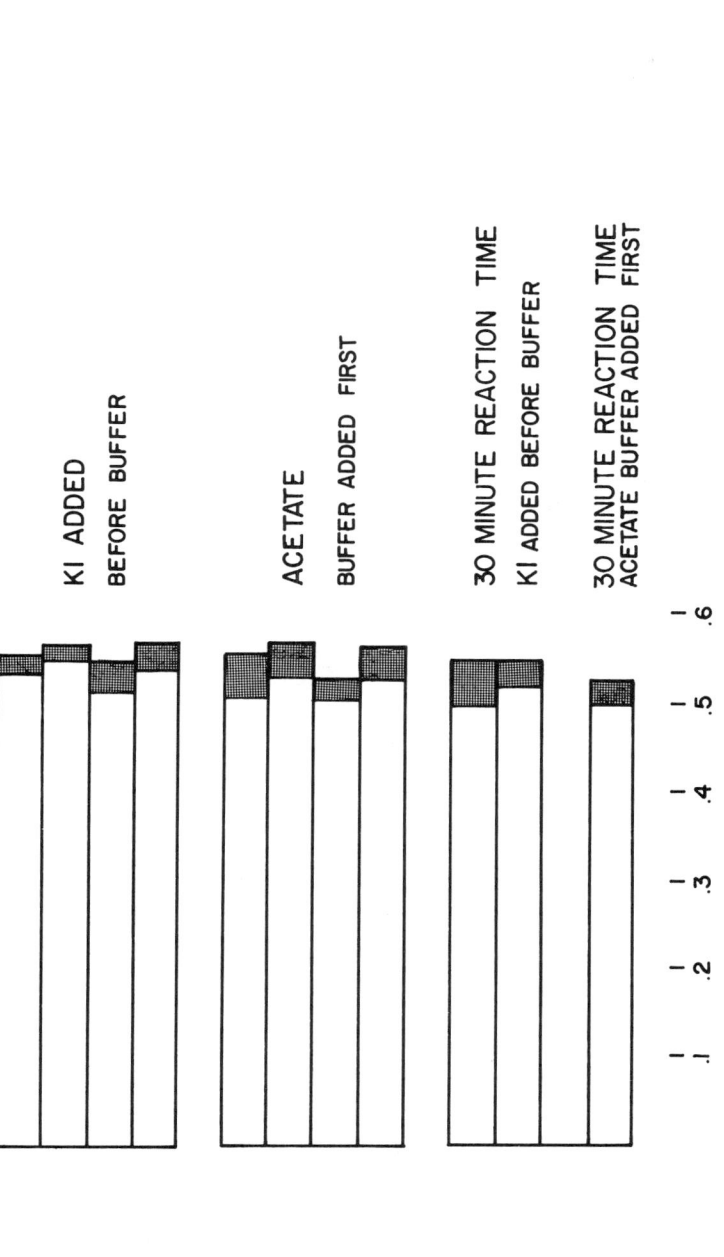

Figure 2. Results of tests for chlorite in the solutions used for chlorination. Acetate buffer addition before potassium iodide would produce lower results if chlorite were present.

We have titrated the iodine in replicate sea water samples, chlorinated to 12 ppm, with thiosulfate and phenylarsine oxide without finding differences that were larger than the reproducibility of the experiments.

DISCUSSION

Our observations are inconclusive but it is clear that the widely used procedure for estimating "residual chlorine" by rapid titration of sea water samples, immediately after adding millimolar potassium iodide and adjusting to pH 4, underestimates the actual "residual oxidizing components" to an extent that the results are misleading. The identification of the chemical species that are formed when chlorine is added to sea water is a prerequisite to designing proper analytical methods. However, the lack of understanding of exactly what occurs when sea water is chlorinated should not distract attention from the significance of the present empirical observations on the responses of aquatic organisms to acute chlorine exposure, as reported in other presentations at this conference. The need for assessment of the effects of chronic exposure to the inorganic and organic products of sea water chlorination seems obvious.

ACKNOWLEDGMENTS

Much of the experimental work was performed carefully by Cynthia Moore. The research was supported by the U.S. Environmental Protection Agency, Grant No. R 803893-01.

REFERENCES

Bellanca, M. A. and D. S. Bailey. 1975. "A Case History of Some Effects of Chlorinated Effluents on the Aquatic Ecosystem of the Lower James River in Virginia." *Abstract, 48th Annual Conference Water Poll. Control Fed.,* Washington, D.C.

Bradbury, J. H. and A. N. Hambly. 1952. "An Investigation of Errors in the Amperometric and Starch Indicator Methods for the Titration of Millinormal Solutions of Iodine and Thiosulfate." *Australian J. Sci. Res.,* Ser A, 5:541-554.

Carpenter, E. J., B. B. Peck and S. J. Anderson. 1972. "Cooling Water Chlorination and Productivity of Entrained Phytophlankton." *Mar. Biol.* 16:37-40.
Carritt, D. E. and J. H. Carpenter. 1966. "Comparison and Evaluation of Currently Employed Modifications of the Winkler Method for Determining Dissolved Oxygen in Seawater; a NASCO Report." *J. Mar. Res.* 24:286-318.
Eppley, R. W., E. H. Renger and P. M. Williams. 1975. "Chlorine Reactions with Seawater Constituents and Inhibition on Photosynthesis of Natural Marine Phytoplankton." *Estuarine Coastal Mar. Sci.* 4:147-161.
Gutmann, H., M. Lewin and B. Perlmutter-Hayman. 1968. "The Ultraviolet Absorption Spectra of Chlorine, Bromine and Bromine Chloride in Aqueous Solution." *J. Phys. Chem.* 72:3671-3673.
Johannesson, J. K. 1955. "Note on the Chlorination of Water in the Presence of Traces of Natural Bromide." *N. Z. J. Sci. Technol.*, 36B:600-602.
Kolthoff, I. M. 1922. *Rec. Trav. Chim.* 41:740, as quoted in *Volumetric Analysis*, Interscience Publishers, Inc., New York, 1957, p. 262.
Kolthoff, I. M., R. Belcher, V. A. Stenger and G. Matsuyama. 1957. *Volumetric Analysis*. Interscience Publishers, Inc., New York, p. 267.
Lewis, B. G. 1966. "Chlorination and Mussel Control. I: The Chemistry of Chlorinated Seawater: a Review of the Literature." *C.E.R.L. Report* No. RD/L/N 106/66.
Morris, J. C. 1966. "The Acid Ionization of HOCl from 5 to 35°C." *J. Phys. Chem.* 70:3798-3805.
Pungor, E., K. Burger and E. Schulek. 1958. "Interhaloid Complexes in Aqueous Solution." *J. Inorg. Nucl. Chem.* 11:56-61.
Riley, J. P. and G. Skirrow. 1975. *Chemical Oceanography*. Academic Press, New York.
Standard Methods for the Examination of Water and Wastewater, 13th ed. 1971. American Public Health Association, Washington, D. C.
Sugam, R. and G. R. Helz. 1975. *Apparent Ionization Constant of Hypochlorous Acid in Seawater*. Research in Aquatic Geochemistry, University of Maryland, College Park, Maryland.
Szabo, Z. G. and L. Csanyi. 1952. *Anal. Chim. Acta* 6:208, as quoted in *Volumetric Analysis*, Interscience Publishers, Inc., New York, 1957, p. 255.
Van der Meulen, J. H. 1931. *Chem. Weekblad.* 28:82, as quoted in *Volumetric Analysis*, Interscience Publishers, Inc., New York, 1957, p. 255.
Weszelszky, von J. 1900. *Z. Anal. Chem.* 39:81, as quoted in *Volumetric Analysis*, Interscience Publishers, Inc., New York, 1957, p. 254.

DISCUSSION

David H. Rosenblatt, U.S. Army Medical Bioengineering Research and Development Laboratory. The formation of chlorite in hypochlorous acid-hypochlorite ion stock solutions is most unlikely in view of the careful study by D'Ans and Freund, *A. Elektrochem.* 61: 10 (1957), on the formation of chlorate from such solutions. The mechansim is:

$$2HOCl + OCl^- = [H_2Cl_3O_3^-] = 2H^+ + 2Cl^- + ClO_3^-$$

This completely bypasses possible formation of ClO_2^-, which those authors looked for and did not find.

J. Donald Johnson, University of North Carolina-Chapel Hill. How did you dismiss bromate, BrO_3^-?

Carpenter. We couldn't find it polarographically after we destroyed the hypochlorous acid with thiosulfate. It makes a beautiful wave at pH 7.

Johnson. Did you have enough sensitivity?

Carpenter. Yes. It is not more than 5×10^{-5} M.

Johnson. Well, I don't know anything about this polarographic method you mentioned, but I like bromate.

Carpenter. Well I did too. I was very frustrated when I didn't find it.

Walter J. Blogoslawski, National Marine Fisheries Service. In some recent work we have found the presence of a long-lived oxidant (measured by iodometric titration) after ozonization of sea water. This compound may be a bromite or hypobromite as suggested by the work of McIlhenny at Dow Chemical Company, Freeport, Texas. Chlorine when added to sea water may produce a similar effect.

Carpenter. Did you say hypobromite or bromite?

Blogoslawski. We think it might be hypobromite.

EXTENDED DISCUSSION

Joseph E. Draley, Argonne National Laboratory. Are there comments concerning the subject of the symposium thus far, namely the chemistry of chlorine in water? It occurs to me to remark that we have a lot of confusion about what is needed. I ought to define what I mean by what is needed. Let's eliminate temporarily the pure research that we do because we are engaged in fundamental research, and ask the question from the point of view of the needs for the control of the addition of substances to our environment or to our drinking water that would be harmful either to people or to our ecology. What things do we need? We've found or mentioned a number of uncertainties and unknowns. I'd like to ask you to offer comments on what our problems are that we could consider in a kind of a focused way.

Robert S. Ingols, Georgia Institute of Technology. Chlorinated organic pesticides apparently concentrate in the biota. Therefore I believe that data should be obtained on the accumulation of chlorinated organics in the biota below wastewater discharges. This may be a very important point.

Draley. Are there others who would like to relate to that particular topic? Do we know that if you accumulate this material in plankton or in a higher species that it is authomatically bad if it contains chlorine? Is there a tenet that if you get chlorinated organics in these systems that it will be bad for them or that it will be bad for us? Is there a threshold level below which you don't have to worry and above which you do have to worry? Do we know these things or are we speculating? If we are speculating, do we need to know?

William H. Glaze, North Texas State University. Obviously, I can't answer that question. But I would like to tell you about some results, and this sort of anticipates tomorrow. I am reporting the results of a study that was done this summer in my laboratory. There's a real neat system close by to us, namely a power plant lake, which is almost dead-end. That is to say, except at flood stage, there's not any water getting out except by evaporation, but there is water being brought in from a river that rivals the Ohio, and Trinity. So there's a good deal of organic material in the lake both from natural and unnatural sources. The point of this is that I had a student who did look at accummulation of chlorinated organics, not specific compounds, but as measured by TOCl in fish, in perch, in that lake as compared to a control lake—that is, a lake of about the same size, about the same terrain, which was only getting agricultural runoff as far as we could tell. There was significant higher concentration as measured by TOCl, the way I talked about it a minute ago, in the power plant lake. That lake being such that it takes water through the cooling towers and chlorinates periodically, and so it's a natural system for the buildup of these compounds that we've been talking about. What the concentration is in the lake, I don't know. We are now in the process of looking at the lake water and the sediments thereof. We are going to look at specific compounds, and also the TOCl values, in the lake and the biota thereof.

Draley. I have a comment here. Bob Baker, I am going to ask you about the measurement of chlorine in sea water in a little while. You may be thinking about how terrible a question it is going to be.

H. Sikka, Syracuse University. I think the possibility for bioaccumulation of the compounds mentioned by Dr. Jolley is going to be much less compared to the chlorinated hydrocarbon pesticides. This is because they are much more polar chlorocompounds than the chlorinated pesticides. As a result the bioaccumulation by fish and animals will not be much of a concern. Of course, this does not exclude the possibility of doing research in that area.

Draley. No. The reason I asked the question is you can't do everything you can possibly think of. Somehow we have to find a way to guide those who fund and those who do research for, shall we say, practical benefit, in what needs to be done.

Glaze. I think you're quite right. Nevertheless, there is some significant chance for accumulation. But another thing to understand now is what I think Jolley pointed out also, that it's not just accumulation that we're interested in. We are also interested in other specific effects. When you look at 5-chlorouracil and think about the implication there, and when you think about chlorinated amino acids and we know that some of the chlorinated amino acids are antagonists for the unchlorinated amino acids, then it's not necessarily bioaccumulation that one is concerned about. One is worrying about some of those specific effects whether they be communication or genetic or whatever, which were referred to earlier.

Sikka. Yes, but these are two separate areas. We have to distinguish between the two areas.

Robert B. Dean, U.S. Environmental Protection Agency. We have been looking a lot at sewage. We have been looking a lot less at what is produced when you chlorinate paper pulp. Chlorinated paper pulp and paper mill wastes can be expected to contain chlorinated terpene compounds similar to known chlorinated pesticides. These would be expected to bioaccumulate.

Unknown. I think bioaccumulation is certainly a thing we need to look at. Another thing relating to bioaccumulation, which is really more broad than bioaccumulation, and that is movement. Movement in the sediment, whether it is facilitated by bacteria or not, should be studied. Maybe some compounds go into storage. This might be important.

Draley. I would like to ask my question now of Bob Baker about the measurement of chlorine in sea water. The reason I brought it up is that in the last talk some problems were discussed with respect to what one can hope to measure by what might be called ordinary or present methods. Do you have any comment to make?

Robert J. Baker, Wallace & Tiernan Division, Pennwalt Corporation. I see you've done a perfect job of putting me on the spot. I've gone through about three phases where I thought we were measuring chlorine, and then I thought we were measuring bromine, and now I haven't the vaguest idea. What actually occurs is that the results we get are simply recorded as if they were chlorine. That doesn't really answer your question. In some of the work we have done, obviously, the results don't seem to be comparable as far as the reaction is concerned. This comment on the flash back—I've spotted it in some cases, and in some cases I have not. That's all I can offer.

Draley. Other comments?

Robert L. Jolley, Oak Ridge National Laboratory. Relative to chlorination of paper waste effluents, Dr. Larry Keith at the Southeastern Environmental Research Laboratory, Athens, Georgia, has done some work. He has analyzed several chlorinated phenolic ethers.

Glaze. The Fisheries Research Board of Canada has done a whole series of publications where they have looked for chlorinated compounds in paper effluents and are beginning to look at specific toxicities. In one of the last editions there was a good bibliography for those interested in chlorination at paper mills. This is a bleaching process. There is some good research and they are getting into it in the same way that you have done here.

9

DECISION-MAKING IN THE REGULATION OF CHEMICALS

Edward M. Brooks
   Chief, Policy Review Staff
   Office of Toxic Substances
   U.S. Environmental Protection Agency
   Washington, D. C. 20460

INTRODUCTION

I'm delighted to have the opportunity to talk to you this evening. I must confess that, being neither scientist nor engineer, I approach the opportunity with no little trepidation. I am somewhat consoled by the fact that my topic has to do with decision-making in the regulation of chemicals, since most decision-makers in this area are also not scientists—a fact which must no doubt frustrate many of you from time to time. Of course there is that body of opinion which holds that the regulation of toxic substances is too important to leave to toxicologists. In any event, the information exchanged—or lack thereof —between scientists and the decision-maker is what I want to discuss tonight.

REGULATIONS

When he approached me to give this talk, Dr. Jolley was kind enough to suggest a number of questions I might address, among which I found four particularly provocative:

1. What type of toxicity data are considered in making regulations?
2. Can cost-benefit analyses be applied in developing regulations?
3. What is the mechanistic procedure or protocol for developing regulations?
4. What can data generators do to make decision-making easier?

All four questions touch upon a more general proposition I now put to you, namely: *regulations to control serious chronic toxicants are not developed within a consistent logical framework.* I will first demonstrate the truth of this general proposition, and then explore a few reasons why it is the case and what, if anything, ought to be done. Before embarking let me define, for purposes of discussion, two key phrases in Brooks' proposition. By "serious chronic toxicants" I mean those that are of concern because they may cause death or illness to humans—either after many years of continuous exposure, or after a latency period of many years—at levels markedly below the dosage at which the tolerable levels of risk can be detected in laboratory test animal experiments. Such toxicants include, but are not in my opinion limited to, human chemical carcinogens.

The term "consistent logical framework" refers to a system that imposes a *stipulated* set of values, principles and rules upon the manner in which data are evaluated and exploited to reach decisions—such that different players, operating independently with the same information, or lack of information, reach essentially the same conclusions. A major function of such a system is to compel consideration of costs, risks and benefits across the range of available regulatory options.

The value of such a rational approach to regulatory decision-making is generally appreciated. The National Academy of Sciences (NAS) recently completed a study entitled *Decision Making for Regulating Chemicals in the Environment* (NAS, 1975) the fruit of which is a series of 34 recommendations—from four of which I quote in part.

17. The quality of chemical regulatory decisions is dependent largely upon the adequacy of the available information. To develop an adequate data base, research efforts in basic clinical and environmental toxicology and epidemiology and in economic analysis must be strengthened . . .

30. Highly formalized methods of benefit-cost analysis can seldom be used for making decisions about regulating chemicals in the environment. However, benefit-cost and decision frameworks can be useful in organizing and summarizing relevant data on regulatory alternatives which the decision maker must review.

31. Value judgments about noncommensurate factors in a decision such as life, health, aesthetics, and equity should be explicitly dealt with by the politically responsible decision makers and not hidden in purportedly objective data and analysis.

32. The decision process should require the agency's technical staff to present a full set of options with a corresponding range of cost-benefit-hazard data and explicit statements on the confidence limits of each analysis.

The Academy thus expects staff scientists to identify several regulatory alternatives and objectively estimate and present to the politically accountable decision-maker the costs, risks and benefits associated with each—together with explicit probability statements regarding the reliability of those estimates. From these analyses the decision-maker selects a regulatory option and, in proposing and promulgating the decision, explicitly sets forth the value judgments he applied. 'Tis a consummation devoutly to be wished.

It will be instructive to look at a few proposed or promulgated regulations to see how closely they approach this ideal. I emphasize two points. First, nothing that follows is intended as criticism of any given regulatory agency or decision. Further, where discrepancies are found between different regulations controlling the same substance, I offer, in these remarks, no opinion regarding their relative merits. My purpose here is to examine what has been done to assess possible weaknesses in the *process* rather than in any particular regulation. Secondly, I trust it goes without saying that any weaknesses found hardly constitute grounds for not continuing to aggressively implement our diverse authorities as best we can. I will now discuss four proposed or promulgated standards—two for asbestos and one for vinyl chloride and aflatoxins.

Clean Air Act—Sec. 112—Asbestos:

In April of 1973, the U.S. Environmental Protection Agency (EPA) promulgated a regulation (EPA 1973) to control asbestos emissions at a level designed to protect human health with an "ample margin of safety." In the preamble to this standard the Agency stated that no numerical concentration or mass emission limit was practicable because it is "impossible to estimate even roughly the quantitative relationship between asbestos-caused illness and the doses that cause those illnesses." Although we concluded that there are no levels known at which asbestos does not involve risk, and that the effects of inhaling it are cumulative, we did not ban the substance outright because to do so would prohibit many extremely important activities. Accordingly, the standard, in major part, simply banned "visible emissions" of asbestos.

OSHA—Sec. 6—Asbestos:

One week ago Thursday last (*i.e.,* October 9, 1975), the Occupational Safety and Health Administration (OSHA) of the Department of Labor proposed a new standard (OSHA, 1975) to regulate work-place exposure to asbestos. The initial OSHA asbestos standard (OSHA, 1971a), promulgated in May of 1971, was 12 fibers/cm^3, with no fibers longer than 12 μm. In December of 1971 (OSHA, 1971b) this was reduced to 5 fibers, no longer than 5 μm. The current standard (OSHA, 1972), promulgated in 1972, established the level at 5 fibers, but added the provision that this would automatically go down to 2 fibers on July 1, 1976. Now the October 9 proposal would reduce the level, still further, to 0.5 fibers. The proffered rationale is that (1) sufficient evidence has accumulated to warrant designating asbestos a human carcinogen, (2) a "no effect" level has not been demonstrated, and (3) in the absence of evidence to establish a safe level, employee exposure must be reduced "as low as feasible."

Thus both EPA in its promulgation under Sec. 112 of the Clean Air Act and OSHA in its promulgation under the Occupational Safety and Health Act concluded that there is no known exposure level for asbestos at which adverse human health effects do not occur. Within that framework EPA concluded that no numerical concentration limit is possible because the dose-reponse relationship is unknown, while OSHA not only established a numerical limitation, but has systematically reduced it over the years.

The significant point is that no data have been provided to the decision-maker in *either* agency regarding the dose-response relationship. Relative hazard assessment is therefore out of the question. In this regard it is worth noting that, although the rationale for the OSHA standard is to reduce exposure levels to "as low as feasible," no evidence is provided to suggest that 0.5 fibers is either feasible or safe. In point of fact the rationale and evidence used to justify the 1972 5-fiber standard could just as well have been used to justify the 1975 0.5-fiber proposal, and vice versa. The evidence to establish a "no effect" or "safe" level was just as absent when OSHA promulgated the 5-fiber standard to protect against asbestosis as it was a week ago Thursday last when it proposed the 0.5-fiber standard to protect against mesothelioma—and this lack of evidence apparently has nothing to do with any special attributes of carcinogens. In neither instance were "no effect" levels or dose-response relationships established.

I now want to contrast the OSHA vinyl chloride standard with the tolerance for aflatoxin proposed by the Food and Drug Administration (FDA).

OSHA—Sec. 6—Vinyl Chloride:

In October of 1974 OSHA promulgated a standard (OSHA, 1974) for vinyl chloride based on (1) the fact that 31 vinyl chloride workers had died of angiosarcoma of the liver, (2) Maltoni's experiments inducing angiosarcoma in rats at 250 parts per million (ppm), and (3) Industrial Bio-Test Laboratories' studies inducing angiosarcoma in rats *and* mice at 50 ppm. OSHA concluded in accordance with the 1970 Report of the

Surgeon-General's Ad Hoc Committee on the Evaluation of Low Levels of Environmental Chemical Carcinogens that, on the basis of the demonstration of cancer in two animal species, vinyl chloride posed a carcinogenic hazard to man. In further accordance with that report, OSHA took the position that "safe exposure levels for carcinogenic substances can not be scientifically determined"—a position supported at the Hearings by both NIOSH and NCI. On the grounds that vinyl chloride is a carcinogen and "safe" levels for carcinogens cannot be established, OSHA promulgated a "no detectable level" standard as measured by methods sensitive to 1 ppm.

FD&CA—Sec. 406—Aflatoxins:

The Food and Drug Administration's proposed December, 1974, tolerance for aflatoxins (FDA, 1974) in shelled peanuts and peanut products provides an interesting contrast. The Federal Register NPRM notes that 25 rats, fed aflatoxins at 15 parts per billion (ppb), *all* developed liver cancer, as did monkeys fed aflatoxin. In addition, epidemiological studies in Southeast Asia and Africa indicated a correlation between the incidence of liver cancer in humans and exposure to aflatoxins. FDA concluded from this that human exposure should be held to the "lowest level possible."

FDA survey data indicated that 4% of the U.S. peanut butter exceeded the 20 ppb level, 7% exceeded 15 ppb, 11% exceeded 10 ppb, and 25% approximated one ppb. Thus, about 4% of the U.S. production fell between 10 and 15 ppb, and 3% between 15 and 20 ppb. FDA proposed a 15 rather than 10 ppb tolerance to "avoid causing significantly increased losses of food."

Thus, in quite comparable situations—with substances demonstrated to be carcinogenic in two animal species and with strong epidemiological evidence implicating them as human carcinogens —OSHA promulgated a "no detectable level" standard (for vinyl chloride) well below the lowest levels at which any adverse effects have been found in any species, while FDA proposed a tolerance (for aflatoxin) at the same level at which 25 of 25 mice developed liver cancer. The two Federal Register

Notices reflect the different philosophies. First the OSHA text regarding vinyl chloride:

> There is little dispute that vinyl chloride is carcinogenic to man and we so conclude. However, the precise level of exposure which poses a hazard and the question of whether a 'safe' exposure level exists cannot be definitively answered on the record. Nor is it clear to what extent exposures can be feasibly reduced. We cannot wait until indisputable answers to these questions are available, because lives of employees are at stake. Therefore, we have had to exercise our best judgment on the basis of the best available evidence. These judgments have required a balancing process in which the overriding consideration has been the protection of employees, even those who may have regular exposures to vinyl chloride throughout their working lives.

And the counterpart passage from the FDA proposal regarding aflatoxins:

> In addition, because there is no direct evidence that aflatoxins cause cancer in man or of what may be the level of no effect, the Commissioner cannot conclude that there is any tangible gain from lowering the permissible level to either ten or five ppb. Such uncertain benefit to the public health must be weighted against the clear loss of food that would result.

Again the point is not who, if anyone, is right or wrong, but rather that no data were provided to either decision-maker to permit a reasoned analysis of the risks incurred at various possible exposure levels. Without such information the levels can and will be set almost anywhere from zero up to the levels that obtain without any regulation at all. Any consonance between such permitted exposure levels and a balanced tradeoff between health and economic impact considerations will be purely fortuitous.

The inadequacies of these regulations warrant a moment's reflection. Most remarkable is the fact that none makes any attempt to specify an "acceptable" level of risk. Instead they offer analytically meaningless platitudes about the need to reduce exposures "as much as possible" or as is "feasible." There are no estimates of the extent of the adverse human health effects presumed to be caused by these substances, much less of the extent to which this incidence is expected to be reduced by the regulation. Not only is it impossible to evaluate how well these regulations achieve their objectives—we can't even define the objective. In the most fundamental sense, then, it is impossible to assess their worth.

Even if "acceptable" levels of risk had been established, however, there are still no health effects data available to indicate the exposure levels at which those risk levels would be exceeded. In this situation it is fatuous to speak of cost-risk-benefit analyses, judicious tradeoffs, balancing competing factors or any other phrase that connotes a reasoned application of useful information.

One must ask how this came to pass? How is it we write regulations in such an important area with so little comprehension of what we are about? From whence comes the pressure to promulgate such regulations? Why is the scientific documentation such a paltry product? What impels decision-makers to act upon such tenuous evidence? Aside from our ignorance, what accounts for the striking inconsistencies found in these regulations—not only in their stringency but, more basically, in their underlying philosophy?

While there are obviously many reasons, I would like to briefly mention two contributing factors. First, I believe that we may sometimes be moved to precipitate an unwarranted action in response to public pressure. In this regard, incidentally, I have just read a perceptive article in the fall issue of *The Public Interest* (Nisbet, 1975) which reflects my concern precisely. I commend it to your attention and will attempt to whet your appetite with the following quote:

> Of all the heresies afloat in modern democracy, none is greater, more steeped in intellectual confusion, and potentially more destructive of proper governmental function than that which declares the legitimacy of government to be directly proportional to its roots in public opinion—or, more accurately, in what the daily polls and surveys assure us is public opinion. It is this heresy that accounts for the constantly augmenting propoganda that issues forth from all government agencies today—the inevitable effort to shape the very opinion that is being so assiduously courted—and for the frequent craven abdication of the responsibilities of office in the face of some real or imagined expression of opinion by the electorate.

This tendency to precipitate action in the face of uninformed but aroused popular opinion is reinforced by the wide and increasing discrepancy between our ability and disposition to detect the *presence* of *potential* toxic substances and our ability to assess the degree of risk they pose. EPA's recent and continuing concern with organics in drinking water illustrates this phenomenon. Surveillance and analytical technology now yields impressively precise quantification of very low levels of organics in water, while the state-of-the-art of health effects assessment apparently permits only crude qualitative estimates of the human health hazards posed at these levels. In 1969 the Federal Water Pollution Control Administration found chloroform, benzidine and *bis*-chloromethyl ether in the New Orleans drinking water (EPA, 1972). In 1974 EPA found 66 organic chemicals in the New Orleans drinking water (EPA, 1974). This past year we completed a survey of the water in 80 cities and found at least one of the six organics for which we sampled—and most particularly chloroform—in every location (EPA, 1975a). At various stages in the course of these events substantial pressure was brought to bear on EPA to "do something." But what do we really know? When asked to review the findings of the 80-city National Organics Reconnaissance Survey, EPA's Science Advisory Board concluded that:

> . . . there may be some cancer risk associated with consumption of chloroform in drinking water. The level of risk, estimated from consideration of the worst case and for the expected cancer site for chloroform (the liver) might be extrapolated to account for up to 40% of the observed liver cancer incidence rate. A more reasonable assumption, based upon current water quality data which show much lower levels than the worst case in the majority of U.S. drinking water supplies, would place the risk of hepatic cancer much lower and possibly nil. Further, it is emphasized that both the experimental carcinogenicity data and the mathematical and biological extrapolation principles used to arrive at the upper estimate of risk are extremely tenuous. Epidemiologic studies do not, thus far, support the conclusion of an increased risk of liver cancer; although hypothesis formulating studies in southern Louisiana suggest the possibility of an association with contaminated water and overall high cancer incidence (EPA, 1975b).

There is no obvious solution to this problem. It is clearly important for a public agency to be responsive to the public; this notwithstanding, it is at least equally important to allocate resources and conduct the public's business in an orderly and reasoned manner.

There is also apparently no immediate solution to our inability to quantify the human health risks associated with low levels of chronic toxicants. As I understand it, neither epidemiology nor test animal experiments provide a really acceptable solution. The uncertainty regarding exposure level, the long latency or exposure periods, and the confusion created by multiple exposures all diminish the utility of epidemiology. On the other hand, the problem of translating from test animal to human response and, perhaps more importantly, the high dose/low dose extrapolation problem, seriously limit the utility of test animal experiments. This problem is so serious that Messrs. Hoel, Gaylor, Kirschstein, Saffiotti and Schneiderman, in a recent article in the first edition of the *Journal of Toxicology and Environmental Health* (Hoel et al., 1975) flatly state that:

> There is no adequate method for determining the best estimate of risk for a given dose and the best estimate of dose for a given risk. Because of model dependency there does not appear to be a reliable method for obtaining such direct estimates and their required confidence limits.

I have been given to understand that the National Center for Toxicological Research was established, at least in part, precisely in order to illuminate this problem by using very large numbers of test animals to generate experimentally derived points much lower on the dose-response curve. I certainly hope my understanding is correct—and that NCTR is soon successful in this endeavor—for until we can quantify the human health risks associated with very low levels of serious chronic toxicants I see no hope of materially improving the standard-setting process.

REFERENCES

Food and Drug Administration. 1974. "Aflatoxins in Shelled Peanuts and Peanut Products Used as Human Foods, Proposed Tolerance." *Federal Register* 39(236):42748. December 6.

Hoel, David G., David W. Gaylor, Ruth L. Kirschstein, Umberto Saffiotti and Marvin A. Schneiderman. 1975. "Estimation of Risks of Irreversible Delayed Toxicity." *J. Toxicol. Environ. Health* 1:133-151.

National Academy of Sciences. 1975. *Decision Making for Regulating Chemicals in the Environment*, Washington, D. C.

Nisbet, Robert. 1975. "Public Opinion Versus Popular Opinon." *The Public Interest* 41:166.

Occupational Safety and Health Administration. 1971a. "Occupational Safety and Health Standards, National Concensus Standards and Established Federal Standards." *Federal Register* 36(105):10466. May 29.

Occupational Safety and Health Administration. 1971b. "Emergency Standard-for Exposure to Asbestos Dust." *Federal Register* 36(234):23207. December 7.

Occupational Safety and Health Administration. 1972. "Standard for Exposure to Asbestos Dust." *Federal Register* 37(110):11318. June 7.

Occupational Safety and Health Administration. 1974. "Standard for Exposure to Vinyl Chloride." *Federal Register* 39(194):35890. October 4.

Occupational Safety and Health Administration. 1975. "Occupational Exposure to Asbestos, Notice of Proposed Rulemaking." *Federal Register* 40(197):47652. October 9.

U.S. Environmental Protection Agency. 1972. *Industrial Pollution of the Lower Mississippi River in Louisiana.* Region VI, Dallas, Texas, April.

U.S. Environmental Protection Agency. 1973. "National Emissions Standards for Hazardous Air Pollutants—Asbestos, Beryllium, and Mercury." *Federal Register* 38(66):8820. April 6.

U.S. Environmental Protection Agency. 1974. *New Orleans Area Water Supply Study.* Lower Mississippi River Facility, Slidell, Louisiana and Region VI, Dallas, Texas. November.

U.S. Environmental Protection Agency. 1975a. *National Organics Reconnaissance Survey.* Office of Research and Development, Cincinnati, Ohio. April.

U.S. Environmental Protection Agency. 1975b. *Assessment of Health Risk from Organics in Drinking Water*, p. ix. Science Advisory Board. April.

SECTION II
BIOMEDICAL EFFECTS OF CHLORO-ORGANICS

Robert B. Cumming, Session Chairman
    Biology Division
    Oak Ridge National Laboratory
    Oak Ridge, Tennessee 37830

    Yesterday's sessions of this conference to a large degree dealt with chlorination technology and the science involved in understanding the chemistry of the chlorination process. We perhaps gained some appreciation, particularly in Dr. White's talk, of the tremendous economic and social stake that this society has in chlorine and in chlorination. The data are probably available to calculate in purely economic terms the costs and benefits of water chlorination. If one neglects the difficult areas of the impact on man himself and upon the environment in which he must live, there is a clear benefit. The benefit from chlorine vastly outweighs the cost.

    Today, it is time to take a look at what we know and what we can know about the other side of the ledger. With what precision can we add the costs to human health and the costs of the potential environmental degradation into the cost-benefit calculation? It is clear that chlorination of water does add to the load of man-made chemicals which humans encounter in their environment.

    This morning we are going to focus on the very difficult area of the biomedical effects. What we are really dealing with is exposure of a large population to extremely low concentrations of a variety of potentially biologically active compounds. The nature of the problem is such that we can almost immediately eliminate from our consideration acute toxicity. We do have to consider two areas of toxicity which are particularly difficult to handle. One is carcinogenesis and the other is mutagenesis. This morning we have assembled a group of people who can approach these toxicological endpoints from several different angles. Two methods which will be explored, and perhaps they're not the only two, are (1) extrapolation of data from experimental animal models and (2) the epidemiological approach.

10

HALOGENATED ORGANICS IN TAP WATER: A TOXICOLOGICAL EVALUATION

Robert G. Tardiff

    Health Effects Research Laboratory
    U.S. Environmental Protection Agency
    Cincinnati, Ohio 45268

Gary P. Carlson

    Department of Pharmacology and Toxicology
    Purdue University
    West Lafayette, Indiana 47907

Vincent Simmon

    Microbial Genetics Program
    Stanford Research Institute
    Menlo Park, California 94025

ABSTRACT

Evaluation of hazard of halogenated hydrocarbons in drinking water requires the consideration of many factors, including degree of exposure, intrinsic toxicity of the agents, interactions among compounds and with other environmental factors, and species sensitivity.

Halogenated hydrocarbons are ubiquitous in the public water supplies of the United States. Of the compounds that have been identified in the nation's drinking water, approximately 34% are halogenated. A recent survey of five U.S. cities revealed that approximately 50% of the volatiles in tap water are halogenated. Although the number of volatile halogenated organics is relatively large, they constitute only a small percent of the total organic concentration in drinking water. Generally, chloroform is present in the highest concentration (approximately 100 ppb), and the majority of the other compounds are present in 1 ppb or less.

Assessment of the toxicity of the organics in tap water is following two lines of investigation. Mixtures of organics from tap water are being bioassayed for mutagenic, carcinogenic and teratologic activity. Specific compounds present in tap water are being subjected to in-depth toxicity evaluations. Among the compounds being investigated are the chloroethers, the chlorobenzenes and the bromobenzenes. Compounds of interest for future research include the halomethanes and the chloroethanes. Toxicologic questions being addressed using these compounds include questions of extrapolation from experimental animals to man through investigations of comparative metabolism and predictability of synergistic interactions through studies of alterations of basic metabolic pathways.

INTRODUCTION

Publicity of the past 12 months has focused attention on the presence in drinking water of some of the more toxic organic compounds. In more recent months, research findings indicated that halogenated organic compounds having carcinogenic properties were actually synthesized during chlorine disinfection—a fact that has given way to claims that chlorination of drinking water may pose a cancer threat or hazard.

That experience reiterates the frequently observed misconception that toxicity alone is equated with hazard—or more specifically, that a toxic property such as experimental carcinogenesis is, of necessity, identical with a threat to human health and a human cancer risk. Evaluation of safety and hazard, the subject of this discussion, is the cornerstone to understanding the impact on human health of chemicals in our surroundings.

Evaluation of safety involves the conceptual integration and interpretation of the physical, chemical and biological properties or effects of a particular compound or product for the purpose of assessing the safety of that agent under conditions of its intended use.

The evaluation of safety and hazard of halogenated hydrocarbons in drinking water requires consideration of many factors, including the degree of exposure (including both concentration and duration), intrinsic toxicity of the agents, interactions among compounds as well as between compounds and other environmental stimuli, and sensitivity of the exposed population and its subsets.

EVALUATION

Halogenated hydrocarbons are ubiquitous in the water supplies of the United States (Symons *et al.*, 1975). Of the compounds that have been identified in the nation's drinking water, approximately 38% or 111 of 289 compounds are halogenated, as noted in Table I (U.S. Environmental Protection Agency, 1976). A recent survey of five U.S. cities (U.S. Environmental Protection Agency, 1975) revealed that approximately 50% of the volatile hydrophobic compounds in tap water were halogenated. The data in Table I also categorized the organic compounds in drinking water according to structure; thus it can be seen that 66% of the compounds are aliphatic and the remainder are aromatic in structure. Although the number of volatile halogenated compounds is relatively large, these agents constitute only a small fraction by weight of the total organic concentration in tap water. The term "volatile" refers to those compounds that are purged from water at 25°C with an inert gas (*e.g.*, helium). Examples of these compounds include the halomethanes such as chloroform and bromoform, and the chloroethenes such as vinyl chloride and vinylidine chloride. The "nonvolatile" fraction, by contrast, is operationally defined as that mixture of compounds selectively extracted from water by reverse osmosis (cellulose acetate and nylon membranes). Generally, one volatile compound, chloroform, is present in the highest concentration—approximately 100 ppb (μg/l) on the average—and the majority of the other compounds are present at 1 ppb or less.

Recently, our laboratory initiated the analysis of the nonvolatile fraction of one sample of organics from tap water. Preliminary data indicate that the molecular weights of the compounds in this fraction vary from 200 to 500 as contrasted to the predominantly lower-molecular-weight species in the volatile fraction (*i.e.*, less than 100 molecular weight, MW). Approximately 200

Table I. Occurrence of Organic Compounds in
Tap Water of the United States

Type	Number	Percent
Halogenated	111	38
Nonhalogenated	178	62
Total	289	
Aliphatic	190	66
Aromatic	99	44
Total	289	

compounds have been separated in the nonvolatile fraction, only a few of which have been identified. Several of the approximately 200 compounds appear to contain from 4 to 10 halogen atoms each, in contrast to the occurrence of between 1 and 6 halogen atoms in the compounds of the volatile fraction. Work is proceeding to identify the structures of the nonvolatile agents by mass spectrometry.

Since dosage is a primary factor in the evaluation of safety and hazard, the concentrations to which man is and can be exposed require definition. Based on general parameters such as total organic carbon (TOC), the concentrations of organics in tap water range from less than 1 mg/l (*i.e.*, ppm) to at least 6 mg/l. Assuming that carbon comprises on the average 50% of the total weight of the compound, then the concentrations of the mixtures would range from 1 ppm to 12 ppm, respectively. In view of the relatively large numbers of these compounds in these mixtures, individual compounds are expected to be present in μg/l (ppb) quantities. Quantitative analyses of volatiles in tap water are in agreement with this hypothesis. Although those doses are seemingly minute, exposure to these agents via drinking water is chronic or continuous in nature, whereas other forms of exposure are believed to be intermittent. Because of the length of the half-life of these chlorinated compounds (usually longer than 12 to 24 hr), it is highly likely that continued exposure (not only on a daily basis but throughout the day) may lead to a bioaccumulation of these compounds. Consequently, the relationship between the equilibration concentrations and the critical levels in soft tissues necessary to initiate toxic responses must be determined. An additional modifiying factor is the fact that exposure occurs during certain life stages which may be more critical than others with regard to susceptibility to the toxic properties of these chemical agents (*i.e.*, increased sensitivity). One example of the effect of dosage regimen was recently noted with chloroform, in which the target organ changed when going from low to high exposures. Acute toxicity was observed in the kidney after low-dose exposure, whereas higher-dose exposure led initially to kidney pathology followed by hepatic centrolobular necrosis (Hill *et al.*, 1975).

The toxic properties of the compounds can be defined by employing a battery of bioassays. The array of assays ranges from organ function test, to histopathologic evaluation, to tissue enzyme assays, to determinations of molecular alterations of genetic material. No single battery of tests is guaranteed of predicting all possible pathologic lesions in the whole animal. Exhaustive testing with all conceivable test systems would be economically prohibitive and wasteful. Nevertheless, there is a real need to obtain sound and relatively comprehensive toxicity data on the compounds to which man is exposed (as for example in his drinking water).

The evaluation of the toxicity of organic compounds, including halogenated hydrocarbons in drinking water, follows two avenues of investigation. First, an attempt is made to screen the toxic potential of mixtures of organic compounds that are actually obtained from municipal drinking water. Second, individual compounds are subjected to in-depth toxicological evaluation in animal models that are predictive of human responses. Although not a part of this discussion, note must be taken of the fact that epidemiologic investigations are also pursued and that both the toxicologic and epidemiologic work are complementary to one another.

For the purposes of toxicological assessment as well as chemical characterization, samples of organic mixtures can be concentrated and recovered from tap water through the use of reverse osmosis technology (cellulose acetate and nylon membranes) (Kopfler *et al.*, 1975). Such procedures require the separation of inorganic salts from organic compounds in order to avoid any additional injury due simply to high salt concentrations. The utilization of solvent extraction and certain forms of dialysis can be used effectively to generate samples of organic mixtures. At present, recovery of organics from tap water is approximately 30 to 40% by utilizing these techniques.

A program has been developed to ascertain the types of water supplies which most likely present some health hazard to the consumer. The program requires the sampling and bioassay of six samples or organic mixtures from water supplies representing each of the major sources of raw water. Table II lists the cities that have been selected for this study, the type of water supply utilized by each city, and the type of raw water used in generating drinking water. Consequently, a matrix of both ground and surface sources as well as waters containing predominantly industrial wastes, agricultural runoff, or municipal wastes are also being separately sampled and assayed along with presumably uncontaminated ground and surface sources. One duplication is immediately obvious: the inclusion of two cities both having surface sources containing predominantly industrial wastes. Cincinnati was originally selected as the prototype city, the one in which all of the analytical developments and modifications would be made. Subsequently, the drinking water from the City of New Orleans was selected to obtain additional information for use with standardized methodology, as were Miami, Ottumwa, Philadelphia and Seattle.

The screening of organic mixtures or concentrates from tap water for toxicologic activity is summarized in Table III. With the exception of the range-finding assay, the assays attempt to screen for irreversible toxic phenomena. Such endpoints include mutagenesis, teratogenesis, carcinogenesis and other forms of chronic intoxication. Measurement of mutagenic potential is performed by the histidine reverse mutation system in five strains of *Salmonella typhimurium* (TA1535, TA1537, TA1538, TA98 and TA100) (Ames *et al.*, 1973; McCann *et al.*, 1975).

Because of the relatively small amount of material required for the *in vitro* assays, it is entirely possible that an active parent mixture can be subfractionated with various solvents and other techniques to yield fractions for further bioassay.

Table II. Cities Selected for Bioassay of Organics in Drinking Water

City	Type of Water Supply	Type of Raw Water
Cincinnati, OH	Surface	Industrial Wastes
Miami, FL	Ground	Uncontaminated[a]
New Orleans, LA	Surface	Industrial Wastes
Ottumwa, IA	Surface	Agricultural Runoff
Philadelphia, PA	Surface	Municipal Wastes
Seattle, WA	Surface	Uncontaminated[a]

[a]Refers to no known contamination from municipal, agricultural and industrial wastes—but contamination presumably from decomposition products of natural origin.

Table III. Toxicologic Screens of Organic Concentrates from Tap Water

Assay	Sample/City at 2-Month Intervals[a]					
	1	2	3	4	5	6
Range-finding (LD50 Mouse)	x					
Mutagenesis (*Salmonella*)		x-f			x-f	
Mammalian cell transformation		x-f			x-f	
In vivo carcinogen bioassay (Neonate)			x			?
Teratogen assay (rat)				x	?	
Chemical characterization (GCMS)	?	?	?	?	?	x-?

[a]x = analysis of sample; x-f = analysis of sample and subfractions; ? = possible analysis.

Consequently, the active ingredients can be narrowed down in specific fractions which can be analyzed chemically at a later time.

Initial results using the aforementioned mixtures from the City of Cincinnati's tap water have yielded data which indicate that the two mixtures used with TA100 were both mutagenic (Simmon and Tardiff, 1976). Subfractionation of these parent mixtures sequentially with petroleum ether, diethyl ether and acetone yielded results indicating that the mutagenic components were located predominantly in the diethyl ether fraction. Chemical characterization of this fraction is currently ongoing.

As a measure of possible carcinogenic activity, an *in vitro* mammalian cellular transformation assay (3T3) is being utilized to determine the ability of the mixtures to produce malignant transformations in culture (DiPaolo *et al.*, 1972). The malignancy of transformed cells is confirmed by the development of carcinoma in whole animals injected with transformed cells. Both the mutagenesis assay in bacteria and the transformation assay in mammalian cells are presumed to be indicators of a high degree of suspicion of carcinogenesis. An additional carcinogen/chronic toxicity bioassay is also utilized. Several studies have indicated that the neonatal animal is far more susceptible to a chemical carcinogen than is the adult animal (Kelly and O'Gara, 1961; O'Gara *et al.*, 1965; Pietra *et al.*, 1959). Consequently, such a model using a neonatal rat is being employed, first, to determine whether the mixtures have carcinogenic activity and, second, to ascertain whether other endpoints of irreversible chronic intoxication with subchronic exposures can be observed. The assay includes the oral administration of the test mixtures to rats from day 1 to day 21 after birth with subsequent observations throughout the majority of the animals' lifespans and terminal histopathologic examinations. Additionally, the mixtures are assayed for teratogenic potential by administration of the materials to pregnant rats during the period of organogenesis with subsequent gross morphologic observations of the progeny at birth.

If the outcome of any or all of the screening test is positive, an archival sample of the mixture is analyzed for its chemical composition in order to obtain an indication of the compounds which may be responsible for the observed effects. Suspect agents are then retested in the appropriate bioscreens. Separation of the individual components is performed with glass capillary columns in a gas chromatograph, and identifications are confirmed by mass spectrometry.

The determination that certain mixtures of organic compounds derived from drinking water have the potential for irreversible toxicity inevitably leads to a list of compounds for which in-depth evaluation of safety and hazard must be conducted. In addition to the stated criteria, compounds can be selected for safety/hazard evaluation on the basis of predominant occurrence in potable waters, as in the case of the haloforms. The primary objectives of the toxicology program on halogenated organics in drinking water are listed in Table IV. The identification of pathologic lesions and the identification of target organs in progressive sensitivity leads to experimental designs aimed at defining the dose-response relationships as well as to the definition of those pathologic phenomena which are likely to occur in man after exposure. The elucidation of the pathogenesis of subchronic and chronic intoxication may yield valuable clues to enable epidemiologists to conduct informative clinical and subclinical investigations of possible effects in the human populations. Investigations of toxicokinetic properties of the agent may lead to accurate predictions of accumulation and of possible toxic metabolites and may yield valuable hints to determine possible interactions with other environmental factors. Synergism or antagonism as related to the exposure to a specific hydrocarbon or class of hydrocarbons is approached from the point of view of comparative metabolism as well as from the systematic evaluation of various environmental factors including host susceptibility.

Table IV. Primary Objectives of Toxicology Program on Halogenated Organics in Drinking Water for Extrapolation to Man

1. Definition of pathologic lesions and target organs
2. Determination of dose-response relationships
3. Identification of pathogenesis
4. Elucidation of toxicokinetics
5. Investigation of factors influencing toxic manifestations (*i.e.*, interactions)
6. Elucidation of differential response and sensitivity

The classes of chlorinated compounds under active investigation for their toxicologic activity are listed in Table V. The three main classes include the chloroethers, which are chemical congeners of the human carcinogen *bis*-chloromethyl ether; the halobenzenes, which are known to occur in potable waters;

Table V. Classes of Chlorinated Compounds in Drinking Water
under Toxicological Investigation

   I. Chloroethers
      1. *Bis*-(2-chloroethyl) ether
      2. *Bis*-(2-chloroisopropyl) ether
  II. Halobenzenes
      1. Chlorobenzenes
      2. Bromobenzenes
      3. Chlorobromobenzenes
 III. Haloforms
      1. Chloroform
      2. Bromodichloromethane
      3. Chlorodibromomethane
      4. Bromoform

and the haloforms, which are synthetic by-products of disinfection of tap water with chlorine.

One of the greater concerns associated with the organic contaminants in drinking water is the possibility that the population genome may be affected such that adverse effects may be seen in future generations. Consequently, it is of interest to determine which of the compounds in tap water may be mutagenic hazards to man. As indicated above, almost 300 organic compounds have already been identified from tap waters from the United States. Of this number, approximately one-third have been spot tested for mutagenic activity in *Salmonella typhimurium* TA100; and twelve of these compounds have demonstrated mutagenic activity in this strain, which is a measure of base-pair substitutions with some overlap to frameshift mutations. The compounds that demonstrated mutagenic activity in this test system are listed in Table VI.

The extrapolation of experimental data from animal to man is one of the most critical factors in the evaluation of the safety and hazard of chemicals to man by the use of experimental animals. Three main parameters affecting this problem are: (a) the biotransformation of the compound with respect to both activation and detoxification of the chemical, (b) the toxicokinetics of the agent, and (c) the sensitivity of the target organism (man vs experimental animal). The biotransformation of compounds has been under active investigation since it was originally

Table VI. Compounds Tested and Found to be Mutagenic in *S. typhimurium* TA100[a]

Bromoform	Hexachloro-1,2-butadiene
Chlordane (technical grade)[b]	Isopropylbenzene
1,2-bis-Chloroethoxyethane	Methylchloride
m-Chloronitrobenzene	Methylbromide
1,2-Dichloroethane	Nitroanisole
1,1-Dichloroethylene	Vinyl chloride

[a]Base-pair substitutions with overlap to frameshift mutations.
[b]Pure chlordane is negative in same system.

learned that organophosphate pesticides were initially activated to the toxic form prior to detoxification (Brooks, 1972). Subsequent studies demonstrated that the activity of chemical carcinogens often follows the same route through oxidative pathways in which epoxides may be formed (Arcos and Argus, 1974). Consequently, if man metabolizes an agent or several agents in such a fashion that the pathway is through the activation-inactivation process, it is of practical interest to find a species which metabolizes the compound in the same manner. Thus, the active chemical form responsible for the pathologic lesions is present in the experimental animals as well as in man. Not only is it important to have the same metabolic pathways in the experimental animals as in man, but the rates of reactions for activation and inactivation (detoxication) should also be as similar as possible. One of our research programs is currently focusing on this problem by attempting to study the metabolism of specific chlorinated hydrocarbons in seven species of experimental animals, both *in vivo* and *in vitro*, and in man *in vitro*. The test compounds utilized in these investigations are principally the chlorinated ethers; however, future plans involve the measurement of the metabolism of the haloforms in experimental animals and possibly in man *in vivo*. The species utilized in the comparative metabolism program are listed in Table VII. The experimental species were selected on the basis of their reasonable accessibility as well as on the basis of their extensive utilization for toxicity experimentation. Confirmation that one or several species of experimental animals metabolizes the subject

Table VII. Species Employed in Comparative Metabolism and Toxicokinetics

Mouse (Carworth CD1)
Rat (Charles River CD)
Hamster (Syrian Golden)
Guinea Pig (Hartley)
Rabbit (NZW)
Dog (Beagle)
Monkey
Man

compounds similarly to man will lead to formulation of protocols for in-depth chronic toxicity investigations in the predictive animal models.

While interactions are being investigated indirectly in the studies utilizing mixtures from organic materials taken from drinking water, more extensive and systematic investigations with known compounds are also being conducted. Of the many approaches that can be employed to study toxic interactions, our laboratory has selected the mechanism by which one compound alters the metabolism of other compounds or classes of compounds. It has been demonstrated, for example, that compounds in the organophosphate class can have their toxicity potentiated by an alteration of the biochemical processes that metabolize the organophosphates (Murphy and DuBois, 1957). Additional work has demonstrated that metabolic alterations influencing toxicity are also observed with the chlorinated hydrocarbon insecticides (Deichmann and Keplinger, 1970) and the polynuclear aromatic hydrocarbon contaminants (Uchiyama et al., 1974). For this work, the homologous series of chlorobenzenes has been selected to determine their influence upon the metabolism of various classes of xenobiotics (Carlson and Tardiff, 1976). Relatively little is known about the simple halogenated benzenes, despite their widespread use. Chlorobenzene, p-dichlorobenzene, 1-bromo-4-chlorobenzene, 1,2,4-trichlorobenzene and hexachlorobenzene were administered orally for 14 days to adult male rats. The parameters measured included hexabarbital sleeping time, cytochrome c reductase, cytochrome P-450, EPN detoxication, glucuronyltransferase, benzpyrene hydroxylase and azoreductase. The results indicated that all of the compounds, with the exception of chlorobenzene, decreased hexabarbital sleeping time immediately following and/ or 14 days after treatment, at doses of 600 to 2000 mg/kg/day. At levels ranging from 10 to 40 mg/kg/day, much lower doses than those necessary to cause hepatotoxicity or mortality, all but chlorobenzene caused alterations in the various parameters measured. Administration of p-dichlorobenzene or 1,2,4-trichlorobenzene at these levels for 90 days resulted in increases in EPN detoxification and benzpyrene hydroxylase, and azoreductase. Despite cessation of administration of the compound for as much as 30 days, the EPN detoxification and benzpyrene hydroxylase were still altered in animals having been exposed subchronically to either p-dichlorobenzene or 1,2,4-trichlorobenzene. The more potent inducer of these metabolic pathways and electron transfer components we judged to be 1,2,4-trichlorobenzene. It is concluded that even simple chlorinated benzenes can induce the metabolism of foreign organic compounds with effects continuing substantially beyond the period of exposure. Thus, some compounds in this class of chlorobenzenes may significantly compromise the ability of the organism to properly respond to the ambient gram aimed at the toxicologic definition of the chlorinated hydrocarbons in drinking water. Two specific approaches were mentioned: (a) that dealing with bioscreen of mixtures or organic compounds for mutagenesis, chronic toxicity/carcinogenesis and teratogenesis; and (b) that dealing with specific chlorinated hydrocarbons or classes of these compounds with specific emphasis on comparative metabolism for prediction of human responses and on interactions for predictions of synergism and antagonism.

ACKNOWLEDGMENTS

The authors gratefully acknowledge the technical assistance of W. Emile Coleman and Judith L. Mullaney and the clerical assistance of Shirley A. Tenhover.

REFERENCES

Ames, B. N., F. D. Lee and W. E. Durston. 1973. "An Improved Bacterial Test System for the Detection and Classification of Mutagens and Carcinogens." *Proc. Nat. Acad. Sci.* 70:782-786.

Arcos, Joseph C. and Mary F. Argus. 1974. *Chemical Induction of Cancer: Structural Bases and Biological Mechanisms.* Academic Press. New York. pp. 135-183.

Brcoks, G. T. 1976. "Pathways of Enzymatic Degradation of Pesticides," p. 106. In F. Coulston and F. Korte (Ed.) *Environmental Quality and Safety: Global Aspects of Chemistry, Toxicology and Technology as Applied to the Environment.* Volume 1. Academic Press. New York.

Carlson, Gary P. and Robert G. Tardiff. 1976. "Effect of Chlorinated Benzenes on the Metabolism of Foreign Compounds." *Toxicol. Appl. Pharmacol.* 36:383-394.

Deichmann, W. B. and M. L. Keplinger. 1970. "Protection Against Acute Effects of Certain Pesticides by Pretreatment with Aldrin, Dieldrin and DDT. p. 121-123. In *Pesticides Symposia.* Halos and Associates, Inc. Miami, Florida.

DiPaolo, J. A., K. Takano and N. C. Popescu. 1972. "Quantitation of Chemically Induced Neoplastic Transformation of BALB/3T3 Cloned Cell Lines." *Cancer Res.* 32:2686-2695.

Hill Richard N., Thomas L. Clemens, Dai Kee Liu and Elliot S. Vesell. 1975. "Genetic Control of Chloroform Toxicity in Mice." *Science* 190:159-161.

Kelly, Margaret G. and Roger W. O'Gara. 1961. "Induction of Tumors in Newborn Mice with Dibenz (a,h) anthracene and 3-Methylcholanthrene." *J. Nat. Cancer Inst.* 26(3):651-673.

Kopfler, F. C., R. G. Melton, J. L. Mullaney and R. G. Tardiff. 1975. "Human Exposure to Water Pollutants." Presented at the Division of Environmental Chemistry Meeting, American Chemical Society. Philadelphia, Pennsylvania. April 6-11.

McCann, J., N. E. Spingarn, J. Kobori and B. N. Ames. 1975. "Detection of Carcinogens as Mutagens: Bacterial Tester Strains with R Factor Plasmids." *Proc. Nat. Acad. Sci.* 72:979-983.

Murphy, Sheldon D. and Kenneth P. DuBois. 1957. "Quantitative Measurement of Inhibition of the Enzymatic Detoxification of Malathion by EPN." *Proc. Soc. Exp. Biol. Med.* 96:813-818.

O'Gara, Roger W., Margaret G. Kelly, Jewel Brown and Nathan Mantel. 1965. "Induction of Tumors in Mice Given a Minute Single Dose of Dibenz(a)anthracene or 3-Methylcholanthrene as Newborns. A Dose-Response Study." *J. Nat. Cancer Inst.* 35:1027-1042.

Pietra, Giuseppe, Kathyrne Spencer and Philippe Shubik. 1959. "Response of Newly Born Mice to a Chemical Carcinogen." *Nature* 183(4676):1689.

Simmon, Vincent F. and Robert G. Tardiff. 1976. "Mutagenic Activity of Drinking Water Concentrates." Presented at the Annual Meeting of The Environmental Mutagen Society. Atlanta, Georgia. March 11-15.

Symons, James M., Thomas A. Bellar, J. Keith Carswell, Jack DeMarco, Kenneth L. Kropp, Gordon G. Robeck, Dennis R. Seeger, Clois J. Slocum, Bradford L. Smith and Alan A. Stevens. 1975. *National Organics Reconnaissance Survey for Halogenated Organics in Drinking Water.* Water Supply Research Laboratory, U.S. Environmental Protection Agency, Cincinnati, Ohio. April. (Prepublication Print).

Uchiyama, Mitsuri, Takako Chiba and Kiichiro Noda. 1974. "Cocarcinogenic Effects of DDT and PCB Feedings on Methylcholanthrene Induced Chemical Carcinogenesis." *Bull. Environ. Contam. Toxicol.* 12(6):687-693.

U.S. Environmental Protection Agency. 1975. "National Organics Reconnaissance Survey: Analysis of Tap Water from Five U.S. Cities for Volatile Organic Compounds, a Staff Report." Health Effects Research Laboratory, Cincinnati, Ohio.

U.S. Environmental Protection Agency. 1976. "List of Organic Compounds Identified in Drinking Water in the United States, January 1." Health Effects Research Laboratory, Cincinnati, Ohio.

DISCUSSION

Max Eisenberg, Maryland State Department of Health. Are you doing any epidemiological studies? In particular, have you considered following a specific fraction of the population at risk, such as pregnant women, to observe any anatomical malformations?

Tardiff. I'll answer your question very briefly by saying that, yes, our program is involved in epidemiologic studies. They will be discussed by Dr. Cantor later in the program.

11

ORIGIN, CLASSIFICATION AND DISTRIBUTION OF CHEMICALS IN DRINKING WATER WITH AN ASSESSMENT OF THEIR CARCINOGENIC POTENTIAL

Herman F. Kraybill

National Cancer Institute
Bethesda, Maryland   20014

ABSTRACT

Of the wide array of chemical contaminants identified in potable waters some have carcinogenic activity referenced to studies with experimental animals. Some appear to be universally distributed both nationally and internationally. Some carcinogenic chemicals may be traced back to point source contamination while others may be formed or magnified to levels above those in raw water supplies during the chlorination process. Some carcinogenic chemicals fall into use classes such as pesticides, industrial chemicals, drugs and other categories. Not all these chemicals classified as having carcinogenic potential or activity can be assessed as to their equivalent hazard. Differentiation is necessary to identify those that are well recognized as classical carcinogens and those that are of equivocal nature when referenced to experimental animal studies and may thus be termed suspect carcinogens. Some chemicals may be characterized as potential carcinogens on the basis of structural relationships or ancillary studies on mutagenicity. Many chemicals remain to be characterized for their carcinogenic activity. The integrated insult from multiple carcinogens in the water supply may have additive or inhibitory properties. This aspect of the problem remains to be qualified or quantified in terms of human risk assessment.

INTRODUCTION

The interaction of man with his environment and the health consequences therefrom are well recognized, but not fully comprehended. Environmental stresses of biological, physical and chemical nature may entail both benefits and risks. More and more attention is being given to assess these stresses in a benefit-risk equation. The spectrum of biomedical responses to environmental stresses imposes a cytotoxic effect which includes carcinogenic, mutagenic or teratogenic activities. Our concern in this presentation is oriented toward carcinogenic or neoplastic events.

Environmental cancer in its fullest context implies not the examination of singular insults as a traditional approach, but rather examination of the multiple insults that impact on man from air and water pollutants, diet contaminants, drugs and occupational exposures. Water contaminants, in the form of inorganic or organic carcinogens, add to the total carcinogenic environmental load encountered from all stresses.

Relevant to hazard—carcinogenicity or noncarcinogenicity as components of the overall toxic stress—there is not now and probably never will be an ideal state of absolute safety or zero exposure. Cytotoxic agents abound in nature from plants, bacteria and geologic origin. These environmental stressors on a global basis are omnipresent, as background below the levels added by man through his technological developments. Thus, conceptually, even if one could biologically and pragmatically estimate a safe dose or threshold for a carcinogen, one should view this problem in terms of a potentially "added risk;" that is, above the risk that would prevail naturally. Boyland (1969) believes that 90% of all cancers are due to chemicals, and Higginson (1969) maintains that 90% of cancers are "theoretically preventable" because they may be environmentally related. These views are shared by Epstein (1970).

With approximately four million chemicals already in the world, and the introduction of many new ones into the environment, it becomes prohibitively expensive, if not logistically impracticable, to prove out the safety or risk of each compound (Lewin, 1974). Thus, one sees the necessity to have new methods for rapid evaluation of the carcinogenic potential of whole classes of compounds.

Experimental studies in animals are at best only presumptive measures for assessment of a carcinogenic risk in man. Obviously, epidemiologic pursuits are of high validity; but, it is socially, legally and medically unacceptable to subject man to tests unless he is inadvertently receiving such challenges, as is the case with water contaminants. However, animal studies are used for predictive purposes. The determination of carcinogenic and/or tumorigenic risk to chemicals at ambient concentrations is a difficult task. One can never feel secure as to whether or not a chemical is carcinogenic in man, since an animal system may reveal false positives or false negatives. Indeed, there is a variable response in man which leads one to ask "to what man is this effect relevant?" Thus, the perfect model will probably never be attained.

Beyond the considerations of species variation is the evaluation of the potency of the chemical for tumor induction and the time frame for tumor induction. The question arises whether very low levels, on a dose-response relationship for a defined population, would require over 100, 200 or 300 years to induce tumors; which, of course, is beyond man's life span.

Carcinogenic contaminants in the drinking water supply may originate from various sources. Since many municipal water supplies are derived from rivers, lakes, streams or ground waters, they carry pollution from agricultural runoff, industrial effluents and accidental or deliberate dumping. Some organic chemicals are formed from treatment processes such as chlorination. Many of the carcinogens are in tap water not from treatment or processing alone but because of failure to adequately remove them from raw water supplies. Carbon tetrachloride and chloroform do not need to be formed by chlorination if the rivers carry these chemicals originally.

For purposes of this presentation, the matter of carcinogens in water will be referenced to those chemicals identified in drinking water. An analysis of a list of 221 organic chemicals provided by the Water Supply Research Laboratory of the Environmental Protection Agency in Cincinnati, Ohio, on June 1, 1975, is used in the categorization of carcinogens. Since this listing, a revised list with 14 more chemicals was provided on September 1. This listing was generated from an exhaustive search of the literature and from reports on analysis of some municipal water supplies. The listing is, therefore, not all-inclusive and, on a quantitative basis, probably reflects only a small percentage of the total organics in the water (Tardiff *et al.*, 1975).

As a result of concern in 1974 over certain organic chemicals (chalomethanes) in drinking water, the Environmental Protection Agency instituted the National Organics Reconnaissance Survey. This survey of 80 cities had three major objectives. One of these was to deal with the trihalomethanes; the second was to ascertain the effect of water treatment processes on the formation of these chemicals; and the third was to characterize and possibly quantify the organic content of finished water (Symons, 1975).

GENERAL CLASSIFICATION OF ORGANIC CHEMICALS IN DRINKING WATER: REFERENCES TO TECHNOLOGICAL USE

The selection process of chemicals for carcinogenic activity assessment by animal bioassay is based on the uses of that chemical, probable exposures, structure-activity relationships and available toxicity data. These criteria provide a priority scheme for investigating those chemicals which may impose the maximum hazard to man in his environment. The production estimate and extent of probable exposure are most significant. Accordingly, classifying these chemicals according to their use pattern is most important. In Table I, 221 organic compounds identified in drinking water as of June 1975, have been grouped by classes, and the number in each class and the percentage are shown.

Because of uses in industry, agriculture and by consumers as food chemicals, drugs, and for other applications, these chemicals may be introduced into water supplies which are the sources of municipal tap water.

Table I. Representation of Organic Chemicals in Drinking Water by Technological Use[a]

Technological Use	Number of Chemicals in Use Class	Percent of Total in Class
Industrial chemicals	100	45.2
Pesticides	28	12.6
Food chemicals—derivatives	16	7.3
Drugs—perfumes	13	5.8
Decomposition products	10	4.5
Natural product or toxin	9	4.2
Laboratory chemicals	3	1.3
Tobacco products	1	0.4
Miscellaneous or unknown	41	18.5
Total	221	100.0

[a]For reference purposes, grouping of chemicals for bioassay origin by use patterns is important. Chemicals could fall into several classes—major use adopted (Furia and Bellanca, 1971; *Merck Index*, 1960).

CRITERIA USED IN CLASSIFICATION OF CARCINOGENS

Attempts have been made to categorize carcinogens under such terms as potent, weak, recognized, suspect or potential on the basis of toxicological studies. Some prefer not to classify and simply indicate that a carcinogen is a carcinogen, but such a rigid viewpoint provides no basis for risk assessment when considering the host of environmental agents which man encounters in his life. Potent carcinogens such as aflatoxin (Kraybill and Shimkin, 1964) and vinyl chloride (Maltoni and Lefemine, 1974) produce in a variety of experimental animals a high incidence of cancers, even at low levels in a diet or in the environment. Some of the potent experimentally proven (animals) carcinogens, such as benzidine, *bis*-chloromethyl ether, betanaphtylamine and vinyl chloride, have been well established, epidemiologically, as occupational carcinogens in man. A precedent has been established for classification of carcinogens by listing chemicals and mixtures that cause cancer in man by direct observations on exposed populations (National Cancer Institute, 1975a). There are 32 substances on that list.

Professor Maltoni (1973) in discussing occupational carcinogenesis classified occupational carcinogens into three groups: (a) definite carcinogens, (b) suspect carcinogens and (c) potential carcinogens. Goldsmith (1975) also developed a taxonomic approach to environmental carcinogenesis and provided, from epidemiological experience and analysis, eight classes (including such descriptors as likely and possible) for development of a decision scheme for the different classes that would help in regulatory interpretation. Kraybill (1976) also developed a classification scheme for carcinogens in water which, in 1975, was adopted in a classification used by the Water Supply Research Laboratory of the Environmental Protection Agency. This classification scheme and criteria are presented in Table II.

Table II. Criteria for Development of Classification of Aquatic Carcinogens

   I. Recognized and classical
      A. From epidemiological observations
      B. Response in many species, strains, and repeatedly confirmed
      C. Tumors of rare type, tumor incidence increase in short time, exposure route relevant, tumors transplantable
  II. Suspect
      A. Response only in one species, strain or sex
      B. Tumor incidence increase with exposure
      C. Evidence from *in vitro* studies
 III. Potential
      A. Suggestive evidence, basis of structure activity
      B. Mutagenic data
  IV. Promoters
      A. Will increase tumor incidence of known carcinogens
   V. Fragmentary data—inadequate tests
      A. Inadequate number of test animals, observation time, nonrelevant exposure, improper species and strain
      B. Question of contaminants—cocarcinogens
      C. Overt toxicity and metabolic overloading
  VI. Noncarcinogenic
      A. Negative in exhaustive testing
      B. Physiological and cellular constituents
      C. Negative findings—human observations—long time
      D. Noncarcinogenic when contaminants are removed

Pragmatically, such a classification provides a working scheme for decision-making processes as to relevant hazard. For example, one would not develop the same type of guidelines for exposure situations relevant to vinyl chloride as for DDT, *bis*-2-chloroethyl ether, or other chemicals where experimental evidence is tenuous.

Although no formal agreement has been reached on criteria and a taxonomic classification of the activity of chemicals proclaimed or suspected as being carcinogenic, Hueper, as early as 1961, recognized the need for such a classification. He believed that the concept of "chemical group carcinogens" had restricted applicability and would be valid for only two classes of carcinogens, that is, radioactive chemicals and chemicals with estrogenic properties (Hueper, 1961).

The first delineation of carcinogens by Hueper was by site of action or target organ. His classification contained four groups: (a) primary contact point, *i.e.*, skin, lungs, etc.; (b) at site of excretion, *i.e.*, bladder or skin; (c) at site of secondary retention and storage, *i.e.*, radium, arsenic, Thorotrast, etc. and effect on bones, bone marrow, liver and skin; and (d) at site of special affinity, *i.e.*, hormones and organotropic, goitrogenic and estrogenic chemicals, etc.

Another classification was by species specificity. Here, Hueper uses subclassifications such as recognized, suspect, potential and experimental. His explanation of varying biological activity is comparable to our previous assessment. In the experimental category, he lists those chemicals which produce cancers in experimental animals but which are not important contaminants in the environment yet may be of scientific significance, for example, 2-acetylaminofluorene (2AAF). Hueper's final classification is according to carcinogenic potency and carcinogenic completeness. In the former case, he specifies the variation in activity in different species, strains, sexes and specific test conditions. In the latter class, he alludes to multistage mechanisms for a chemical in the induction of cancer; he discusses the fact that some chemicals are only "initiators" or "promoters;" and some chemicals derive their activity at the cell level predominantly as cocarcinogens. He also lists some chemicals that are "incomplete" carcinogens which are conditioned by other factors for eventuating in neoplasia.

CONSIDERATION AND ASSESSMENT OF THE CARCINOGENIC PROPERTIES OF ORGANIC CHEMICALS IN WATER

Some lists on the status of organic compounds identified in drinking water have been published, while others are unpublished. For purposes of this presentation, we have confined our review mostly to the listing of organic compounds identified in drinking water in the United States, made available on June 1, 1975, by the Water Supply Research Laboratory of the Environmental Protection Agency. This listing, over the months, has been expanded when new organic chemicals were identified. As mentioned before, the June 1975 listing contained 221 organic chemicals; 235 chemicals were listed by September 1975.

The evaluation of carcinogenic activity, whether classified as recognized, suspect or potential, is based on reports in the literature or recent reports on preliminary findings from the National Cancer Institute Carcinogenesis Program. Many of the chemicals appearing on lists of biorefractories in drinking water have been suggested for bioassay in the National Cancer Institute Bioassay Program. Some of the chemicals appearing as biorefractories had been on bioassay at the National Cancer Institute prior to the realization that they were water contaminants. Data on these latter compounds have only recently become available.

There are further studies planned on the haloethers, originally reported by Van Duuren and co-workers (1972) who studied the structure-activity relationships of analogs of *bis*-chloromethyl ether. One of the haloethers appears to be a contaminant in many water supplies in this country and in Europe (Piet *et al.*, 1973). This haloether, the *bis*-2-chloroisopropyl ether, is currently under bioassay. Haloethers, products of glycol synthesis in industrial processes, are introduced into waterways as industrial effluents. This class of compounds is interesting in that the carcinogenic activity of the chemical is dependent on structure; namely, the location of the halogen atom in the alpha or beta position. The alpha compound appears to be highly active, that is, for such compounds as the chloroethyl ethers. As the chlorine atom is positioned further away from

the ether linkage, the chemical has less activity. Van Duuren
et al. (1972) reported that chloromethyl methyl ether (CMME),
bis-chloromethyl ether (BCME), and bis-α-chloroethyl ether
have significant carcinogenic activity.

As previously indicated, the National Organics Reconnaissance
Survey of the Environmental Protection Agency had as one
assignment the study of six halogenated organic compounds to
determine their presence in eighty water supplies surveyed. The
six chemicals of concern are (1) chloroform, (2) carbon tetrachloride, (3) bromodichloromethane, (4) dibromochloromethane,
(5) 1,2-dichloroethane and (6) bromoform. Of these six, chloroform, carbon tetrachloride and dichloroethane have carcinogenic
activity. The National Cancer Institute data base indicates that
carbon tetrachloride, showing a carcinogenic response in several
species and strains in several laboratories, would be a recognized
carcinogen. There are some reports on carcinogenic activity in
man (Tracey and Sherlock, 1968). Data on chloroform indicate
that this chemical may be suspect as a carcinogen (International
Agency for Research on Cancer, 1972, p. 61-65; Eschenbrenner
and Miller, 1945). Chloroform is obviously of interest in that
it appears in raw water supplies, and in all samples from 80
locations surveyed. The concentration ranged from 0.1 to 311
$\mu g/l$ with 50% of the finished waters containing 25 $\mu g/l$ of
chloroform or less (Environmental Protection Agency, 1975).
While chloroform appears in raw water supplies (range of 0.1
to 0.9 $\mu g/l$), its concentration was considerably higher in treated
(chlorinated) or finished water (Bellar and Lichtenberg, 1974).
Since chloroform is widely distributed at significant levels, has
carcinogenic activity (hepatoma production), and occurs in
other ingested products (drugs), its health risk cannot be ignored.
Further support for concern on chloroform may be reinforced
because of the ancillary data on carbon tetrachloride and the
potential additive effect of these two chemicals in the environment.

Out of the list of 221, or now 235, chemicals in drinking
water, only 21 were characterized as having carcinogenic activity.
Four of the chemicals listed are recognized carcinogens; the
remaining are classified as suspect. This is not to imply that
there may not be more, since some remain to be identified in
future analytical works (Table III). There is also the possibility

Table III. Organic Carcinogenic Chemicals in Drinking Water

	Chemical	Concentration[a] (μg/l)	Carcinogenesis Reference
1.	Aldrin	–	Davis and Fitzhugh (1962)
2.	Benzene	50	Ishifaru et al. (1971)
3.	Benzo(a)pyrene[b]	0.0002 - 0.002	IARC (1973a, pp. 91-137)
4.	Bis-2-chloroethyl ether	0.07 - 0.16	Innes et al. (1969)
5.	Bis-chloromethyl ether[b]	–	Thiess et al. (1973)
6.	BHC (Lindane)	–	Thorpe and Walker (1973)
7.	Carbon tetrachloride[b]	5	Tracey and Sherlock (1968)
8.	Chlordane	–	NCI (1975b)
9.	Chloroform	0.1 - 311	Eschenbrenner and Miller (1945)
10.	1,2-Dibromoethane (EDB)	–	Olson et al. (1973)
11.	1,1-Dichloroethane (EDC)	–	NCI (1975b)
12.	Dieldrin	0.05 - 0.09	Davis and Fitzhugh (1962)
13.	DDT	–	Innes et al. (1969)
14.	DDE	–	Tomatis et al. (1974)
15.	Endrin	0.004	Deichmann et al. (1970)
16.	Heptachlor	–	IARC (1974, pp. 173-191)
17.	1,1,2-Trichloroethane	0.35 - 0.45	NCI (1975b)
18.	Trichloroethylene	–	NCI (1975b)
19.	Tetrachloroethane	10	NCI (1975b)
20.	Tetrachloroethylene	0.4 - 0.5	NCI (1975b)
21.	Vinyl chloride[b]	–	Maltoni and Lefemine (1974)

[a] Determinations from EPA Water Quality Program.
[b] Recognized carcinogens.

that those classified as potential carcinogens may prove to be positive carcinogens on further testing. Not to be overlooked is the possibility that some chemicals may act as promoters or cocarcinogens, giving an added insult to those present as recognized or suspect carcinogens (Table IV). The characterization of inorganic and organic carcinogens in water supplies (raw or untreated) deserves careful scrutiny since they may not be completely removed in some municipal water supplies. Six recognized carcinogens have been identified in various water supplies (Table V).

Table IV. Potential Carcinogens and Promoters in Drinking Water[a]

1. Chloromethyl ethyl ether	Kraybill 1976
2. *Bis*-2-chloroisopropyl ether	Kraybill 1976
3. Decane (promoter)	Horton *et al.* 1965
4. Dodecane (promoter)	Horton *et al.* 1965
5. Octadecane	Sice 1966
	Horton *et al.* 1957

[a]Promoters or cocarcinogens demonstrated in tests with recognized carcinogens. Potential on basis of structural analogs.

Table V. Other Carcinogenic Chemicals in Raw Water Supplies

	Chemical	Reference on Activity
1.	Arsenic[a]	IARC (1973b, pp. 48-73)
2.	Asbestos[a]	IARC (1973b, pp. 17-47)
3.	1,2-Benzanthracene	Public Health Service (1958-59)
4.	Cadmium[a]	IARC (1973b, pp. 74-99)
5.	Benzidine[a]	IARC (1972, pp. 80-86)
6.	Dibenz (a,h) anthracene	IARC (1973a, pp. 178-196)
7.	Ethylene thiourea[a]	Innes *et al.* (1969)
8.	Chromium (hexavalent)[a]	IARC (1973b, pp. 100-125)
9.	3-Methylcholanthrene	Kelly and O'Gara (1973)
10.	Mirex	Innes *et al.* (1969)
11.	Polyurethane	Hueper (1961)
12.	Strobane	Innes *et al.* (1969)

[a]Recognized carcinogen.

DISCUSSION: RISK ASSESSMENT

Biorefractories in municipal water supplies represent a wide spectrum of organic and inorganic chemicals. The ability to detect the presence of these organic compounds is exemplified by the ever-increasing numbers identified in the surveillance program. Among the organic chemicals, the chlorinated hydrocarbons and ethers, and a few polycyclic aromatic hydrocarbons represent thus far the major concern because of their carcinogenic properties. Some of the halogenated compounds are ubiquitous, both in raw and treated waters, reflecting the extent of pollution of waterways, the extent of formation through chlorination, or the failure to remove them in filtration processes. Certainly, from the standpoint of preventive health measures, a more comprehensive understanding of this problem is needed.

The chlorination process may not only produce chlorinated hydrocarbons or augment the levels already present, but also produce other compounds, such as chlorinated phenols, which may have more toxic and biological properties than their parent phenols. Some aromatic compounds would fall in this class as well as the aliphatics. Of significant interest is the fact that molecular species appear in drinking water that were not previously identified in raw water. Ozonation, as another choice for disinfecting water in the place of chlorine, could also offer some problems since this process could yield such products as ozonides, peroxides, epoxides and aldehydes. Chemicals with some of these structures have demonstrated carcinogenicity, but many of them would have to be evaluated for their carcinogenic activity.

While major attention may be given to the organic chlorine compounds, some more attention will also have to be given to inorganic chemicals. Certainly, the asbestos in drinking water as a potential hazard is a problem that remains to be assessed. Some studies have been accomplished on the trace metal contaminants, such as arsenic, beryllium, cadmium, chromium, cobalt, iron, lead and nickel (Berg and Burbank, 1972) in an attempt to correlate these trace metal concentrations in drinking water with cancer mortality. There appeared to be some correlation between nickel concentrations and cancer death rates from mouth and intestinal cancer, and between arsenic and cancer of the eye and larynx, and myeloid leukemia. Beryllium appeared to be correlated with bone, breast and uterine cancer; and lead was associated with kidney, stomach, ovarian and intestinal cancers. Cancers from arsenic traced to water supplies in Taiwan and Argentina have been reported (Tseng et al., 1968). Some trace metals are necessary for physiological function. An absence of the metal can lead to nutritional deficiencies. Beyond these low levels, however, an opposite effect may be noted where there could be enhancement of toxicity through overloading or interactions.

The relevance of epizootics in aquatic animals with respect to neoplasia (fish and shellfish tumors) associated with water pollution, as claimed by Brown and co-workers (1973), is a warning not only on potential hazard but could also serve as a clue to potential association of point source contamination with increased incidence of human cancers in certain geographical areas.

Some of the halogenated organic compounds found in water have carcinogenic activity and because of this there could be a cancer risk associated with consumption of such compounds. This hypothesis, advanced for a probable cause-and-effect relationship in the case of the New Orleans water supply, might be confounded by other variables. Since these compounds are predominantly hepatocarcinogenic (from experimental studies), one might expect an increased liver cancer rate among certain populations. Epidemiologic studies thus far have not revealed this to be the case. Most of the attention has been focused on chloroform and carbon tetrachloride, which could react additively in hepatoma induction. Another additive effect could come from the haloethers and polynuclear hydrocarbons, but their exposure levels to man are lower in drinking water than those of the aforementioned halogenated methanes, ethanes and olefins.

The levels of the carcinogenic organic contaminants are in the parts-per-trillion and parts-per-billion range. Collectively, they could have a greater impact on a target organ through continuous exposure than the insult from a single chemical exposure. From mathematical and biological extrapolation factors used to estimate an upper level of risk, the evidence thus far does not support a conclusion of increased risk. More critical and definitive data need to be set forth to test the hypotheses currently advanced. Because of the low-level exposures to man, and in some locations an infrequent occurrence even at higher concentrations, *e.g.*, 300 µg/l, it would appear that further studies are needed. This would require more extensive experimental studies on the carcinogenicity of lifetime exposures to provide clues on the potential human health hazard. Beyond this, because of limited epidemiologic studies on these contaminants, more definitive monitoring and surveillance data must be acquired delineating possible groups at risk where there may be a continuous exposure at high levels from single agents or multiple stresses. Until then no conclusive assessment of any health risk can be made.

ACKNOWLEDGMENTS

The author wishes to acknowledge the technical assistance given by Dr. Kirtland McCaleb and Mr. Arthur McGee, Stanford Research Institute, Menlo Park, California; and Dr. Sidney Siegel, Carcinogenesis Program, National Cancer Institute, Bethesda, Maryland.

REFERENCES

Bellar, T. A. and J. J. Lichtenberg. 1974. *The Occurrence of Organohalides in Chlorinated Drinking Waters.* EPA-670/4-74-008. U.S. Environmental Protection Agency, Washington, D. C.

Berg, J. W. and F. Burbank. 1972. "Correlations Between Carcinogenesis Trace Metals in Water Supplies and Cancer Mortality." *Ann. N. Y. Acad. Sci.* 199:249-264.

Boyland, E. 1969. "The Correlation of Experimental Carcinogenesis and Cancer in Man." *Prog. Exp. Tumor Res.* 11:222-234. Basel/New York: S. Karger AG.

Brown, E. R., J. J. Hazdra, L. Keith, I. Greenspan, J. B. G. Kwapinski and P. Beamer. 1973. "Frequency of Fish Tumors Found in a Polluted Watershed as Compared to Non-Polluted Canadian Waters." *Cancer Res.* 33:189-198.

Davis, K. J. and O. G. Fitzhugh. 1962. "Tumorigenic Potential of Aldrin and Dieldrin for Mice." *Toxicol. Appl. Pharmacol.* 4:189.

Deichmann, W. B., W. E. MacDonald, E. Blum, M. Berilacqua, J. Radomski, M. Keplinger and M. Balkus. 1970. "Tumorigenicity of Aldrin, Dieldrin and Endrin in the Albino Rat." *Ind. Med. Surg.* 39:426.

Environmental Protection Agency. 1975. "Report to the U.S. Congress on Suspect Carcinogens in Water Supplies." U.S. Environmental Protection Agency, Washington, D. C. June 17.

Epstein, S. S. 1970. "Control of Chemical Pollutants." *Nature* 228: 816-819.

Eschenbrenner, A. B. and E. Miller. 1945. "Induction of Hepatomas in Mice by Repeated Oral Administration of Chloroform with Observations on Sex Differences." *J. Nat. Cancer Inst.* 5:251.

Furia, T. E. and N. Bellanca. 1971. *Fenaroli's Handbook of Flavor Ingredients.* Chemical Rubber Co., Cleveland, Ohio.

Goldsmith, J. 1975. "Proposed Terminology and Classification for Environmental Cancer." National Cancer Institute, Division of Cancer Cause and Prevention. Unpublished Report.

Higginson, J. 1969. "Present Trends in Cancer Epidemiology." *Proc. Can. Cancer Conf.* 8:40-75.

Horton, A. W., D. T. Denham and R. P. Trosset. 1957. "Carcinogenesis of the Skin. II. The Accelerating Properties of Aliphatics and Related Hydrocarbons." *Cancer Res.* 17:758-766.

Horton, A. W., P. A. Van Dreal and E. Bingham. 1965. "Physiochemical Mechanism of Acceleration of Skin Carcinogenesis." *Adv. Biol. Skin* 7:165-181.

Hueper, W. C. 1961. "Carcinogens in the Human Environment." *Arch. Pathol.* 71:237-267 and 355-380. March.

Innes, J. R. M., B. M. Ulland, M. G. Valerio, L. Petrucelli, L. Fishbein, E. R. Hart, A. J. Pallota, R. R. Bates, H. L. Falk, J. J. Gart, M. Klein, I. Mitchell and J. Peters. 1961. "Bioassay of Pesticides and Industrial Chemicals for Tumorigenicity in Mice: a Preliminary Note." *J. Nat. Cancer Inst.* 72(6):1101-1114.

International Agency for Research on Cancer. 1972. "IARC Monographs on the Evaluation of Carcinogenic Risk of Chemicals to Man." 1:61-65, 80-86. Lyon, International Agency for Research on Cancer.

International Agency for Research on Cancer. 1973a. "IARC Monographs on the Evaluation of Carcinogenic Risk of Chemicals to Man: Certain Polycyclic Aromatic Hydrocarbons and Heterocyclic Compounds." 3:91-136, 178-196. Lyon, International Agency for Research on Cancer.

International Agency for Research on Cancer. 1973b. "IARC Monographs on the Evaluation of Carcinogenic Risk of Chemicals to Man: Asbestos,

Inorganic and Organometallic Compounds." 2:17-47, 48-73, 74-99, 100-125. Lyon, International Agency for Research on Cancer.

International Agency for Research on Cancer. 1974. "IARC Monographs on the Evaluation of Carcinogenic Risk of Chemicals to Man: Some Organochlorine Pesticides." 5:173-191. Lyon, International Agency for Research on Cancer.

Ishimaru, T., H. Okada, T. Tomiyasu, T. Tsuchimoto, T. Hoshino and M. Ichimaru. 1971. "Occupational Factors in the Epidemilogy of Leukemia in Hiroshima and Nagasaki." *Am. J. Epidemiol.* 93:157-165.

Kelly, M. and R. O'Gara. 1973. "Induction of Tumors in Newborn Mice with Dibenz (a,h) anthracene and 3-Methylcholantrene." *J. Nat. Cancer Inst.* 26:651-679.

Kraybill, H. F. and M. B. Shimkin. 1969. "Carcinogenesis Related to Foods Contaminated by Processing and Fungal Metabolites." *Adv. Cancer Res.* 8:191-248. Academic Press, New York.

Kraybill, H. F. 1976. "Distribution of Chemical Carcinogens in Aquatic Environments," *Prog. Exp. Tumor Res.* 20:3-34.

Lewin, R. 1974. "Environmental Search for the Source of Cancer." *SR/World–Science Section* 1974:50-51. April 20.

Maltoni, C. and G. Lefemine. 1974. "Carcinogenicity Bioassay of Vinyl Chloride Current Results. Part II. Carcinogenesis Associated with Vinyl Chloride." *Ann. N. Y. Acad. Sci.* 246:195-218.

Maltoni, C. 1973. "Occupational Carcinogenesis." In *Advances in Tumor Prevention, Detection and Characterization. Vol. 2. Detection and Characterization.* Preprint of International Congress Series. No. 322.

Merck Index of Chemicals and Drugs. 1960. P. Stecher, M. J. Finkel, O. H. Siegmund and B. M. Szafranski (Eds.). Merck and Company, Rahway, New Jersey.

National Cancer Institute. 1975a. "List of Chemicals that Cause Cancer in Man," p. 85. *House of Representatives Hearings, Subcommittee of the Committee on Appropriations, March 1975.* Department of Labor, Health, Education and Welfare. U.S. Government Printing Office, Washington, D. C.

National Cancer Institute. 1975b. "Preliminary Findings on Bioassay of Chemicals for Carcinogenic Activity." Unpublished data from Bioassay Program.

Olson, W. A., R. T. Haberman, E. K. Weisburger, J. M. Ward and J. H. Weisburger. 1973. "Brief Communication: Induction of Stomach Cancer in Rats and Mice by Halogenated Aliphatic Fumigants." *J. Nat. Cancer Inst.* 51:1993-1995.

Piet, G. J., B. C. T. Zoeteman, A. H. Nettenbreiger and G. T. H. Ruijgrok. 1973. "*Bis*(2-chloroisopropyl) Ether in Surface and Drinking Water in the Netherlands." Orientation Report of *Rijksinstituut voor Drinkwatervoorziening*, S-Gravenhage, Parkweg 13, The Netherlands.

Public Health Service. 1958-59. "Survey of Compounds Which Have Been Tested for Carcinogenic Activity." PHS-149, Series Entry 393. DHEW Publ. No. (NIH) 72-35. U.S. Government Printing Office, Washington, D. C.

Sice, J. 1966. "Tumor-Producing Activity of N-Alkanes and 1-Alkanols." *Toxicol. Appl. Pharmacol.* 9:70.

Symons, J. M. 1975. "Suspect Carcinogens in Water Supplies." Interim Report to Congress. Appendix A. Character and Extent of Contamination. Section 1. National Organics Reconnaissance Survey. EPA Progress Report. Cincinnati, Ohio. April.

Tardiff, R. G., G. F. Craun, L. J. McCabe, P. E. Bertozzi. 1975. "Suspect Carcinogens in Water Supplies." Interim Report to Congress. Appendix B. Health Effects Caused by Exposure to Contaminants. EPA Progress Report. Cincinnati, Ohio. April.

Thiess, A. M., W. Hey and H. Zeller. 1973. "Zur Toxikologie von Dichloromethylether—Verdacht auf kanzerogene Wirkung auch beim Menschen." *Zentralbl. Arbeitsmed. Arbeitschutz.* 23:97.

Thorpe, E. and A. I. T. Walker. 1973. "The Toxicology of Dieldrin (HEOD). II. Comparative Long-Term Oral Toxicity Studies in Mice with Dieldrin, DDT, Phenobarbitone, β-BHC and γ-BHC." *Food Cosmet. Toxicol.* 11:433.

Tomatis, L., V. Turusov, R. T. Charles and M. Boiocchi. 1974. "The Effect of Long-Term Exposure to 1,1-Dichloro-2,2-*bis*(p-chlorophenyl) ethylene(p,p'-DDE), to 1,1-Dichloro-2,2-*bis*(p-chlorophenyl) ethane (p,p'-DDD) and to the Two Chemicals Combined on CF-1 Mice." *J. Nat. Cancer Inst.* 52:883-891.

Tracey, J. P. and P. Sherlock. 1968. "Hepatoma Following Carbon Tetrachloride Poisoning." *N. Y. State J. Med.* 68:2202.

Tseng, W., H. Chu, S. How, J. Fong, C. Lin and I. Shuyeh. 1968. "Prevalence of Skin Cancer in an Endemic Area of Chronic Arsenicism in Taiwan." *J. Nat. Cancer Inst.* 40:453-463.

Van Duuren, B. L., C. Katz, B. M. Goldschmidt, K. Frenkel and A. Sivak. 1972. "Carcinogenicity of Haloethers. II. Structure Activity Relationship of Analogs of *bis*(chloromethyl) ether." *J. Nat. Cancer Inst.* 48:1431-1439.

DISCUSSION

David Friedman, Food and Drug Administration. On your last slide, Hueper has shown that polyurethane was in water and was also carcinogenic. Could you indicate the type of urethane this was and what type of carcinogenicity?

Kraybill. At this point we should correct some impressions gleaned from identification of carcinogens (see Table V) in water. This table does not necessarily mean that these chemicals were tested in water and then found to be carcinogenic. The reference to Hueper's work, for example, on polyurethane, means that Hueper and others tested this chemical separately and reported on the carcinogenicity. Most of these polymeric materials have been tested by subcutaneous injection in which instance they induced sarcomas (Hueper, *J. Nat. Cancer Inst.* 33:1005, 1964). Other sites of tumor induction for this plastic and medicinal agent is the cecum and connective tissue.

Friedman. Was this chemical shown to be actually in water?

Kraybill. This chemical appeared on lists of chemicals monitored in water. Polyurethane (spandex) appeared on a list of organics in the Kanawha River—a report made available (unpublished) by Dr. Edward Light, Research Director of Campaign Clear Water, Box 567, Charleston, West Virginia 25322.

12

THE POTENTIAL FOR INCREASED MUTAGENIC RISK TO THE HUMAN POPULATION DUE TO THE PRODUCTS OF WATER CHLORINATION

Robert B. Cumming

    Biology Division
    Oak Ridge National Laboratory
    Oak Ridge, Tennessee 37830

ABSTRACT

    The chlorination of water in which there are organic materials produces stable chlorine-containing compounds which may have substantial biological activity. Such compounds are released into surface waters by chlorination processes at very low concentrations, and the assessment of their potential for producing genetic damage in the human population presents a formidable technological problem. The approach which seems most productive is to select several from the array of chlorine-containing compounds known to be produced by water chlorination and to study their behavior in well established genetic test systems including those which have been used to estimate levels of genetic risk for other agents. If risk estimates can be determined for several compounds and some factor added for potential interaction, then a genetic risk estimate can be made for the entire water chlorination process. Such an estimate would always be imperfect and subject to revision as additional data were obtained.

    In the model studies described in this paper, 5-chlorouracil (5-ClU) has been tested in several mammalian and submammalian genetic test systems. The incorporation of 5-ClU into the DNA of mice which had been exposed to this compound in their drinking water has been measured. Both a specific-locus mutation test and a dominant-lethal mutation test have been performed on mice which have been similarly exposed. The compound has been tested for mutagenicity in several types of bacteria and has been found to be highly mutagenic in *E. coli*. The data from 5-C1U incorporation studies into the DNA of mice, together with the specific-locus mutation data, allow the calculation of the upper 95% confidence limit for mutations induced in the human population at environmental exposure levels. This calculation demonstrates that 5-C1U, by itself, does not pose a significant genetic hazard to humans at current release levels. The more significant question of whether, in sum, all of the chlorine-containing organic compounds produced by water chlorination pose a significant hazard is yet to be determined.

INTRODUCTION

Our primary biomedical concern related to water chlorination and human exposure to chlorinated organics has been the potential for increasing the incidence of cancer among exposed individuals. Various aspects of this problem are dealt with by the other three speakers in this session. I will deal with a different but related problem, that is, the potential for increasing the amount of induced heritable genetic damage in exposed human populations.

There are several important differences in the techniques and the data used for the estimation of mutagenic risk and carcinogenic risk. These are two biomedical endpoints with very different implications for the human population and very different problems associated with their assessment. Although the medical consequences of genetic disease may be altogether as devastating as the consequences of cancer, identifying the environmental factors that lead to the damage is very much more difficult. The expression of genetic disease is usually far removed in time from the induction of the mutations which caused it and, thus, from the toxic agent which lead to the mutations. The consequences of genetic damage would not be expected to fall upon the individual in which the mutations were induced but upon his progeny or descendants—perhaps several or many generations later. In carcinogenesis the target is the individual, and the cancer statistics are compilations of many individual tragedies. The real target for environmental mutagens is not the individual, or even the small group, but the human gene pool, and the human gene pool is not only the source of all our admirable and less than admirable attributes, it is also the key to the continued existence of our species.

There are other differences between the assessment for mutagenicity and carcinogenicity. A person who has cancer knows it. He is easy to identify medically. He knows or can quickly estimate the consequences of his disease. It is not very difficult for him or others to come up with some estimate of the cost of his condition to himself and to the society, and to the individual it is frequently catastrophic. One can make cost estimates for induced cancer by simply adding up the number of affected individuals and recognizing that for each individual we are dealing with a very serious problem. Genetic effects may be very much more subtle and the estimation of the cost of genetic damage is very much more difficult. Genetic disease ranges from things which are as totally incapacitating as cancer —lethal conditions—to things which are much less serious to the individual, but which in their sum total, since they may affect very large populations, may have as great a cost to society.

Another difference between cost assessments of mutagenicity and carcinogenicity in human populations is that there is a straightforward relationship between carcinogenicity and cancer incidence, whereas the relationship between increased mutation frequency and increased genetic disease is not straightforward. The way in which increased mutation frequency is related to increased medical costs to the society is presently the subject of much controversy and debate. This makes the assessment of mutagenic risks—risks of increased mutation frequency—very much easier than estimating the increased medical costs to the society.

These differences in the nature of the two major public health problems which may be associated with exposure to chlorinated organics demand different approaches to risk assessment. Ironically, simple *in vitro* mutational tests may be more useful in dealing with the problem of carcinogenesis than they are for predicting genetic hazards (McCann and Ames, 1976).

What are the characteristics of the problem we are facing? The problem is estimating the potential public health hazards from water chlorination and from the products released therefrom. First, as has been amply documented by the first two speakers in this session, we are not dealing with a single compound, or even a few, but with many compounds, some of which are poorly characterized. Secondly, these compounds are individually present in very low concentrations. Third, the exposure would be expected to be continuous throughout the lifetime of the individual or, more importantly for genetic damage, throughout the reproductive lifetime of the individual, which for humans is generally taken to be thirty years. Fourth, very large populations would be exposed so that even small effects might involve significant numbers of people. So we have large populations exposed for long periods of time to low levels of very many compounds.

Some of the compounds identified from this array would be expected to have rather potent biological effects. For others we would not be able to even guess what their effects might be. Some of the identified compounds would be expected to induce, at least in some test systems, a class of genetic damage which we can loosely refer to as point mutations, though the induction of gross chromosomal abnormalities is not excluded. This means that test systems should be provided which will measure point mutations as well as chromosome aberrations.

What kinds of test systems and other means of obtaining data relevant to this problem are available? The epidemiological approach may, in certain selected instances, yield valuable information about carcinogenicity. The next speaker will discuss this approach in greater detail. The epidemiological approach is not likely to provide useful information on potential genetic effects of chlorinated organics. This is true because of the long time lag between exposure and observed effect, the tremendous genetic diversity of the human population, the lack of discrete large populations with identifiably different exposure conditions over the required long periods of time, and several other factors.

Several *in vitro* tests for mutagenicity are available, mostly involving prokaryotes. A high correlation has been claimed (McCann and Ames, 1976) between tests for mutagenicity involving reversion to prototrophy of histidine auxotrophs of *Salmonella* and carcinogenicity in *in vivo* tests in mammals. Thus, some simple bacterial tests may serve as indicators of compounds with mutagenic or carcinogenic potential, but these tests do not provide data which can be used to estimate human risk. The usefulness of bacterial tests as preliminary screens may be enhanced by the addition of enzyme systems which mimic mammalian metabolism in the activation of certain pro-carcinogens or promutagens, but *in vitro* metabolic activation systems are complex and unpredictable. Therefore, these tests may leave us with data which are difficult to interpret with confidence.

Historically there has not been good correlation of the results of mutagenicity testing in whole mammals with the data derived from mutagenicity tests in other systems (Russell, 1972). For risk assessments, data are needed from genetic tests involving intact mammals. In addition, these tests must be able to measure both point mutations and transmitted chromosomal effects. Clearly it is impossible to deal with the entire range of chlorinated organics in a definitive way. Simple tests and structural considerations must be used to identify potentially hazardous compounds among the many present and we must concentrate initially on these. There will be many compromises, but it is essential to start gathering genetic data on a problem of such immense potential public health importance as water chlorination.

STRATEGY FOR AN EXPERIMENTAL APPROACH TO MUTAGENIC RISK ASSESSMENT

We have adopted the following procedure as a strategy in investigating the potential for genetic damage from chlorinated organics. We select from the many compounds known to be produced by water chlorination a few and try to understand the mutagenicity of these in mammalian systems. We concentrate on basic mechanisms so that the information derived from the studies may have some predictive value for a much larger group of compounds. We also employ the selective use of nonmammalian and *in vitro* test systems to clarify points about the mechanism of action of our models. Only data from intact animals are used to quantify probable risks for the compounds we study. Of course, this is a big order. We will never have all of the pieces of the puzzle in place, but I think we will generate some data which will help to put the problem into perspective.

I would like to illustrate this approach with a particular model compound. This is a compound which we have worked on and which has also received the attention of some of the environmental scientists in this laboratory. It is a compound which has been identified from the chlorinated sewage effluents in Oak Ridge and, therefore, has some local interest. The compound is 5-chlorouracil.

5-Chlorouracil has several attributes which commend it. First, on theoretical grounds, it would be expected to be mutagenic in some test systems. If we look at the whole list of compounds known to be present in chlorinated sewage effluents, from structural considerations we would expect 5-chlorouracil to be one of the bad actors.

The mechanism of action of 5-chlorouracil (5-ClU), at least by analogy with its big brother, 5-bromouracil (5-BrU), is thought to be fairly well understood. Sometimes, however, when something is thought to be well understood, and it is looked into deeply enough, certain discrepancies become apparent. 5-Bromouracil is mutagenic in a number of prokaryotic systems. It is mutagenic in phage (Litman and Pardee, 1956; Terzaghi, Streisinger and Stahl, 1962). It is mutagenic in *E. coli* (Witkin and Parisi, 1974). It appears not to be mutagenic in certain other bacterial systems, but we do not know enough about why it is not mutagenic in those systems to say very much about it. We know that in many of these systems it is incorporated into DNA where it replaces the base thymine (Dunn and Smith, 1954). Bacteria, and particularly thymine-requiring bacteria, will take up this base analog and incorporate it into their DNA; and incorporation into the DNA of the halogen-containing base appears to be necessary for mutagenesis to occur.

The mechanism for mutagenesis by 5-bromouracil, and by analogy 5-chlorouracil, as mentioned earlier is thought to be well understood. It results from what molecular geneticists call transitions. Transitions are the substitution of one pyrimidine for another or one purine for another in the DNA. This results in a change in the coded information in the gene involved. Halogenated uracils could produce this type of mutation by a mistake in pairing at DNA replication.

Since mutagenesis by these compounds requires incorporation into the DNA, we have a biochemical handle to work with in mammals and other test organisms. We can measure the amount of incorporation of the compound into DNA and compare that with the mutation frequency we observe. The work with phage has demonstrated (Terzaghi, Streisinger and Stahl, 1962) that one gets mutations from 5-bromouracil after the compound is no longer present in the environment of the organism as long as it has been previously incorporated into the DNA. This indicates that once it is in the DNA it can continue to make mistakes. We need to know if 5-chlorouracil, like 5-bromouracil, is incorporated into DNA in place of thymine, if it can produce mutations in the same way, and what sort of effects we can observe in mammals.

EXPERIMENTAL

The first question that needed to be answered was whether 5-chlorouracil, like 5-bromouracil, is mutagenic in bacteria. We used the same bacterial strain, *E. coli* strain WP-1, used to study 5-bromouracil mutagenesis by Witkin and Parisi (1974) and the same mutational test. Table I shows the results of a typical experiment. Mutations per survivor are given for bacteria grown in the presence of the natural base thymine, 5-chlorouracil or 5-bromouracil. The controls (thymine) show about two mutants per 10^8 viable bacteria. There is about a 300-fold increase in mutation frequency after the bacteria have grown in the presence of either halogenated base at 50 μg/ml for 1 hr. The mutation frequency drops for longer exposure times, and we think that at higher levels of incorporation of the base analogs the bacteria lose their ability to express the mutations. It is clear that 5-chlorouracil is mutagenic to bacteria and that its mutagenic behavior is very similar to 5-bromouracil.

Now we will look at 5-chlorouracil in mammals. The first question is: how toxic is 5-chlorouracil to mice? The answer is that it isn't toxic. Male mice, (101 x C_3H) F_1, were put on

Table I. Mutations Induced in *E. coli* by
Halogenated Pyrimidine Base Analogs

Time (min)	Mutants/Survivor x 10^8		
	Thymine[a]	5-ClU	5-BrU
0	3.22	2.01	5.78
30	3.00	628.92	318.88
60	1.89	635.13	641.12
90	0.93	414.93	337.12
120	0.46	13.19	60.12
150	1.77	6.56	24.31

[a]Mean for all thymine points, 1.88 ± 1.09.

drinking water containing 5-chlorouracil at near the maximum soluble concentration (1 g/l). They drank the water well and there were no adverse health effects noted for a period of greater than one year. This concentration is more than one million times the estimated human environmental level. The weight of the animals was near that of controls kept on regular water. Reproduction was normal in every respect and the young appeared normal. The next question is: do the mice incorporate 5-chlorouracil into their DNA? The answer, reported in preliminary form elsewhere (Cumming, Pal, Walton and Russell, 1975), is yes; DNA was extracted from the livers and testes of mice which had been on water containing 5-chlorouracil for various periods of time. This DNA was hydrolyzed by successive incubation with DNase I, snake-venom phosphodiesterase and alkaline phosphatase to produce a mixture of deoxynucleosides. 5-Chlorodeoxyuridine was identified and quantified by chromatography of the DNA hydrolysate on Aminex A-6 (cation exchange resin). It was absent from the DNA of animals which had not been exposed to 5-chlorouracil. About 1.4% of the thymine residues were replaced by 5-chlorouracil in those animals which had been exposed to this compound at a level of 1 g/l in their drinking water. This amounts to about one 5-ClU for every 250 nucleotides (averaged throughout the genome) or about 2.2×10^7 5-ClU's per genome or about four 5-ClU's per structural gene. At this level of incorporation no physical effects on the mice were noted. It is clearly necessary to look for genetic effects in mammals. The meaning of these experiments appears to be that if mammals, presumably including humans, drink water containing 5-chlorouracil, some of it goes into their DNA.

A small dominant lethal mutation experiment was performed to check the possibility that the incorporated 5-chlorouracil was leading to increased chromosome breakage. The procedure for doing the dominant lethal test is essentially as described by Ehling (Ehling, Cumming and Malling, 1968) except that a modification was made to allow for continuous exposure of the males to water containing 5-chlorouracil while not exposing the females. The way this test was performed all male germ cell stages were exposed. Females were checked each morning for vaginal plugs as an indication of mating, and mated females were removed and replaced. At 13 to 15 days of gestation the females were sacrificed and the uterine contents were scored. The results are shown in Table II.

Table II. Dominant Lethal Mutation Experiment with 5-Chlorouracil

	Number of Females	Total Implants (Female)	Live Embryos (Female)	Dead Embryos (%)	Dominant Lethal Effect
Control	36	7.53	6.67	11.44	-
5-Chlorouracil	42	6.90	6.33	8.27	5%

The 5% dominant lethal frequency shown in the table is not significant and thus, within the limits of resolution of this test, no dominant lethal effect is demonstrated at about one million times the maximum human exposure level. But these data still cannot be used to calculate a maximum level of human genetic risk.

There are two general ways of estimating genetic risk to a population exposed to a particular insult.

1. To estimate the overall damage: This is done by estimating the total number of new mutations which will be induced by the agent at a given exposure level and the medical consequences of these mutations. This would be a good way to estimate the effect of mutagens except that there is presently no way to do it very effectively in mammals.
2. To express the increased mutation frequency at particular loci in treated animals in terms of the spontaneous frequency at the same loci: The assumption is made that man has a spontaneous mutation frequency with which he can live, and that if we do not alter it much the compound will be relatively safe.

For the purpose of estimating risk by the second method, we have used the specific-locus test of Russell (Russell, 1951; Russell and Cumming, 1975).

The specific-locus test measures recessive visible mutations at seven marked loci in the mouse. The test is highly quantitative and simple to perform. It has been very important historically as the source of data which have been used to set maximum allowable radiation exposure levels for man. The test may be very large and has gained a reputation for being cumbersome and expensive. However, as I shall show, under certain conditions useful information can be obtained with a very small test. The specific-locus test has the advantages of producing highly quantitative data on a very relevant genetic endpoint.

The specific-locus test has been used on a very small scale to estimate the upper limit of genetic risk that might occur at the human exposure levels of 5-chlorouracil (Russell and Cumming, 1975). We did a mini-specific-locus test. Eleven males were continuously exposed to 1 g/l 5-chlorouracil and mated to T-stock females. No mutations were observed in 314 offspring. The mice had been exposed to a concentration of 5-chlorouracil in the drinking water at least one million times higher than the human exposure level. Their offspring were conceived, on the average, three months after a steady-state value for incorporation of 5-chlorouracil had been reached. This is 1/120 of the 30-yr generation time exposure in man, but the mouse still gets about 8000 times the human exposure ($10^6/120$). With this exposure factor and the observation of zero mutations in 314 offspring we can calculate an estimate of the upper level of genetic risk that might occur at the human exposure level. Taking 3.3 as the upper 95% confidence limit of the observed zero mutation frequency in the 314 offspring, subtracting the unknown spontaneous frequency of 28 mutations in 531,500 offspring and dividing by the exposure factor of $10^6/120$, we come out with an estimated induced mutation rate, at the human exposure level, that is only 2% of the spontaneous frequency:

$$\left(\frac{3.3/314 - 28/531,500}{10^6/120} \right) \bigg/ \left(\frac{28}{531,500} \right) = 0.02$$

We can see from this calculation that 5-chlorouracil by itself does not seem to be a particular genetic hazard to man even though it is mutagenic in bacteria and other lower organisms. We are 95% confident that at human exposure levels we have not increased the mutation frequency by more than 2% of the spontaneous. That is a small number. There are few compounds in our environment for which we have that kind of information.

What is an acceptable increase in the mutation frequency is a political and not a scientific question. We really do not know what a given increase in mutations means in terms of increased human suffering and economic expense. We assume that any increase in the human mutation frequency will not be without its costs.

We have looked at one compound from the many present and find that by itself it is relatively safe. But we have not addressed the more difficult problem of additive effects of many compounds and interactions. Thus we have just started to scratch the surface of a large area of legitimate concern.

ACKNOWLEDGMENTS

I would like to acknowledge my collaborators on the various projects mentioned in this paper. They are: W. L. Russell, Bimal Pal, Marva F. Walton, Barbara J. Elmhorst, Donna L. George and David L. Sultzer. This work is sponsored by the United States Energy Research and Development Administration under contract with Union Carbide Corporation.

REFERENCES

Cumming, R. B. and B. J. Elmhorst. 1975. "A Test for 5-Chlorouracil-Induced Dominant-Lethal Mutations in the Mouse." *Biol. Div. Annu. Prog. Rep.*, ORNL-5072, p. 135.

Cumming, R. B., D. L. George, M. F. Walton and B. J. Elmhorst. 1975. "Mutations Produced by 5-Chlorouracil and 5-Bromouracil in *Escherichia coli*." *Biol. Div. Annu. Prog. Rep.*, ORNL-5072, pp. 135-136.

Cumming, R. B., B. C. Pal, M. F. Walton and W. L. Russell. 1975. "Studies on Potential Genetic Effects of 5-Chlorouracil in Mammals." *Biol. Div. Annu. Prog. Rep.*, ORNL-5072, pp. 134-135.

Dunn, D. B. and J. D. Smith. 1957. "Effects of 5-Halogenated Uracils on the Growth of *Escherichia coli* and Their Incorporation into Deoxyribonucleic Acids." *Biochem. J.* 67:494-506.

Ehling, U. H., R. B. Cumming and H. V. Malling. 1968. "Induction of Dominant Lethal Mutations by Alkylating Agents in Male Mice." *Mutation Res.* 5:417-428.

Litman, R. M. and A. B. Pardee. 1956. "Production of Bacteriophage Mutants by Disturbance of Deoxyribonucleic Acid Metabolism." *Nature* 178:529-531.

McCann, J. and B. N. Ames. 1976. "Detection of Carcinogens as Mutagens in the *Salmonella*/Microsome Test: Assay of 300 Chemicals: Discussion." *Proc. Nat. Acad. Sci. U.S.A.* 23:950-954.

Russell, W. L. 1951. "X-Ray-Induced Mutations in Mice." *Cold Spring Harbor Symposia on Quant. Biol.* 16:327-336.

Russell, W. L. 1972. "Radiation and Chemical Mutagenesis and Repair in Mice," p. 239-247. In Roland F. Beers, Jr., Roger M. Herriott and R. Carmichael Tilghman (Eds.) *Proceedings of Miles Fifth International Symposium on Molecular Biology: Molecular and Cellular Repair Processes*, Johns Hopkins University, Baltimore, Maryland, June 3-4, 1971.

Russell, W. L. and R. B. Cumming. 1975. "An Example of Conditions that Make the Mouse Specific-Locus Test Highly Efficient at Low Expense." *Biol. Div. Annu. Prog. Rep.*, ORNL-5072, pp. 126-127.

Terzaghi, B. E., G. S. Streisinger and F. W. Stahl. 1962. "The Mechanism of 5-Bromouracil Mutagenesis in the Bacteriophage T-4." *Proc. Natl. Acad. Sci. U.S.A.* 48:1519-1524.

Witkin, E. M. and E. C. Parisi. 1974. "Bromouracil Mutagenesis: Mispairing or Misrepair?" *Mutation Res.* 25:407-409.

DISCUSSION

A. D. Venosa, U. S. Environmental Protection Agency. Since uracil is a base in RNA, would you expect 5-chlorouracil to be incorporated into the RNA as opposed to DNA and thereby elicit a translation effect rather than transcriptional effect on protein synthesis?

Cumming. I would have no *a priori* expectation, but in fact, when you measure it, you find that 5-chlorouracil goes only into DNA and not into RNA. The 5-chlorouracil molecule is shaped more like thymine than uracil since the chlorine atom is in the same position as a methyl group on thymine. Apparently the enzymatic machinery which assembles nucleic acids recognizes 5-chlorouracil and 5-bromouracil only as thymine. 5-Fluorouracil is, however, incorporated into RNA.

13

THE EPIDEMIOLOGIC APPROACH TO THE EVALUATION OF WATER-BORNE CARCINOGENS

Kenneth P. Cantor

U.S. Environmental Protection Agency
Washington, D. C. 20460

and

National Cancer Institute
Bethesda, Maryland 20014

ABSTRACT

The classic mid-19th century London study of John Snow, in which cholera was linked to water-borne contamination, is reviewed to provide a context for discussing the epidemiologic approach to evaluation of carcinogens in drinking water. The value of epidemiologic studies arises from the fact that observations are made directly on human populations so that extrapolation from animal models and/or unrealistically high doses is not necessary to predict effects in humans. Limitations of the approach include the long latent period for most cancers, difficulties in estimating dose, the definition of at-risk populations, and the relatively low exposure levels to carcinogenic agents. Recent studies of a preliminary nature are reviewed.

INTRODUCTION

Previous sessions of this conference have raised the issue of how one should weigh the findings of various scientific disciplines which help shape environmental regulatory policy. In this discussion, I will focus on one of these fields—epidemiology—and dwell on problems of methodology central to epidemiologic investigation of chronic diseases as related to drinking water contaminants. An understanding of methodology leads to an appreciation of the utility of epidemiologic studies as well as knowledge of their limitations.

The concern for methodology is related to the philosophical problem of what we accept as evidence of a cause-and-effect relationship in disease etiology. When, for example, should we demand more than a high degree of statistical association between presumed cause and effect before taking action on the findings of epidemiologic investigations? In addition, an understanding of methodology is crucial in deciding how to approach the study of possible carcinogenic effects in human populations exposed to very low levels of substances which are known to cause cancer in laboratory animals, induce mutations in microorganisms, or are carcinogenic to persons who were knowingly exposed in occupational settings to far higher levels.

Critical examination of the methodology used in epidemiologic investigations is of paramount importance when we are faced with acting on the results of such studies. To put these problems in context, I will review one of the first and most definitive epidemiologic investigations into health aspects of drinking water conducted in the mid-19th century by Dr. John Snow. Snow's study of cholera in London—well known to students of epidemiology and biostatistics—will serve as a basis from which we can examine aspects of today's epidemiologic approach to water-borne carcinogens.

CHOLERA AND DRINKING WATER

In 1855, Snow, an English physician, published the second edition of "On the Mode of Communication of Cholera" (Snow, 1855). This work was the culmination of seven years of intensive investigation in which Snow personally looked into conditions surrounding thousands of cholera deaths. The conclusions he drew regarding the cause and spread of this disease are recognized today as being essentially correct.

Snow commenced his investigation in 1848 with the hypothesis that the disease is caused by some characteristic of particular drinking waters—in his words, some "poison." His reasoning grew from the observation that in cholera, initial symptoms localize in the alimentary canal and then proceed to more general systemic involvement. With other diseases studied at that time, symptoms such as headaches, fever and a higher pulse rate appeared first. Snow argued that this pattern of symptom development indicated that the causative agent in cholera must first come into contact with the alimentary mucosa before reaching other organs. In addition to the observation of disease progression, there were also a large number of anecdotal accounts of the occurrence and local spread of cholera which tended to support the water-borne "poison" hypothesis. Drinking water was therefore implicated as a prime suspect as the carrier of the "morbid material."

Snow was able to test his hypothesis owing to the unique arrangement of London's water supply system. The system was operated by two private companies, each of which had intakes in the Thames River as well as its own distribution system within the city. London suffered a series of cholera epidemics in the 1840's and 50's, two major ones occurring in 1849 and 1854. In the interval between these outbreaks, in 1852, one of the suppliers (the Lambeth Company) moved its intake from the highly polluted part of the Thames, which was serving both companies, to a less contaminated stretch of the river. Intakes of the other utility (the Southwark and Vauxhall Company) remained in place.

In examining mortality rates from cholera, Snow observed that areas served by the Lambeth Company, with its new water supply, fared much better in the epidemic of 1854 than in the 1849 episode. Snow's compilation of deaths and rates is shown in Table I. Cholera death rates in districts served by the Lambeth Company in 1849, before the intakes were moved, were comparable to those in districts served by the Southwark and Vauxhall Company, the death rate in Southwark and Vauxhall districts being 1.2 times that in Lambeth Company districts. In the 1854 epidemic, after the change in supply for the

Table I. Cholera Deaths in London, 1849 and 1854[a], in Districts Served by Different Water Companies[b]

Water Supplier of District	Population (1851)	Cholera Deaths				Risk Relative to the Lambeth Co. District	
		Number		Rate/1000			
		1849	1854	1849	1854	1849	1854
Southwark and Vauxhall Co.	167,654	2,226	2,458	13.5	14.7	1.2	8.6
Both Companies	300,149	3,905	2,547	13.0	8.5	1.2	5.0
Lambeth Company	14,632	162	25	11.1	1.7	1.0	1.0

[a]In 1849 data were collected for the 13-week period ending Nov. 19. In 1854 data were collected for the 14-week period ending Oct. 14.
[b]From John Snow. 1855. *On the Mode of Communication of Cholera*, 2nd edition.

Lambeth Company, Southwark and Vauxhall districts suffered death rates 8.6 times those of Lambeth Company service areas. The part of the city served by both companies had intermediate rates. The data necessary to make these observations consisted of group statistics—cholera mortality rates for different districts of the city and knowledge of the water distribution patterns for these same districts. Snow used no information regarding the personal backgrounds, demographic characteristics, economic class and so on, of the populations he studied to make his observations. While there was a clearly observable difference in death rates in districts served by the water companies, Snow was concerned that the fact of association of cholera deaths and water supply was not sufficient to make the causal argument incontrovertible. After all, one could argue that differences in geographically separated populations other than water supply—what we today would call confounding causes—could have led to his observation. He therefore carried out a detailed house-to-house survey of cholera deaths in the 1854 epidemic in districts served by both companies. In his words:

> ... the intermixing of the water supply of the Southwark and Vauxhall Companies, over an extensive part of London, admitted of the subject being sifted in such a way as to yield the most incontrovertible proof on one side or the other ... The pipes of each company go down all the streets, and into nearly all the courts and alleys ... In many cases a single house has a supply different from that on either side. Each Company supplies both rich and poor, both large houses and small; there is no difference either in the condition or occupation of the persons receiving the water of the different companies.

Snow thus had available for observation exposed and nonexposed populations which were almost perfectly matched with respect to all other possible differences which might have had a bearing on the outcome. Table II shows the result of his investigation. For the first 8 weeks of the 1854 epidemic, houses served by the Southwark and Vauxhall Company experienced a cholera death rate of 4.76 per 1000 population, over 7 times the rate of 0.67 per 1000 in houses supplied by the Lambeth Company. The rates from Table II are not directly comparable with those on Table I, since cholera deaths for different lengths of time were counted in each case.

Table II. Cholera Deaths in London Houses Served by Two Water Companies, 1854[a,b]

Water Supply for Individual Houses	Population (est.) of Houses Served	Cholera Deaths Number	Cholera Deaths Rate/1000	Risk Relative to Houses Served by the Lambeth Co.
Southwark and Vauxhall Co.	106,309	506	4.76	7.1
Lambeth Company	139,472	94	0.67	1.0
Total	245,781	600	2.44	

[a] Data for the 8-week period through August 26.
[b] John Snow. 1855. *On the Mode of Communication of Cholera*, 2nd edition.

Snow's successful investigation was performed before the germ theory of disease had been formally proposed, and thus modern methods of bacteriology were not available to him. His achievement, however, was aided by the characteristics of the disease itself. Cholera is highly specific in that it is caused only by one offending agent—*Vibrio cholera*—which enters the body via an oral route. The disease, morever, has a brief incubation period. A cholera patient shows symptoms within a few days of exposure and occasionally within hours. A close relationship between cause and effect, both in specificity and in elapsed time between one and the other, aided his analysis.

CHRONIC DISEASE EPIDEMIOLOGY

The incompletely understood causative links in most chronic disease development stand out in contrast to cholera. Rather than a few hours or days, chronic diseases such as heart disease, cancer and chronic respiratory disease develop over a period of years or decades. While many factors would seem to influence whether or not an individual contracts and dies of cholera, the presence of the infective agent is a necessary precondition for development of the disease. In the case of many chronic diseases, any of a number of environmental, genetic, metabolic or other factors, acting alone or in concert, can give rise to similar disease outcomes in different individuals.

There are a few exceptions to this general rule. Mesothelioma, for example, a cancer of the pleural or peritoneal lining, has been closely linked to exposure to asbestos fibers (Newhouse and Thompson, 1965). Angiosarcoma of the liver has been associated with occupational exposure to vinyl chloride (Selikoff and Hammond, 1975a) but may also be caused by a drug, Thorotrast, and probably by certain arsenicals. In choosing populations for study, one must make an effort to obtain information on every possible known variable which might affect the outcome as expressed by morbidity or mortality rates. In practice, this can be difficult or impossible to accomplish.

An epidemiologic determination of possible carcinogenic effects of water-borne chlorinated organics will, no doubt, be performed under conditions somewhat less ideal than those enjoyed by Snow in his cholera study. The great value of the epidemiologic approach for Snow and for us resides in the fact that human beings are the subjects of the study. To evaluate results, there is no need to extrapolate from microbe or laboratory animal to man, from high- to low-level exposure, or from both. In a large number of cases, of course, when suggestive data from other lines of inquiry are available, it would be unethical to await the results of an epidemiologic study before taking preventive action. But if effects are observed in human populations, such action can proceed with great confidence.

Aspects of carcinogenesis which must be carefully weighed when designing epidemiologic studies include the long latent period for development of cancer, influence of the age at first exposure to the presumed carcinogen, the magnitude of risk posed by the carcinogenic agent and the presence of confounding factors.

Table III lists a number of substances which have been identified as carcinogens and strongly linked to cancer of specific sites. Indicated in the table are the so-called "latent" periods for most of the listed carcinogens as well as the relative risk where it has been estimated. The concept of "latent period" is used in different ways. In some studies, it refers to the time after a single exposure or after the start of a continuous exposure when a peak in the mortality rate is observed. In other investigations, for example the ongoing studies in Japan by the Atomic Bomb Casualty Commission, the term refers to the average time between a single exposure and appearance of cancer in affected members of the population. In cases where a specific carcinogenic effect has been demonstrated, the latent period before a tumor is observed is on the order of decades, not days or weeks. In most epidemiologic investigations of carcinogenic effects of environmental contaminants, estimation of dose—even of relative dose—is difficult, as it usually must be made many years after the offending exposure.

In both animal studies and the few epidemiologic investigations where the question has been raised, it appears that the length of the latent period is inversely related to dose. In the study of Japanese atomic bomb survivors, for example, Bizzozero and co-workers (1966) observed that acute leukemia patients under

Table III. Latent Period and Relative Risks for Cancers of Known Etiology

Carcinogen	Anatomical Site or Type	Latent Period (yr)	Risk Relative to Unexposed Populations	Reference
Mustard gas	Respiratory tract (various sites)	24		Wada et al. (1968)
Radiation	Leukemia	8-9		Bizzozero et al. (1966)
Vinyl chloride monomer	Liver angiosarcoma	17-19	400-3000	Selikoff and Hammond (1975a)
B-Naphthylamine	Bladder	18		Case et al. (1954)
Benzidine	Bladder	18		Case et al. (1954)
Asbestos and smoking	Lung	20	90	Selikoff and Hammond (1975b)
Asbestos	Pleural mesothelioma	20	250	Rall et al. (1973)
Cigarettes	Lung		5-20	U.S. Public Health Service (1964)

15 years of age who were within 1,500 m of the hypocenter developed the disease 8.6 yr, on the average, after exposure; while those who were between 1,500 and 10,000 m away from the blast center experienced a latent period of 11.6 yr. The data base is not yet extensive enough to feel fully confident that the latent period is universally an inverse function of dose, but most evidence is pointing in this direction.

The sensitivity of the human organism to carcinogens is a complex function of age, with the age at exposure often an important determinant of cancer development many decades later. Fraumeni and Hoover (1975), in reviewing the epidemiologic literature on age at first exposure and environmental carcinogenesis, report that the risk for some cancers appears to be elevated among persons exposed at older ages (*e.g.*, bladder cancer in dye workers, nasal sinus cancer in nickel refiners, lung cancer in asbestos workers) while in other situations risk is greatest in persons exposed at younger ages (various cancers in atomic bomb survivors, lung cancer in cigarette smokers).

In designing studies to examine the effects of chlorinated organics in water, the long latent periods for carcinogenesis and their possible dependence on dose impose major boundary conditions. How should we estimate dose, when the exposures of greatest significance to today's disease experience took place 20, 30 or 40 years ago? To what extent are present environmental measures of value? Are there reliable methods for extrapolating present analyses back in time, given that we know something about the industrial mix and pollution potential in years gone by? Can we assume that the relative contamination of water between regions in the past was similar to today's situation and that only the absolute values have changed?

In addition to the changes in water supply which might have occurred, migration of the observed populations must be carefully evaluated and statistically controlled for. People who die of cancer were not necessarily exposed to effective doses of carcinogens at their place of death but may have moved from the area of crucial exposure many years before. The cancer mortality patterns in Los Angeles, for example, resemble those of the Midwest. The rates for many different cancers in Miami,

Florida, are similar to those in the New York/New Jersey area. In indirect studies, where the primary unit of observation is the group, the migration statistic must be considered. In direct studies, where individuals are followed, residential histories are of prime importance.

The problems inherent in dealing with diseases having long latent periods are illustrated by a study of cancer mortality in Duluth. Mason and co-workers (1974) did not observe cancer mortality rates consistent with the hypothesis that water-borne asbestos fibers, as they are found in the Duluth water supply, cause cancer. The asbestos fibers in Duluth's drinking water originate from taconite ore tailings which have been emptied into Lake Superior since 1955. Mason examined the cancer mortality data through 1969, 14 years after exposure to waterborne asbestos had started. He concludes: "The period of observation is short relative to the latent period for occupationally induced carcinogenesis from asbestos. A longer period of follow-up than was possible in this study will be necessary before one can conclude that there is no cancer hazard related to the drinking water supplies of Duluth and neighboring communities." In this case the limits of information from an epidemiologic study preclude a definitive result. A positive result from an epidemiologic study, in a case such as this, usually means that we have waited too long. In the case of chlorinated organics in drinking water, it is possible that adverse health effects will be shown to have occurred in human populations, but if unambiguous epidemiologic evidence is not forthcoming, a negative result cannot be interpreted as proof of no effect.

Table III lists the cancer risk of exposed persons relative to the risk in the general population. In general, a high relative risk makes the work of the epidemiologist relatively easy, since a high relative risk implies a strong cause-effect relationship and it is unlikely that other differences between groups (confounding factors) are responsible for the observation. Recall that John Snow observed relative risks of 7 or 8 for cholera. As indicated in Table III, workers exposed to high levels of vinyl chloride monomer have a risk of contracting liver angiosarcoma much greater than 400 times the general population—perhaps about 3,000 times as great (Selikoff and Hammond, 1975a). This carcinogen first became suspect when only a few workers in a single vinyl chloride plant developed this rare disease and the suspicions of the plant physician were thereby aroused. Pleural mesothelioma due to asbestos exposure, with a relative risk well in excess of 250, belongs in the same class (Rall et al., 1973). Lung cancer related to asbestos exposures or to cigarette smoking have much lower relative risks. This is not an indication of the relative public health significance of these diseases, since this is determined not only by the relative risks but also by the frequency of the disease and exposure of the general population. Lung cancer, for example, is not an uncommon disease, and the relative risk for smokers of 8 or 10 is a reflection of tens of thousands of premature deaths.

Developing evidence for a cause-effect relationship, especially with chronic diseases having multifactorial etiologies, becomes difficult when the relative risk is lower than 2.0. When this is the case, the need for specific data on exposure and disease incidence in large populations becomes ever greater, and obtaining the needed information becomes exceedingly difficult. At this time, we do not know the relative risk of cancer in populations exposed to chlorinated or other organics found in drinking water, but if it is much below 2.0, epidemiologic answers to the questions posed by this conference may not be forthcoming.

Let us now turn briefly to a few studies which have examined the possible relationship between organics in drinking water and cancer rates. The studies cited below use the descriptive approach. They examine characteristics of populations with regard to their exposure to suspect carcinogens and their cancer mortality experience. The best known are the investigations by Harris and co-workers of the Environmental Defense Fund, in which multiple regression techniques were used to examine this relationship (Harris, 1974; Page, Harris and Epstein, 1975).

In a short paper which has not been widely circulated, Mr. Lee McCabe of the EPA Health Effects Research Laboratory at Cincinnati, Ohio, performed a simple linear regression analysis of age-adjusted cancer mortality rates against chloroform concentration in 50 U.S. cities which were included as part of the EPA 80-City National Organics Reconnaissance Survey (McCabe,

1975; U.S. Environmental Protection Agency, 1975). Cancer death rates were not available for the cities not included in this analysis. McCabe recognized several problems with this study and concluded that "these preliminary analyses only provide enough in the way of results to stimulate more study of the problem." A third indirect study was conducted by Dr. Thomas Mason at the Epidemiology Branch of the National Cancer Institute (Mason, 1975). Mason used county cancer mortality data for the 10 counties which are coterminous with cities for which data on the chloroform content of drinking water were available. Cancer mortality rates were calculated for these counties and compared with standard rates adjusted for several potentially confounding variables.

These studies are valuable in that they represent the first steps taken to evaluate the possible role of low-level carcinogens in drinking water. All of the studies are lacking with respect to the pollution indicator that was used. The Harris study employed a variable indicating the percentage of each Louisiana parish supplied with drinking water from the Mississippi River, a source of presumably highly contaminated water. The McCabe and Mason studies used the amount of chloroform in the drinking water of the cities they analyzed, based on measurements made on a one-time grab sample from an EPA nationwide survey of the drinking water of 80 cities. In one of the studies, cancers of all anatomical sites were grouped together, and there was no consideration of other possible factors contributing to the observed cancer rates (McCabe, 1975). In another, possible confounding of the results occurred because populations with large exposure differences also differed ethnically and religiously (Harris, 1974). The remaining study used a small number of cities and only an extremely large effect would have been observed as giving a statistically significant result. The authors of these studies recognize several of these shortcomings and point out the need for better exposure information and data on the health status of exposed populations.

We should keep in mind that epidemiologic studies are comparative studies. An initial task of the epidemiologist is to define populations for comparison, either on the basis of differential mortality or morbidity rates, or of gradients in exposure to known or suspected toxic substances. Both approaches require information of three types: measures of response—that is, disease incidence and prevalence data; estimates of dose; and information on as many other demographic and exposure variables as may reasonably be thought to influence the development of the disease or diseases of concern. Of these areas, we are probably on weakest ground in the availability of good dose estimates. EPA is currently expanding the 80-City survey of organics in drinking water by including additional supplies for analysis and performing year-round measurements to determine the influence of temperature, flow fluctuations and other variables on the concentration of these chemicals. This effort will enable us to develop somewhat better estimates of dose to populations presently exposed to the measured drinking water supplies and hopefully provide information to allow assessment of historical exposures.

In the future, it will be necessary to further coordinate activities between epidemiologists, demographers and water supply experts so that dose estimates over time, measures of disease frequency and potential confounding variables are collected for the same group of people. Only with such cooperation can we develop the necessary data base to conduct meaningful epidemiologic studies of the health effects of low levels of chlorinated organics in drinking water.

ACKNOWLEDGMENTS

This work was performed while the author was detailed as a visiting scientist to the National Cancer Institute from the U.S. Environmental Protection Agency. Thanks are due to all in EPA and NCI who helped make this arrangement possible. The author is especially grateful to Dr. Robert Hoover of the National Cancer Institute, who provided several helpful suggestions during development of the presentation and in clarification of ambiguities in the manuscript.

REFERENCES

Bizzozero, O. J., Jr., K. G. Johnson and A. Ciocco. 1966. "Radiation Related Leukemia in Hiroshima and Nagasaki, 1946-1964. I. Distribution, Incidence, and Appearance Time." *New England J. Med.* 274: 1095-1101.

Case, R. A. M., M. E. Hosker, D. B. McDonald and J. T. Pearson. 1954. "Tumours of the Urinary Bladder in Workmen Engaged in the Manufacture and Use of Certain Dyestuff Intermediates in the British Chemical Industry." *Brit. J. Industr. Med.* 11:75-104.

Fraumeni, J. F. and R. Hoover. 1975. "Immunosurveillance and Cancer: Epidemiologic Observations." In *Proceedings of the Symposium on Epidemiology and Cancer Registries in the Pacific Basin.*

Harris, R. H. 1974. "The Implications of Cancer-Causing Substances in Mississippi River Water." Environmental Defense Fund, Washington, D. C.

Mason, T. 1975. Personal communication.

Mason, T., F. W. McKay and R. W. Miller. 1974. "Asbestos-Like Fibers in Duluth Water Supply." *J. Am. Med. Assoc.* 228:1019-1020.

McCabe, L. J. 1975. "Association Between Trihalomethanes in Drinking Water and Mortality." U.S. Environmental Protection Agency. Unpublished paper.

Newhouse, M. L. and H. Thompson. 1965. "Mesothelioma of Pleura and Peritoneum Following Exposure to Asbestos in the London Area." *Brit. J. Industr. Med.* 22:261.

Page, T., R. H. Harris and S. S. Epstein. 1975. "Relation Between Cancer Mortality and Drinking Water in Louisiana." Unpublished paper.

Rall, D., J. Churg, E. C. Hammond, A. M. Langer, W. J. Nicholson, I. J. Selikoff and Y. Suzuki. 1973. *Proceedings of Conference on Biological Effects of Asbestos.* National Institutes of Health. February 1.

Selikoff, I. J. and E. C. Hammond (Eds.). 1975a. "Toxicity of Vinyl Chloride—Polyvinyl Chloride." *Ann. N. Y. Acad. Sci.* 246:1-337.

Selikoff, I. J. and E. C. Hammond. 1975b. "Multiple Risk Factors in Environmental Cancer." In J. F. Fraumeni, Jr. (Ed.). *Persons at High Risk of Cancer.* Academic Press.

Snow, J. 1855. "On the Mode of Communication of Cholera," 2nd ed. Churchill, London. Reproduced in *Snow on Cholera.* 1936. Commonwealth Fund. New York. Reprinted by Hafner. 1965. New York.

U.S. Environmental Protection Agency. 1975. "Preliminary Assessment of Suspected Carcinogens in Drinking Water." April.

U.S. Public Health Service. 1964. "Smoking and Health: Report of the Surgeon General of the Public Health Service." Publication No. 1103.

Wade. S., M. Michihiro, Y. Nishimoto, S. Kambe and R. W. Miller. 1968. "Mustard Gas as a Cause of Respiratory Neoplasia in Man." *Lancet* 1969:1161-1163.

DISCUSSION

Nancy Stroup, Environmental Defense Fund. I want to know if EPA or anyone else has been trying to develop retrospective studies on chloro-organic dosages in drinking water.

Cantor. Retrospective in the sense that we are trying to evaluate historical doses? Gordon Robeck might be doing this. Do you want to answer?

Gordon G. Robeck, U.S. Environmental Protection Agency. We are not doing this specifically. We are trying to get grantees to make proposals, however.

Jerome R. McKersie, Wisconsin Department of Natural Resources. You asked if there was any city that chlorinated part of their water system and not other parts. I might suggest that for years Minneapolis, Minnesota, had a ground water supply while its adjacent city, St. Paul, obtained their drinking water from the Mississippi River. It might be interesting to compare health statistics for these two cities. My question is, have there been any studies of bioaccumulation of chloro-organic compounds in human fat between those who have cancer vs noncancer deaths?

Cantor. To my knowledge there has been no such study to date. At EPA we are looking into the possibility of correlating organic residues in human fat with organic constituents in drinking water supplies.

SECTION III.
ENVIRONMENTAL TRANSPORT AND EFFECTS

William A. Brungs, Session Chairman
>Environmental Research Laboratory
>U.S. Environmental Protection Agency
>Duluth, Minnesota 55804

I saw and/or met many of you about two weeks ago at the Water Pollution Control Federation Meeting in Miami Beach and at that time I stewed all week because I had to present something. This week I have really enjoyed myself because I know I don't have to say anything. At least nothing that takes preparation. So I'm going to get right into the program. The papers this afternoon are, of necessity, general in nature because we're not trying to give you a thorough understanding of the toxicity of residual chlorine to aquatic life, but to give you a summary of it and some of the detailed research that is being undertaken.

14

THE TOXICITY OF CHLORINE TO FRESHWATER ORGANISMS UNDER VARYING ENVIRONMENTAL CONDITIONS

Arthur S. Brooks and Gregory L. Seegert

    Center for Great Lakes Studies
    The University of Wisconsin-Milwaukee
    Milwaukee, Wisconsin 53201

ABSTRACT

    Chlorine enters freshwater systems from many sources. It enters waters of variable chemical quality in a variety of chemical forms over wide ranges of temperature and for varying periods of time. Each of these factors is important in determining the toxicity of chlorine to aquatic life.

    This paper reviews studies which have been conducted under a myriad of environmental conditions in attempts to quantify the toxicity of chlorine to freshwater organisms. Included in the review are experiments run in relatively clean waters and sewage effluents. Studies are also included which involved continuous and intermittent chlorine applications and tests conducted under a wide range of temperatures. Data from these studies are reviewed in light of the conditions under which the tests were conducted. A synthesized view is presented of the toxicity of chlorine to freshwater biota in terms of the level and duration of exposure, temperature and the chemical nature of the water in which the experiments were conducted.

INTRODUCTION

Chlorine is used in numerous industrial and water treatment processes as a biocide for undesirable organisms. Chlorine is a very effective biocide, not only for undesirable organisms, but for other forms of aquatic life as well. Because of this lack of specificity, great concern has been expressed over the impact of chlorine on freshwater ecosystems.

Chlorine enters freshwater systems under a wide variety of environmental conditions. These conditions include wide ranges of temperature, water quality, and variable application times and dose rates. Each of these variables must be carefully evaluated in assessing the toxicity of chlorine to freshwater organisms. This paper reviews the literature in an attempt to evaluate the influence of the above-mentioned environmental variables with respect to the toxicity of chlorine to freshwater organisms. The review is not intended to be a complete compilation of the chlorine literature. It draws only on studies that have presented sufficient data to permit meaningful comparisons. Brungs (1973, 1976) has compiled two excellent literature reviews to which the reader may refer for more complete coverage.

FRESHWATER FISH

Species-Related Factors

Species Composition

One of the most important factors in determining the impact of chlorine on a fish community is the species composition of that community. Cold water species (salmonids) have generally been considered more sensitive to chlorine than are warm water species (Brungs, 1973; Basch and Truchan, 1974). Recent evidence, however, indicates that this distinction may not always be justified. Several species of minnows have been found to have LC50 (median lethal concentration) values approximating those of salmonids (Ward, 1976, as cited in Brungs, 1976). This new information suggests that in order to accurately determine the impact of chlorine on a particular aquatic environment, representatives from the important families making up the local community should be tested.

Size

There exists a general lack of continuity in the literature on the effects of size on chlorine toxicity. Several authors (Fobes, 1971; Wolf *et al.*, 1975; Eren and Langer, 1973), using a variety of species, have reported that smaller fish are more sensitive to chlorine than are larger individuals of the same species. Rosenberger (1971), however, found the opposite relationship in coho salmon (*Oncorhynchus kitusch*). Others (Wolf *et al.*, 1975; Warren, 1975) observed no effect resulting from size. This lack of agreement reflects species differences and probably differences in analytical techniques between authors.

Effects on Eggs, Larvae and Reproductive Behavior

Perhaps because of the inherent difficulties, few authors have attempted to determine the effects of chlorine on fish eggs, larvae or reproductive behavior. Arthur and Eaton (1971) found that spawning in fathead minnows (*Pimephales promelas*) was practically eliminated at 0.085 mg/l of chlorine and that the number of spawnings per female was significantly reduced at 0.043 mg/l. The highest concentration that had no effect in long-term tests was 0.016 mg/l. Carlson (1975, as cited in Brungs, 1976) found that three-spine sticklebacks (*Gasterosteus aculeatus*) exposed for 3-½ months to chloramine concentrations up to 0.0114 mg/l exhibited no change in reproductive behavior. Hughes (1973, as cited in Brungs, 1973) found that larval striped bass (*Morone saxatilis*) were considerably more sensitive to chlorine than were fingerlings. Gehrs *et al.* (1974) studied the effects of 4-chlororesorcinol and 5-chlorouracil on the hatching success of carp (*Cyprinus carpio*) eggs. A significant inhibition of hatching success was seen at most of the concentrations tested between 0.001 mg/l and 10 mg/l. It has been found for some marine species that the egg membrane affords considerable protection against chlorine (Alderson *et al.*, 1975). This has yet to be demonstrated in freshwater fish.

Avoidance

Avoidance to chlorine by fish has been observed in both lab and field situations. Sprague and Drury (1969) observed that rainbow trout (*Salmo gairdneri*) placed in a test tank offering choices between unchlorinated and chlorinated water clearly avoided total residual chlorine (TRC) concentrations of 0.001, 0.01 and 1.0 mg/l. Unexplicably they showed a preference for the chlorinated side at 0.1 mg/l. In another lab study Borgardus (1975) found that mimic shiners (*Notropis volucellus*), river shiners (*N. blennius*) and bullhead minnows (*Pimephales vigilax*), when maintained in a narrow rectangular tank having a laminar flow of monochloramine down one side actively avoided that side. The lowest concentrations avoided were 0.005, 0.150 and 0.05 mg/l for the mimic shiner, river shiner and bullhead minnow, respectively. Meldrin *et al.* (1975, as cited in Brungs, 1976) found that several species of estuarine fish could avoid chlorine. The avoidance concentrations of these fish were generally inversely related to temperature and light levels.

Several field studies have shown that fish can detect and sometimes avoid chlorinated effluents. Fava and Tsai (1973, as cited in Tsai, 1975) determined that blacknose dace (*Rhinchthys atratulus*) could discriminate and avoid chlorinated sewage effluents containing TRC levels as low as 0.01 mg/l. Several authors (Massey, 1972; Basch, 1971) have reported observing fish leaving discharge channels during chlorination periods at power plants. Conversely, Dickson and Stauffer (1974) reported that most of the resident fish population in the power plant discharge they surveyed remained in the channel during the thrice-daily chlorination periods. They hypothesized that the TRC chlorine levels in the channel were not high enough to evoke an avoidance reaction. The lack of species numbers and diversity repeatedly observed in rivers below waste treatment plants (Tsai, 1968, 1970) probably reflects an avoidance of these areas by fish.

It is apparent that when given a clear choice (as in lab studies) or when chlorine is continuously discharged (as with waste treatment plants) fish can actively avoid low levels of chlorine. However, whether fish can safely avoid intermittently discharged effluents has yet to be demonstrated. Data compiled by Truchan (1977) showed that fish kills have occurred as a result of the intermittent discharge of chlorine. These incidences suggest that either the intermittency of chlorination or a temporary misorientation by the fish can override any avoidance response which they would normally exhibit.

Exposure Time

One of the most important factors determining the toxicity of a given level of chlorine is the exposure time. A comparison of several studies on rainbow trout illustrates this importance. Safe concentrations for continuous exposure are generally considered to be from 0.002 to 0.005 mg/l (Brungs, 1973; Basch and Truchan, 1974; Merkins, 1958). Basch et al. (1971), Esvelt et al. (1971), Sprague and Drury (1969) and Wolf et al. (1975) cite 96-hr LC50 values near 0.1 mg/l. One mg/l was required to kill rainbow trout exposed only 4 hr (Sprague and Drury, 1969) and in ½-hr tests at 10°C an LC50 of 2.0 mg/l was found (Seegert et al., 1977).

Blacknose dace exposed for ½ hr to 0.74 mg/l of free chlorine suffered only 4% mortality but suffered 72% mortality after a 2-hr exposure (Tsai and Tompkins, 1974). When the dace were exposed to 0.15 mg/l of free chlorine for 6 hr only 10% mortality occurred, whereas doubling the exposure time to 12 hr increased the mortality approximately eightfold to 83%. Similar results occurred for dace exposed to chloramines. This study by Tsai and Tompkins (1974) also illustrates the interrelationship between exposure time and chlorine levels. At 0.15 mg/l of free chlorine, a 12-hr exposure caused 83% mortality; while at 1.38 mg/l, a 40-min exposure caused 65% mortality; and at 6.6 mg/l, only 8 min were needed to kill all the dace. Again, similar results were found when chloramines were tested.

Stober and Hanson (1974) tested chinook (*Oncorhynchus tshawytscha*) and pink salmon (*O. gorbuscha*) at exposure times ranging from 7.5 min to 60 min. Chinook salmon exposed to chlorine for 7.5 and 15 min had LC50 values of 0.5 mg/l, with th LC50 values decreasing to 0.25 mg/l for the 30- and 60-min exposures. The LC50 value for pink salmon exposed for 7.5 min was 0.5 mg/l and dropped to 0.25 mg/l for the 15-, 30- and 60-min exposures. Both species had an LC50 of 0.045 mg/l after exposure to fluctuating chlorine concentrations for 2 hr. When both species were subjected to a variety of temperature shocks in addition to the chlorine the LT50 (median lethal time) generally decreased with increasing exposure time at each test temperature. In another study Eren and Langer (1973) found that when the exposure time of *Tilapia aurea* was shortened from 18 to 4 hr the lethal total residual chlorine concentrations decreased about 20%.

Temperature

Cairns *et al.* (1975) concluded that there was little information in the literature regarding the effects of temperature on chlorine toxicity. Recently, however, several papers have appeared indicating a general trend of increasing sensitivity to chlorine of fish at higher temperatures. Two types of studies have emerged. The first type tests a species over a wide range of acclimation temperatures. Seegert *et al.* (1977) found that the 30-min LC50 values for the yellow perch (*Perca flavescens*) decreased from 7.7 mg/l to 1.0 mg/l at acclimation temperatures of 10 and 25°C, respectively. Similarly, Eren and Langer (1973) found that *Tilapia aurea* was more sensitive to chlorine at higher acclimation temperatures. In contrast, Bass and Heath (1975a) observed that the 96-hr LC50 for bluegills (*Lepomis macrochirus*) remained constant at acclimation temperatures from 6 to 32°C. They did find, however, that at the higher temperatures the times to death were shorter. Heath (1974) also found that temperature had little effect on the lethal concentrations of chlorine to bluegills.

The second type of temperature-chlorine study involves exposing the fish to a variety of temperature shocks over a range of chlorine concentrations. Stober and Hanson (1974) found that pink and chinook salmon when subjected to a variety of heat shocks of 3 to 10°C above ambient demonstrated a decreased resistance to chlorine. They concluded, however, that the chlorine level, and not the addition of heat, had acted as the principal lethal agent. Thatcher *et al.* (1976) found that there was no difference in the 96-hr LC50 values of brook trout (*Salvelinus fontinales*) acclimated to 7, 10, 15 and 20°C, when they were tested at 10 and 15°C. However, the LC50 dropped significantly when these fish were tested at 20°C. This indicates that it is the test temperature and not the acclimation temperature that determines the toxicity of the chlorine. Wolf *et al.* (1975) found that rainbow trout must experience thermal stress greater than 10°C before a combined effect with chlorine could be detected.

The above results suggest that fish have some temperature range within which temperature has little effect on chlorine toxicity. However, outside this range, which is undoubtedly species-dependent, test temperatures and/or shock temperatures can increase their sensitivity to chlorine.

Water Quality and Chemistry

The quality of the water in which an organism is exposed to chlorine is very important in determining the response of that organism to chlorine. The pH of the water and the concentrations of ammonia and organic materials are responsible for determining the specific chlorine compounds that will be present.

pH

Although pH plays an important role in the equilibria constants between the various chlorine species (White, 1972; Draley, 1972) little information exists on the effect of pH on chlorine toxicity. Tsai and Tompkins (1974), working over a pH range of 6.8 to 7.6, found that if there was any effect by pH it was masked by much larger effects caused by varying chlorine levels and exposure times. Similarly, Warren (1975, as cited in Brungs, 1976) observed no change in the LC50 values of coho salmon and cutthroat trout (*Salmo clarki*) exposed to chlorine over a pH range of 7.5 to 8.1. Merkins (1958) reported that rainbow trout were somewhat more resistant to chlorine at pH 6.3 than at pH 7.0. However, he attributed this difference not to pH per se but to the relative proportions of free chlorine and chloramines which existed at the two pH values. The primary role of pH in chlorine toxicity is probably related to how it affects the relative proportions of various chlorine species and not to a direct effect on the organisms.

Chlorinated Organic Compounds

Chlorinated organics are found wherever chlorine is added to waters with relatively high concentrations of organics such as in sewage treatment effluents (Jolley, 1973, 1974; Glaze et al., 1973). Jolley et al. (1976) have also found chlorinated organics in the discharge waters of power plants. Despite this apparent ubiquitous occurrence, the study of the toxicity of chloro-organics is still in its infancy.

Recently several authors have attempted to determine the toxicities of various chloro-organic compounds to aquatic life (Gehrs and Jolley, 1975; Leach and Thakore, 1975). A study by the Manufacturing Chemists Association (1972, as cited in Brungs, 1975) tested several chloro-organic compounds and found 96-hr LC50 values for fathead minnows ranging from 0.01 to 10 mg/l. As previously discussed, Gehrs et al. (1974) determined that 4-chlororesorcinol and 5-chlorouracil significantly affected the hatching success of carp eggs.

In view of the lack of specific information on the toxicities of chloro-organic compounds, further bioassay work in this area seems justified.

Synergistic Effects with Chemical Pollutants

Synergistic effects with other chemical pollutants have been largely ignored. Allen (1946) reported that chlorine combined with thiocyanate in industrial wastes to form cyanogen chloride, which was highly toxic to rainbow trout. A similar reaction with potassium thiocyanate was reported by Schaut (1939).

Hoss et al. (1974, cited by Brungs, 1976) reported on the effects of copper and chlorine. As with the case of chlorinated organic compounds, additional study is required in this area, especially in a situation where other toxic substances are known to occur simultaneously with chlorine.

Free vs Combined Chlorine

While the toxicities of free and combined chlorine are considered to be of the same order of magnitude (Brungs, 1973; Merkins, 1958) there is some disagreement regarding the specific toxicities of the two fractions. Most authors (Eren and Langer, 1973; Merkins, 1958; and Rosenberger, 1971) consider free chlorine to be more toxic and swifter acting. However, others (Westfall, 1950; Holland et al., 1960) consider chloramines to be more toxic. Holland et al. (1960) found that dichloramine in particular was considerably more toxic than free chlorine to the coho salmon. However, some questions on the analytical methods they used and the unlikelihood of a solution remaining 100% dichloramine at a pH of 7.6 raise some questions about the exact levels that were toxic. Tsai and Tompkins (1974) found that in the blacknose dace the relative toxicities of free chlorine and chloramines were dependent on the chlorine concentrations being tested. At concentrations greater than about 0.5 mg/l they found chloramines to be more toxic while at concentrations less than about 0.5 mg/l free chlorine was more toxic.

Mechanisms of Toxic Action

Many authors, including Cairns (1975), Dandy (1972), Rosenberger (1971), and Bass and Heath (1975b), have concluded that the gills are the primary site of chlorine toxicity. Bass and Heath (1975b) found that rainbow trout exposed thrice daily to free chlorine exhibited increased mucous production and damage to the respiratory epithelium was apparent in histological sections of gill tissue. Measurement of several physiological parameters (blood pO_2 and pH, heart and breathing rates) indicated that the changes seen in these parameters during chlorination were only partially restored to normal levels between chlorinations. With each succeeding chlorination pulse the amount of recovery became less until death finally ensued. They concluded that the primary mode of chlorine toxicity is gill damage resulting in asphyxiation. Conversely, Fobes (1971) found no change in the respiration rate of gill tissue excised from white suckers (*Catastomus commersonni*) subsequent to the exposure of the suckers to a lethal dose of chlorine. He concluded that death was not from gill damage and that the gills were not the primary site of chlorine toxicity. He hypothesized that chlorine enters through the gills and then somehow affects the nervous system of fishes. Wolf *et al.* (1975) also concluded that the gills were not the primary site of chlorine toxicity. In contrast to the findings of Bass and Heath reviewed above, which had indicated a summation effect with regard to chlorine toxicity, Wolf and co-workers hypothesized that chlorine toxicity is an all-or-none effect.

The answer to the apparent disparity in findings between authors may be that chlorine toxicity is not simply acting in one mode or at one site. Rosenberger (1971) has concluded that free chlorine and monochloramine have different modes of action. Holland *et al.* (1960) also observed differences in toxic patterns between the various chlorine species. Eaton (1973) found that monochloramine caused severe methemoglobinemia in fathead minnows. Free chlorine, however, did not have this effect. Similarly, bluegills exposed to free chlorine had normal levels of methemoglobin (Bass and Heath, 1975a). Lending further support to the multimode or site theory are the previously discussed findings of Tsai and Tompkins (1974), who found that the relative toxicities of free chlorine and chloramines varied with the chlorine concentrations.

FRESHWATER INVERTEBRATES

The literature concerning the effects of chlorine on freshwater invertebrates is not nearly as extensive as is that for fishes. Several recent unpublished reports and theses reviewed by Brungs (1976) have increased our knowledge considerably within the past year. Several of the most recent works were not available to the authors for inclusion here. Even with this new information a comparison of experimental results relating to a given species under varying environmental conditions is difficult.

Exposure Level

In general, the range of 96-hr LC50 values reported in the literature for invertebrates is much more narrow than those reported for fish. With few exceptions, the 96-hr LC50 for invertebrates is in the range of about 0.10 mg/l. Among the more sensitive invertebrates tested to date are the rotifers with 1-, 4- and 24-hr LC50 values of 0.032, 0.027 and 0.013 mg/l, respectively (Grossnickle, 1974). Gregg (1974, as cited by Brungs, 1976) investigated several invertebrate species over a range of chlorine concentrations from 0.01 to 0.1 mg/l. He observed that mayflies were the most sensitive group tested while beetle larvae were the most resistant. Other studies have indicated that crayfish are quite resistant to chlorine. Schneider (1975, as cited by Brungs, 1976) observed 96-hr LC50 values for crayfish of 0.96 mg/l. Arthur (1971, as cited by Brungs, 1973) also indicated that crayfish (*Orcanectes virilis*) were the least sensitive to chlorine among the organisms tested with 7-day TL50 values greater than 0.78 mg/l.

Exposure Time

Chronic exposures of invertebrates to chlorine have been conducted for a number of species. Arthur *et al.* (1975, as cited by Brungs, 1976) reported that the lowest mean total residual chlorine (TRC) concentration having measurable adverse chronic effects were 0.19 mg/l for an amphipod and 0.010 mg/l for *Daphnia magna*. The same study also cites the highest mean TRC concentrations having no measurable effect. These values were 0.021 mg/l and 0.002 to 0.004 mg/l for the amphipod and *D. magna*, respectively. Carlson (1975, as cited by Brungs, 1976) observed no effect on the standing crop of amphipods following a 3-½-month exposure to chloramine at concentrations of 0.001, 0.0043 and 0.0114 mg/l.

A comparison of LC50 values which were determined following exposures to chlorine for varying lengths of time demonstrates the importance of the duration of exposure. Grossnickle (1975) exposed the rotifer *Keratella cochlearis* to monochloramine for periods of 1, 4 and 24 hr at 15°C. He observed decreasing LC50 values of 0.032, 0.027 and 0.0135 mg/l as exposure times increased. Beeton *et al.* (1975) reports a 96-hr LC50 value for the copepod *Cyclops bicuspidatus thomasi* of 0.069 mg/l for a mixture of free chlorine and monochloramine at 15°C. Research conducted in the same laboratory under similar conditions determined that the 30-min LC50 value for *C. b. thomasi* was 15.61 mg/l at 15°C (Latimer *et al.*, 1975). Gregg (1974, as cited by Brungs, 1976) observed that the 4- to 7-day LC50 values determined for mayflies, stoneflies, sowbugs, amphipods, caddisflies, water pennies and snails were intermediate between the LC50 values for intermittent exposures based on maximum or peak concentrations and those determined based on mean concentrations.

Effects of Temperature

Gregg (1974, as cited by Brungs, 1976) reported that temperature exerted an influence on chlorine toxicity although the degree of effect varied for different species and for some there was no observed temperature effect. Latimer *et al.* (1975) reported that the 30-min LC50 value for *Limnocalanus macrurus* was 1.54 mg/l at both 5 and 10°C. *Cyclops bicuspidatus thomasi* was tested at 10, 15 and 20°C by the same authors and was found to have 30-min LC50 values of 14.68, 15.61 and 5.76 mg/l, respectively, indicating a definite temperature effect between 15 and 20°C.

Water Quality

The influence of water quality on the toxicity of chlorine to freshwater invertebrates has not been extensively studied. Beeton *et al.* (1976) reported a 96-hr LC50 value for *C. b. thomasi* exposed to monochloramine of 0.089 mg/l while the LC50 was 0.069 mg/l for a solution which was predominantly free chlorine.

Gehrs and Jolley (1975) have investigated the effect of 5-chlorouracil and 4-chlororesorcinol, two stable compounds found in relatively high concentrations in chlorinated sewage effluents, on survivorship and maturation of *Daphnia magna*. Only a slight decrease in survivorship occurred for 5-chlorouracil over a concentration range between 0.01 and 10.0 mg/l. A definite dose-response relationship was noted between survivorship and the concentration of 4-chlororesorcinol. The greatest increase in mortality was observed between 0.1 and 1.0 mg/l. The authors also report a general delay in the onset of reproduction with 5-chlorouracil and a decrease or total elimination of reproduction with 4-chlororesorcinol.

FRESHWATER ALGAE

Research on the toxicity of chlorine to freshwater algae is sorely lacking. Brook and Baker (1972) conducted studies on the phytoplankton of the St. Croix River, Minnesota/Wisconsin. They noted that photosynthetic rate and respiration were depressed to 50% of control values at a chlorine concentration calculated by serial dilution to be 0.320 mg/l. At the highest concentration actually measured, 2.7 mg/l, both photosynthetic rate and respiration were reduced to zero. Temperatures ranged from 32 to 36°C at the time of chlorination to 23 to 25°C during sample incubation. Although it is not specifically stated in the paper, the exposure time to the chlorine is assumed to be slightly in excess of the 2- to 4-hr incubation period.

Research currently underway by the authors of this report indicates that the photosynthetic rate of Lake Michigan phytoplankton is reduced 25 to 100% following 30-min exposures to chlorine ranging in concentration between 0.01 and 1.375 mg/l. The most significant reductions occurred at concentrations above 0.5 mg/l. No recovery was observed up to 24 hr after the initial chlorine exposure. Reductions in chlorophyll *a* concentrations following the 30-min chlorine exposures increased with increasing chlorine concentrations. As in the case of photosynthetic rates, the greatest chlorophyll *a* reductions were observed at chlorine concentrations above 0.5 mg/l. Judging from the fact that both chlorophyll *a* concentrations and photosynthetic rate are reduced with increasing chlorine concentrations and that no recovery in either parameter was noted, it seems quite possible that chlorophyll *a* destruction is the principal cause of the reduced photosynthetic rates.

Other studies on the effects of chlorine on marine and estuarine phytoplankton have shown photosynthetic rate reductions similar to those discussed above (Hamilton *et al.*, 1970; Hirayama and Hirano, 1970; Carpenter *et al.*, 1972; Brooks *et al.*, 1974; Gentile, 1972, as cited by Brungs, 1973; and Fox and Moyer, 1975). A specific discussion of these reports is not within the scope of this freshwater review; however the results obtained from these studies are helpful in evaluating the overall algae-chlorine picture. It appears that algae are extremely sensitive to chlorine, especially at concentrations exceeding 0.5 mg/l. Evidence further suggests that exposure times on the order of a few minutes may be all that is required to effect irreversible damage.

CONCLUSIONS

The importance of environmental variables such as temperature, water quality, and exposure time and dose rate have been demonstrated for a number of organisms. Unfortunately, many studies have been conducted and reported without sufficient detail regarding experimental conditions to permit complete evaluation. Researchers should be encouraged to include as much pertinent environmental information as possible in research reports so that these important variables may be considered in evaluating the responses of organisms to chlorine.

Based on the information presented in this review, the following may be concluded:

1. The response of freshwater organisms to chlorine is species-dependent. Sweeping statements concerning the response of major taxonomic groups to chlorine should be avoided unless specific information exists for important families and genera of the major taxon.

2. The life stage and size of an organism must be considered in evaluating chlorine toxicity. Due to conflicting information in the literature, no assumptions should be made relative to the responses of different-sized organisms without specific data.
3. The period of time that an organism is exposed to chlorine and the concentration of chlorine are critical in determining the final response of that organism. Exposure times should be carefully noted and chlorine measurements determined by an accepted standard procedure.
4. The effect of temperature on the response of freshwater organisms to chlorine appears to be species-dependent and also dependent on the specific range of temperature concerned. At lower temperatures there appears to be little temperature effect, while at higher temperatures the influence of temperature appears to be increased.
5. The role of chemical water quality in determining chlorine toxicity is important in determining what chlorine species will be present. No definitive statement can be made with regard to which chlorine species are more toxic. Some evidence suggests that different forms of chlorine may affect organisms by different modes of action, depending on the concentration range.
6. The avoidance of chlorine by fish has been demonstrated in laboratory tests. While this is also the case in field situations where the chlorine is discharged continuously, the avoidance of fish to chlorine intermittently discharged is poorly defined and needs further study. Consequently, the mode of chlorine application and the physical constraint of the discharge area should be considered when evaluating the impact of any chlorinated discharge.

REFERENCES

Alderson, R. 1970. "Effects of Low Concentrations of Free Chlorine on Eggs and Larvae of Plaice, *Pleuronectes platessa* L." Report No. FIR: MP/70/E-3. *Technical Conferences on Marine Pollution and Its Effects on Living Resources and Fishing.* Food and Agriculture Organization of the United Nations. Rome, Italy. December 9-18, 1970.

Allen, L. A., N. Blezard and A. B. Wheatland. 1946. "Toxicity to Fish of Chlorinated Sewage Effluents." *Surveyor* 105:298.

Arthur, J. W. 1971. "Progress Report." National Water Quality Laboratory, U.S. Environmental Protection Agency. Duluth, Minnesota.

Arthur, J. W. and J. G. Eaton. 1971. "Chloramine Toxicity to the Amphipod, *Gammarus pseudolimnaeus*, and the Fathead Minnow, *Pimephales promelas*." *J. Fish Res. Bd. Can.* 28:1841-1845.

Arthur, J. W., R. W. Andrew, V. R. Mattson, D. T. Olson and G. E. Glass. 1975. *Comparative Toxicity of Sewage-Effluent Disinfection to Freshwater Aquatic Life*. EPA/600/3-75-012. Ecological Research Series. U.S. Environmental Protection Agency. Washington, D. C. 73 pp.

Basch, R. E. 1971. Memo on Campbell Power Plant Chlorine Measurement. Michigan Water Resources Comm. Unpublished.

Basch, R. E., M. E. Newton, J. G. Truchan and C. M. Fetterolf. 1971. *Chlorinated Municipal Waste Toxicities to Rainbow Trout and Fathead Minnows*. Water Pollut. Control Res. Ser. No. 18050GZZ. U.S. Environmental Protection Agency.

Basch, R. E. and J. G. Truchan. 1974. *Calculated Residual Chlorine Concentrations Safe for Fish*. Tech. Bull. 74-2. Michigan Water Resources Comm., Dept. Natural Resources. 29 pp.

Bass, M. L. and A. G. Heath. 1975a. "Toxicity of Intermittent Chlorine Exposure to Bluegill Sunfish, *Lepomis macrochirus:* Interaction with Temperature." *Assoc. Southeast Biol. Bull.* 22:40.

Bass, M. L. and A. G. Heath. 1975b. "Physiological Effects of Intermittent Chlorination on Fish." *Am. Zool.* 15:818.

Beeton, A. M., P. K. Kovacic and A. S. Brooks. 1976. *Effects of Residual Chlorine and Sulfite Reduction on Lake Michigan Invertebrates*. EPA/600/3-76-036. Ecological Research Series. U.S. Environmental Protection Agency, Washington, D. C. 133 pp.

Bogardus, R. B., D. B. Boies and D. A. Etnier. 1975. *The Avoidance Responses of Selected Wabash River Fishes to Monochloramine*. Technical Report. WAPORA, Inc., Washington, D. C. 15 pp.

Brook, A. J. and A. L. Baker. 1972. "Chlorination at Power Plants: Impact on Phytoplankton Productivity." *Science* 176:1414-1415.

Brooks, A. S., R. A. Smith and L. D. Jensen. 1974. "Phytoplankton and Primary Productivity." In L. D. Jensen (Ed.) *Environmental Responses to Thermal Discharges from the Indian River Station, Indian River, Delaware*. Electric Power Research Institute Pub. No. 74-049-00-3. 205 pp.

Brungs, W. A. 1973. "Effects of Residual Chlorine on Aquatic Life." *J. Water Poll. Control Fed.* 45:2180-2193.

Brungs, W. A. 1976. *Effects of Wastewater Chlorination on Freshwater Aquatic Life*. EPA/600/3-76-098. Environmental Research Laboratory, U.S. Environmental Protection Agency. Duluth, Minnesota. 52 pp.

Cairns, J., A. G. Heath and B. C. Parker. 1975. "Temperature Influence on Chemical Toxicity to Aquatic Organisms." *J. Water Poll. Control Fed.* 47:267-280.

Carlson, A. R. 1975. *Reproductive Behavior to the Threespine Stickleback Exposed to Chloramines*. M.S. thesis. Oregon State University. Corvallis.

Carpenter, E. J., B. B. Peck and S. J. Anderson. 1972. "Cooling Water Chlorination and Productivity of Entrained Phytoplankton." *Mar. Biol.* 16:37-40.

Dandy, J. W. T. 1972. "Activity Response to Chlorine in the Brook Trout, *Salvelinus fontinalis* (Mitchill)." *Can. J. Zool.* 50:405-410.

Dickson, K. L. and J. Stauffer. 1974. "Observations on the Effects of Chlorination on Fish at Appalachian Power Company's Gen Lyn, Virginia, Plant, January 10, 1974." Unpublished Memo. Virginia Polytechnic Institute and State University. Blacksburg, Virginia.

Draley, J. E. 1972. *The Treatment of Cooling Waters with Chlorine*. ANL/ES-12. Argonne National Laboratory, Argonne, Illinois. 11 pp.

Eaton, J. W., C. F. Kolpin, C. W. Kiellstrand and H. S. Jacob. 1973. "Chlorinated Urban Water: A Cause of Dialysis-Induced Hemolytic Anemia," *Science* 181:463-464.

Eren, Y. and Y. Langer. 1973. "The Effect of Chlorination on Tilapia Fish." *Bamidgeh* 25:56-60.

Esvelt, L. A., W. J. Kaufman and R. E. Selleck. 1971. *Toxicity Removal from Municipal Wastewaters. Vol. IV. A Study of Toxicity and Biostimulation in San Francisco Bay-Delta Waters*. SERL Report No. 71-7. Sanitary Engineering Research Laboratory, University of California, Berkeley, California. 224 pp.

Fava, J. A. and Chu-fa Tsai. 1973. "Fish Avoidance of Chlorinated Effluents." Program Report 1973. Water Resources Research Center, University of Maryland, College Park, Maryland.

Fobes, R. L. 1971. *Chlorine Toxicity and its Effect on Gill Tissue Respiration of the White Sucker, Catastomus commersoni* (Lacepede). Thesis. Michigan State University. East Lansing, Michigan.

Fox, J. L. and M. S. Moyer. 1975. "Effect of Power Plant Chlorination on Estuarine Productivity." *Chesapeake Sci.* 16:66.

Gehrs, C. W. and R. L. Jolley. 1975. "Chlorine-Containing Stable Organics: New Compounds of Environmental Concern." *Verh. Internat. Verein. Limnol.* 19:2185-2188.

Gehrs, C. W., L. D. Eyman, R. L. Jolley and J. E. Thompson. 1974. "Effects of Stable Chlorine-Containing Organics on Aquatic Environments." *Nature* 249:675-676.

Gentile, J. H. *et al.* 1972. Unpublished data. National Marine Water Quality Laboratory, U.S. Environmental Protection Agency. West Kingston, Rhode Island.

Glaze, W. H., J. E. Henderson, IV, J. E. Bell and V. A. Wheeler. 1973. "Analysis of Organic Materials in Wastewater Effluents After Chlorination." *J. Chromatogr. Sci.* 11:580-584.

Gregg, G. C. 1974. *The Effects of Chlorine and Heat on Selected Stream Invertebrates*. PhD. thesis. Virginia Polytechnic Institute and State University. Blacksburg, Virginia. 309 pp.

Grossnickle, N. E. 1974. *The Acute Toxicity of Residual Chloramine to the Rotifer*, Keratella cochlearis *(Goose) and the Effect of Dechlorination with Sodium Sulfite*. MS thesis. University of Wisconsin-Milwaukee. Milwaukee, Wisconsin. 39 pp.

Hamilton, D. H., Jr., D. A. Flemer, C. W. Keefe and J. A. Mihursky. 1970. "Effects of Chlorination on Estuarine Primary Production." *Science* 169:197-198.

Heath, A. G. 1974. "A Preliminary Investigation of Chlorine Toxicity to Fish and Macroinvertebrates: Interactions with Temperature." Final Report on Grant with American Electricity Production Service Corp. Virginia Polytechnic Institute and State University. Blacksburg, Virginia.

Hirayama, K. and R. Hirano. 1970. "Influence of High Temperature and Residual Chlorine on Marine Phytoplankton." *Mar. Biol.* 7:205-213.

Holland, G. A., J. E. Lasater, E. D. Newmann and W. E. Eldridge. 1960. "Toxic Effects of Organic and Inorganic Pollutants on Young Salmon and Trout," pp. 198-214. In *Res. Bull.* No. 5. Dept. of Fish. State of Washington.

Hoss, D. E., L. C. Coston, J. P. Baptist and D. W. Engel. 1974. "Effects of Temperature, Copper and Chlorine on Fish during Simulated Entrainment in Power-Simulated Plant Condenser Cooling Systems." Symposium on Physical and Biological Effects on the Environment of Cooling Systems and Thermal Discharges at Nuclear Power Stations. Oslo, Norway. August 26-30.

Hughes, J. S. 1973. "Acute Toxicity of Thirty Chemicals to Striped Bass (*Morone saxatilis*)." Presentation at the Western Assoc. of State Game and Fish Comm. Salt Lake City, Utah.

Jolley, R. L. 1973. *Chlorination Effects on Organic Constituents in Effluents from Domestic Sanitary Sewage Treatment Plants*. ORNL/TM-4290. Oak Ridge National Laboratory. Oak Ridge, Tennessee. 342 pp.

Jolley, R. L. 1974. "Determination of Chlorine-Containing Organics in Chlorinated Sewage Effluents by Coupled ^{36}Cl Tracer—High-Resolution Chromatography." *Environ. Lett.* 7:321-340.

Jolley, R. L., C. W. Gehrs and W. W. Pitt, Jr. 1976. "Chlorination of Cooling Water: A Source of Chlorine-Containing Organic Compounds with Possible Environmental Significance," p. 21-28. In C. E. Cushing (Ed.) *Radioecology and Energy Resources*. Dowden, Hutchinson and Ross, Inc., Stroudsburg, Pennsylvania.

Latimer, D. L., A. S. Brooks and A. M. Beeton. 1975. "The Toxicity of 30-Minute Exposures of Residual Chlorine to the Copepods *Limnocalanus macrurus* and *Cyclops bisucpidatus thomasi*." *J. Fish. Res. Bd. Can.* 32:2495-2501.

Leach, J. M. and A. N. Thakore. 1975. "Isolation and Identification of Constituents Toxic to Juvenile Rainbow Trout (*Salmo gairdneri*) in Caustic Extraction Effluents from Kraft Pulpmill Bleach Plants." *J. Fish. Res. Bd. Can.* 32:1249.

Manufacturing Chemists Association. 1972. *The Effect of Chlorination on Selected Organic Chemicals*. Wat. Poll. Control Res. Series 12020 EXG 03/72. U.S. Environmental Protection Agency. Washington, D.C.

Massey, A. 1972. "A Survey of Chlorine Concentrations in Consumer's Power Company's Big Rock Point Power Plant Discharge Channel." Michigan Water Research Commission Report. Unpublished.

Meldrin, J. W., J.J. Gift and B. R. Petrosky. 1974. *The Effect of Temperature and Chemical Pollutants on the Behavior of Several Estuarine Organisms*. National Technical Information Service, PB-239, 347. 129 pp.

Merkens, J. C. 1958. "Studies on the Toxicity of Chlorine and Chloramines to the Rainbow Trout." *Water Waste Treat. J.* 7:150-151.

Rosenberger, D. R. 1971. *The Calculation of Acute Toxicity of Free Chlorine and Chloramines to Coho Salmon by Multiple Regression Analysis*. Thesis. Michigan State University. East Lansing, Michigan. 33 pp.

Schaut, G. G. 1939. "Fish Catastrophes during Droughts." *J. Am. Water Works Assoc.* 31:771.

Schneider, M. J., C. D. Becker, D. H. Fickeisen, T. O. Thatcher and E. G. Wolf. 1975. "Aquatic Physiology of Thermal and Chemical Discharges." In *Environmental Effects of Cooling Systems at Nuclear Power Plants*. Internat. Atomic Energy Agency, Vienna. IAEA-SM-187/15. Battelle-Pacific Northwest Laboratory. Richland, Washington.

Seegert, G. L., A. S. Brooks and D. L. Latimer. 1977. "The Effects of a 30-Minute Exposure of Selected Lake Michigan Fishes and Invertebrates to Residual Chlorine," pp. 91-99. In L. D. Jensen (Ed.) *Biofouling and Control Procedures—Technology and Ecological Effects*. Marcel Dekker, Inc., New York.

Sprague, J. B. and D. E. Drury. 1969. "Avoidance Reactions of Salmonid Fish to Representative Pollutants," p. 169. In *Advances in Water Pollution Research. Proc. 4th Internat. Conf. Water Pollut. Res.* Pergamon Press. London, England.

Stober, Q. J. and C. H. Hanson. 1974. "Toxicity of Chlorine and Heat to Pink (*Oncorhynchus gorbuscha*) and Chinook Salmon (*O. tshawytscha*)." *Trans. Am. Fish. Soc.* 103:569-576.

Thatcher, T. O., M. J. Schneider and E. G. Wolf. 1976. "Bioassays on the Combined Effects of Chlorine, Heavy Metals, and Temperature on Fishes and Fish Food Organisms. Part I. Effects of Chlorine and Temperature on Juvenile Brook Trout (*Salvelinus fontinalis*)." *Bull. Environ. Contam. Toxicol.* 15:40-48.

Truchan, J. G. 1977. "Toxicity of Residual Chlorine to Freshwater Fish: Michigan's Experience, pp. 79-89. In L. D. Jensen (Ed.) *Biofouling Control Procedures–Technology and Ecological Effects.* Marcel Dekker, Inc., New York.

Tsai, C. 1968. "Effects of Chlorinated Sewage Effluents on Fishes in Upper Patuxent River, Maryland." *Chesapeake Sci.* 9:83-93.

Tsai, C. 1970. "Changes in Fish Populations and Migration in Relation to Increased Sewage Pollution in Little Patuxent Rivery, Maryland." *Chesapeake Sci.* 11:34-41.

Tsai, C. 1975. *Effects of Sewage Treatment Plant Effluents on Fish: A Review of Literature.* Chesapeake Research Consortium, Inc., CRC Pub. No. 36. Center for Environmental and Estuarine Studies. University of Maryland, College Park, Maryland. 229 pp.

Tsai, C. and J. A. Tomkins. 1974. *Survival Time and Lethal Exposure Time for the Blacknose Dace Exposed to Free Chlorine and Chloramine Solutions.* Tech. Report No. 30. Center for Environmental and Estuarine Studies. University of Maryland, College Park, Maryland. 22 pp.

Ward, R. W., R. D. Griffin, G. M. DeGraeve and R. A. Stone. 1976. *Disinfection Efficiency and Residual Toxicity of Several Wastewater Disinfectants. Volume I.* Grandville, Michigan. EPA-600/2-76-156. Environmental Technology Series. U.S. Environmental Protection Agency.

Warren, C. E. 1974. "Laboratory Determination of Chloramine Concentrations Safe for Aquatic Life." Quarterly Progress Report, EPA Grant No. R-802286. Dept. Fish and Wildlife, Oregon State University, Corvallis, Oregon. October 1-December 31.

Westfall, B. A. 1946. "Stream Pollution Hazards of Wood Pulp Mill Effluents." Fishery Leaflet 174. Fish and Wildlife Service, U.S. Department of the Interior.

White, G. C. 1972. *Handbook of Chlorination.* Van Nostrand Reinhold Co. New York. 744 pp.

Wolf, E. G., M. J. Schneider, K. O. Schwarzmiller and T. O. Thatcher. 1975. "Toxicity Tests on the Combined Effects of Chlorine and Temperature on Rainbow (*Salmo gairdneri*) and Brook (*Salvelinus fontinallis*) Trout." Proc. *2nd Thermal Ecology Symp.,* Augusta, Georgia. April 2-5.

DISCUSSION

George C. White, Consulting Engineer. Were the chlorine residuals in the 0.001 to 0.005 part per million range actually measured by amperometric titration or were these numbers extrapolated? From a practical standpoint the equipment operator will not be able to get reproducibility of anything less than 0.05, and it would be more believable to me if it were 0.1. So, somewhere, it ought to be said whether the numbers are extrapolated. If they are extrapolated, anything below 0.05 should be considered zero because that is the way we are operating our dechlorination systems.

Brooks. I think that most of the values cited for the chronic exposures were done in Duluth, and I think the sophisticated amperometric titrator can get down to between 1 and 10 parts per billion.

William A. Brungs, U.S. Environmental Protection Agency. Plus or minus 100%.

Brooks. I am not sure about the plus or minus 100%, and I am not sure exactly what is being measured.

Brungs. In many cases the numbers are also based on the known dilution. You can pick up additional confidence that way. For example, if 0.01 were diluted by half and they get a measurement around 0.005. This would be confirmation.

Brooks. I think most of these studies with the chronic exposure were done with a serial dilution apparatus.

White. We need to adhere to the recognition of significant figures and ability of control equipment and operating personnel to measure these residuals. From a practical standpoint this is about 0.1 mg/l total chlorine. Therefore, somewhere in your reporting you have got to say something about significant figures that will help the person operating the equipment.

Brooks. In our plea for additional information in these reports, I think that is a very important thing that should be included.

Brungs. We don't want to spend all our time on chemistry only. Let us move to another subject.

Alan G. Heath, Virginia Polytechnic Institute and State University. I wonder if you or anyone in the audience has any information on the formation or possible formation of organochlorine compounds in the blood of fish?

Brooks. No, we don't. But I was talking with either Bob Jolley or Carl Gehrs about trying to get some chlorine-36 to see if we can find any taken up by the fish.

Michael L. Bass, Mary Washington College. Do you or anyone at this meeting have any information on the relationship of water hardness and chlorine toxicity?

Brooks. Some of Warren's work cites alkalinity values over quite a range. I believe they determined there was no effect. I don't recall the exact numbers they cited, but it was something like from 100 to 400.

Brungs. Dr. Ingols is in the audience and keeps asking me about the calcium concentration. Would you like to make the point?

Robert S. Ingols, Georgia Institute of Technology. A report has been published on the TLm against minnows of halogenated phenols. The toxicity is dependent upon the number of halogens on the phenol. The toxicity increases with number. Chlorine compounds are more toxic than brominated ones which are more toxic than iodinated ones. An examination of the swimming pool literature may give some information on the chlorination of algae. Blue-green algae may live in a pool for months in spite of shock doses and continuous low levels of chlorine. My personal observation indicates that the presence of calcium (hardness) protects the algae.

Brooks. There is some work by Japanese investigators that show that some marine species can tolerate very high chlorine levels. We looked through the biological literature and even, as a biologist reached out into the engineering literature. But I didn't get into the swimming pool literature.

15

A REVISED REVIEW OF THE IMPACT OF CHLORINATION PROCESSES UPON MARINE ECOSYSTEMS: UPDATE 1977

William P. Davis and Douglas P. Middaugh

    Gulf Breeze Environmental Research
      Laboratory
    Bears Bluff Field Station
    Wadmalaw Island, South Carolina   29487

ABSTRACT

For over 175 years chlorine gas has been used in industrial, biocidal and disinfection applications. The chemistry of chlorine in fresh water is relatively well known, but long-range effects on the organisms and the ecological communities of marine waters have barely been studied. Until recently, the so-called "chlorine demand" of treated or receiving waters has been considered a desirable feature which assured degradation of actively oxidizing states of chlorine to a nontoxic state. With continuing and increased use of chlorine as an antifouling biocide in powerplants, and as a disinfectant of municipal wastes, concern has arisen that resulting by-products, such as induced halogenated hydrocarbons, could potentially reach environmentally harmful levels. For example, in the state of Maryland the quantity of chlorine used, which subsequently reaches the Chesapeake Bay, would have sterilized that body of water were not chemical/biological degradation processes in effect. But, what are the limits of natural degradation systems? What, for one example, are the known environmental costs of our present rates of chlorine applications on renewable fishery resources? What kind and at what rate are persistent halogenated compounds being produced? Where do these go in natural systems? From partial or complete answers to these questions will come meaningful environmental management criteria.

This paper presents a theoretical degradation model of chlorine added to marine waters. Additionally it summarizes literature reporting laboratory or ecological effects of chlorination. It is revised to attempt incorporation of pertinent literature through 1977.

INTRODUCTION

Chlorine gas has been used in industry as a bleaching agent since 1800 and has become one of the most versatile chemicals known. In fresh water, chlorination is used to disinfect drinking and recreational waters, for slime and fouling control, as a pathogen control in municipal wastes as well as a bleach and odor suppressant in the pulp and paper industry. From all these applications which utilize vast quantities of chlorine, chemical by-products are transported through effluents to natural ecosystems. The toxicity desired in disinfection biocide applications goes on to introduce nondesirable effects to wildlife and aquatic ecosystems. Determination of the presence of halogenated organics from chlorination, traceable in drinking water in 80 cities, underscores the need for renewed responsible assessment of the control and regulation of chlorination processes and reexamination of potential damage to the environment.

The state of Maryland is often used by planners as a minimodel for the United States. With respect to the rate of chlorination, some of the most accurate statistics exist for Maryland. Additionally, resulting chlorination by-products from Maryland mostly drain into the Chesapeake Bay. An inventory of chlorine discharge from the Maryland contribution alone into Chesapeake Bay, assuming no degradation, reveals discharges of 27 million pounds per year of chlorine from municipal treatment plants and 2.2 million pounds per year from power generation facilities. To the casual observer it appears that without the action of degradation processes, these amounts could sterilize the bay. Data from Jolley (this conference) suggest that 3% or more of these amounts end up as halogenated organic compounds, which then potentially persist in marine ecosystems. There are 24 other states which border marine waters where chlorinated effluents enter marine ecosystems under various combinations of physical-chemical and environmental conditions.

The purpose of this paper is to compile the scarce data available on chlorination effects upon biota of estuarine and marine ecosystems. Chemical detection of oxidants and by-products in marine ecosystems reported in the literature are discussed to point out needs for future investigations.

DEGRADATION PATHWAYS OF CHLORINE IN MARINE SYSTEMS

A theoretical degradation model prepared by Block and Helz (personal communication) to summarize the suspected pathways of chemical reactions resulting from chlorination of natural marine waters is shown in Figure 1. Obviously missing are any coefficients, rates, or the many effects and influences from physical-chemical conditions and factors which control pathways and end products. Although some data exists on the effects of chlorination and a few of the by-products upon specific organisms, there is virtually no information on transport processes, persistence, bioaccumulation, and fate of the majority of halogenated compounds resulting from chlorination processes.

The reaction occurring between levels I and II of the degradation scheme is a result of chlorine decay from a diatomic gas to hypochlorous acid, hypochlorite ions, and sodium hypochlorite. As pointed out by Moore (1951) and Lewis (1966), this reaction occurs rapidly and goes to completion within seconds after the addition of chlorine. Inclusion of sodium hypochlorite within level II is based on work by Sugam and Helz (1975).

The chemical composition and abundance of products formed from level II to level III is a function of physical and chemical parameters of the water, including, but not limited to, temperature, pH, ammonia, sunlight (UV), salinity or the bromine available as a reaction component. In sea water if NH_4^+ ions are less abundant, the predominant species would be bromamines; since full sea water typically has 60 ppm bromide, bromination predominates as salinity increases. Photolysis may influence the level of bromate produced (Macalady *et al.*, 1977).

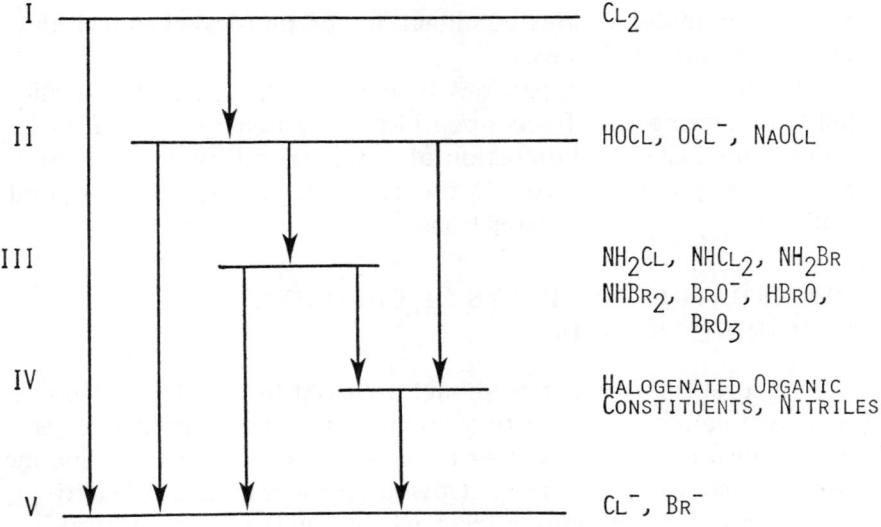

Figure 1. Degradation processes for chlorine in saline waters (modified from Block and Helz, personal communication).

Level IV includes halogenated organic constituents which may be formed by level II or level III species, including chloramines, hypobromite, bromate and bromamines. The stable end products in level V occur through a diverse group of mechanisms operative in steps I through IV.

Charge balance results in one atom of Cl passing from level I to level V to each atom passing from level I to level II. Reduction of hypochlorite by Br^- or Fe^{2+} and Mn^{2+} may release Cl^- from level II to level V. Movement of Cl^- from level III to level V can also occur in a number of ways; the most obvious, suggested by Laubusch (1971), involves destruction of chloramines when the OCl^-/NH_4^+ ratio is large.

Some of the chlorinated organics identified by Jolley (1973) are persistent and the decay from level IV to level V is probably a slower process than decay from levels I through III to level V. Carpenter and Smith (in press), as well as Helz, Sugam and Hsu (in press) demonstrate formation of bromoform, nitriles and other halogenated organics.

CHLORINATION TOXICITY TO MARINE PHYTOPLANKTON

The effects of chlorination and thermal pollution on phytoplankton productivity has been investigated in some detail (Table I). Carpenter et al. (1972) observed an 83% decrease in the productivity of phytoplankton cultured in waters which had passed through the cooling system of a nuclear generating plant on Long Island Sound.

Intake water was chlorinated at a rate of 1.2 mg/l with a chlorination-induced oxidant residual of 0.4 mg/l measured at the discharge. Addition of 0.1 mg/l chlorine at the intake with nondetectable chlorination-induced oxidant residuals at the outfall decreased productivity by 79%. Essentially no decreases in productivity were observed when phytoplankton passed through the cooling system under the same thermal conditions, but without addition of chlorine. Hirayama and Hirano (1970) measured the effect of chlorination on the photosynthetic activity of *Skeletonema costatum* and found that cells were killed when subjected to 1.5 to 2.3 mg/l chlorination for 5 and 10 min.

Gentile et al. (1973, 1976) observed a 55% decrease in ATP content of marine phytoplankton exposed for 2 min to 0.32 mg/l residual chlorination and a 77% decrease after 45 min of exposure to chlorination concentrations as low as 0.01 mg/l. A 50% reduction in growth rates was observed for 10 species of marine phytoplankton exposed for 24 hr to chlorine concentrations that ranged from 0.075 to 0.25 mg/l.

CHLORINE TOXICITY TO INVERTEBRATES

Muchmore and Epel (1973) investigated the effects of chlorination of wastewater on fertilization in marine invertebrates (Table II). Unchlorinated sewage from the Pacific Grove, California, Municipal Treatment Works was a weak inhibitor of fertilization in the sea urchin, *Strongylocentrotus purpuratus*. Exposure of gametes to a 10% unchlorinated sewage-sea water mixture typically reduced fertilization success by 20%. A 0.5% dilution of moderately chlorinated sewage (11 mg/l "total residual chlorine"* undiluted), significantly reduced fertilization.

*Total residual chlorine (TRC) in this case is being reported for marine waters where residual chlorine per se is not likely to be found.

Table I. Summary of Toxic Effects of Chlorinated Wastes and Water on Marine Phytoplankton

Species	Toxicant	Reported Chlorination Level (mg/l)	Duration of Test	Effect	Reference
Phytoplankton	Cl$_2$ injection	0.05- 0.40	12 hr + 4 hr incubation	50-98% loss of productivity	Carpenter et al. (1972)
Chlamydomonas sp.	sodium hypochlorite solution	0.69-12.90	5 min	reduced growth rate	Hirayama and Hirano (1970)
Skeletonema costatum		0.18- 2.40	5 min	none up to 0.29 mg/l; greater amts. inhibited growth	
Phytoplankton	sodium hypochlorite solution	0.32	2 min	55% decrease in ATP	Gentile et al. (1973, 1976)
		0.01	45 min	77% decrease in ATP	
		0.075- 0.25	24 min	55% decrease in growth	
Phytoplankton	Cl$_2$ injection		15 min	91% reduction in photosynthesis	Hamilton et al. (1970)

Table II. Summary of Toxic Effects of Chlorinated Wastes and Water on Marine Invertebrates

Species	Toxicant	Reported Chlorination Level (mg/l)	Duration of Test	Effect	Reference
Strongylocentrotus purpuratus (gametes)	chlorinated sewage effluents	0.02	5 min	none	Muchmore and Epel (1973)
		0.11	5 min	100% inhibition of fertilization	
Urechis caupo (gametes)		0.2	5 min	22% inhibition of fertilization	
		1.0	5 min	100% inhibition of fertilization	
Phragmatopoma californica (sperm)		0.2	5 min	22% loss of motility	
		1.0	5 min	86% loss of motility	
Ostrea edulis	residual chlorine	10.0	48 min + 10°C	none	Waugh (1964)
Eliminius modestus		2.0	10 min	death and inhibited growth	
		5.0	3 min	none	
Balanus sp.	Cl_2 injection	2.5	5 min	80% mortality	McLean (1972, 1973)
Acartia tonsi		2.5	5 min	90% mortality	
Melita nitida		2.5	5 min	near 100% mortality 96 hr after exposure	

Table II. Continued.

Species	Toxicant	Reported Chlorination Level (mg/l)	Duration of Test	Effect	Reference
Palaemonetes pugio		2.5	5 min	near 100% mortality 96 hr after exposure	
Bimaria franciscana		4.5	4 days	none	Turner et al. (1948)
Anemones	residual chlorine	10.0	1,2,4,8 hr/day for 10 days	none	
		2.5	8 days	100% mortality	
		1.0	15 days	100% mortality	
Mussels		10.0	1,2,4,8 hr/day for 10 days	none	
		2.5	5 days	100% mortality	
		1.0	15 days	100% mortality	
Barnacles		10.0	1,2,4,8 hr/day for 10 days	95-100% mortality	
		2.5	4 days	100% mortality	
		1.0	7 days	100% mortality	
Mytilus edulis	Cl_2 injection	0.02-0.05	a few hours	detachment and migration	James (1967)
Homarus americanus	free chlorine	2.02	48 hr	50% mortality	Capuzzo et al. (1976)
	chloramine	0.56	48 hr	50% mortality	

Chlorination had more effect on sperm cells than on eggs. Eggs incubated for 5 min exposed to 0.77 mg/l hypochlorite solution and subsequently washed to remove the oxidant showed no reduction in fertility, whereas incubation of sperm exposed to 0.07 mg/l caused loss of fertilization ability. This loss was attributed to a loss of sperm motility which was not restored by washing to remove the oxidant. Gametes of the echiuroid, *Urechis caupo*, and sperm of the annelid worm, *Phragmatopoma californica*, were not as sensitive to chlorination.

A number of power plant-related studies have been conducted to determine the effect of chlorination of sea water on fouling organisms. Waugh (1964) observed no significant difference in the mortality of unexposed oyster larvae, *Ostrea edulis*, or those exposed to a chlorination rate of 5 mg/l for 3 min at ambient temperature.

Exposure of larvae to thermal stress (10°C above ambient) and a chlorination rate of 10 mg/l for 6 to 48 min also had no significant effect on survival 64 hr after treatment. Nauplii of the barnacle, *Eliminius modestus*, was more acutely sensitive to chlorination. Chlorine concentrations in excess of 0.5 mg/l caused heavy mortality and reduced growth of survivors.

McLean (1973) simulated the conditions encountered by marine organisms passing through a power plant on the Patuxent River, Maryland. Intake chlorination to 2.5 mg/l residual, entrainment for approximately 3 min, and sustained exposure to elevated temperatures for up to 3 hr were used as experimental parameters. While barnacle larvae, *Balanus* sp., and copepods, *Acartia tonsi*, were not affected by a 3-hr temperature stress of 5.5 and 11°C above ambient; exposure to 2.5 mg/l residual chlorination for 5 min at ambient temperatures caused respective mortality rates of 80 and 90%. The amphipod, *Melita nitida*, and the grass shrimp, *Palaemonetes pugio*, showed a delayed death response after exposure to 2.5 mg/l residual chlorination for 5 min. Nearly 100% mortality was observed for both species 96 hr after exposure to the chlorine residual. McLean (1972) showed that established colonies of the euryhaline colonial hydroid, *Bimeria franciscana*, were not greatly affected by 1 and 3 hr of exposure to 4.5 mg/l residual chlorination.

Turner et al. (1948) determined that continuous treatment of sea water conduits with 0.25 mg/l chlorination prevented fouling during a 90-day interval when the flow velocity was 52 cm/sec or less. Intermittent treatment with 10 mg/l residual chlorination for 8 hr/day was ineffective in preventing fouling by anemones, mussels and barnacles.

James (1967), working in Great Britain, observed that residual chlorination concentrations of 0.02 and 0.05 mg/l caused detachment and movement of mussels in the direction of water flow through an aquarium with eventual elimination of the mussels. He concluded that the most effective way to prevent fouling by mussels was not to kill but to discourage settling in cooling water systems by continuous low-level chlorination.

Markowski (1959, 1960) compared the occurrence of marine organisms on concrete slabs placed in the intake and outfall canals of an electric generating plant. Chlorine was injected into cooling condensers for 2 hr/day at a concentration between 1 and 2.5 mg/l. No vegetation was observed to grow in the intake canal where dense animal populations occurred (predominantly invertebrates, Coelenterata, and Polyzoa). The outfall canal contained a prolific growth of algae, *Entermorpha* sp., but fewer invertebrates. The barnacle, *Balanus improvisis,* collected with some regularity from the intake canal, was never observed in the outfall canal. The mollusk, *Eubranchus* sp., was more abundant on the intake slabs than in the outfall.

Gibson et al. (1975) report critical thermal maxima (CTM) for coon stripe shrimp (*Pandalus danae*) and LC_{50} values for chlorine and copper, both singly and in combination. *Pandalus danae* were more resistant to chlorination when acclimated and exposed at 8 to 10°C than when acclimated at 8°C and exposed at 15 or 20°C, or when acclimated and exposed at 15°C which is near their optimum short-term growth temperature (16°C). Residual chlorination at a reported 0.18 mg/l concentration is lethal to 1.0 to 2.0 g coon stripe shrimp at 16°C and reduced growth was observed in shrimp exposed to 0.08 mg/l for one month.

CHLORINE TOXICITY TO ESTUARINE FISH

Tsai (1968, 1970, 1975) observed decreases in abundance and occurrence of brackish water fish species in certain areas of the

Upper and Little Patuxent Rivers that received chlorinated sewage effluent. Tsai suggests that chlorinated sewage effluent may also block the upstream migration of such semianadromous species as white catfish and white perch. He attributed the "blocking effect" to chlorination products rather than to reduced dissolved oxygen or pH resulting from organic decomposition of the effluent (Table III).

Tsai (1973) measured the diversity index of fishes upstream and downstream from 98 sewage treatment plants in Virginia, Maryland, and Pennsylvania. Sewage treatment plants were categorized as Type I engineering facilities (sludge activation, aeration, sedimentation and filtration) with effluent chlorination; Type II engineering facilities with chlorination and effluent holding lagoon; and Type III engineering facilities with lagoon and effluent chlorination at the lagoon outlet. Reductions in number of individual fish, number of species, and the species diversity index were significant downstream from Type I and III plants. These reductions were attributed to total residual chlorination levels and turbidity. No significant changes in diversity indices were found downstream associated with Type II plants.

A study of the effect of chlorinated sewage effluents on sockeye salmon, *Onchorhynchus nerka*, and pink salmon, *O. gorbuscha*, has been conducted by Servizi and Martens (1974). They used three study sites to conduct cage bioassays. The first, Site I, was adjacent to a primary treatment plant with effluents chlorinated following settling and discharged through a 600-ft pipeline directly into the receiving stream. Site II was on a stream receiving wastes from an activated sludge plant in which chlorinated effluents were discharged into a large effluent holding lagoon and retained for 30 to 60 days. Site III was located on a stream receiving effluents which were chlorinated as they left a nonaerated lagoon.

Measured chlorine residuals in the receiving stream at Site I ranged from 0.02 to 0.26 mg/l. These concentrations resulted in 100% mortality of caged sockeye fingerlings placed 30, 60 and 250 ft below the effluent discharge point. Additional tests indicated that the primary effluent without chlorination was also toxic. However, fish exposed to the unchlorinated effluent lived ten times longer than ones exposed when effluents were being

Table III. Summary of Toxic Effects of Chlorinated Wastes and Water on Marine and Freshwater Fishes

Species	Toxicant	Reported Chlorination Level (mg/l)	Duration of Test	Effect	Reference
Fresh Water and Brackish Fishes	chlorinated sewage effluents	0.60 -2.0	long-term	decreased population size and diversity	Tsai (1968, 1970, 1973)
Leiostomus xanthurus	chlorinated sewage effluents	0.07 -0.28	May-June 1973	probable kill 5-10 million fish	Virginia State Water Control Board (1974)
Morone sp.					
Pomatomus saltatrix					
Cynoscion regalis					
Brevoortia tyrannus					
L. xanthurus	sodium hypochlorite	0.09 0.14 0.28	96 hr 24 hr 6 hr	50% mortality 50% mortality 50% mortality	
Oncorhynchus nerka	chlorinated sewage effluents	0.02 -0.26	24 hr	100% mortality	Servizi and Martens (1974)
O. gorbuscha (fresh water)		0.16	72 hr	100% mortality	
O. gorbuscha	residual chlorine	0.50	80 min + 10°C	50% mortality	Stober and Hanson (1974)

Species	Chemical	Concentration	Duration	Effect	Reference
O. gorbuscha		0.50	10 min + 10°C	50% mortality	
Morone americana	residual chlorine	0.08	10 min	avoidance	Meldrim et al. (1974)
Menidia menidia		0.08	10 min	avoidance	
Fundulus heteroclitus		0.03	10 min	avoidance	
Trinectes maculatus		0.03	10 min	avoidance	
Pleuronectes platessa (eggs)	residual chlorine	0.04 -0.08	8 days	none	Alderson (1972, 1974)
		0.62	72 hr	50% mortality	
		0.10	96 hr	50% mortality	
(larvae)		0.62	48 hr	50% mortality	
		0.034	96 hr	50% mortality	
Solea solea (larvae)	residual chlorine	0.03 -0.06	48 hr	50% mortality	
Cyprinus carpio	4-chlororesorcinol 5-chlorouracil (0.001 mg/l)	-	3-7 days	reduced hatch	Gehrs et al. (1974)
Morone americana	residual chlorination	0.20 -0.40	24-76 hr	decreased hatching, larval death	Morgan and Prince (1977)
Morone saxatilis					
Alosa aestivalis (embryos and larvae)					
Menidia menidia					
Menidia beryllina (embryos)		0.20 -0.40	24-48 hr	decreased hatching	

Table III, Continued.

Species	Toxicant	Reported Chlorination Level (mg/l)	Duration of Test	Effect	Reference
Morone saxatilis (larvae)	residual chlorination	0.04 -0.07 0.29 -0.82 0.21 -2.36	48 hr 90 min 48 hr	50% mortality avoidance gill damage	Middaugh et al. (1977a)
Leiostomus xanthurus (juveniles)	residual chlorination	0.06 -0.12 0.05 -0.18 1.57	8 days 90 min 95 min	50% mortality avoidance gill damage	Middaugh et al. (1977b)
Fundulus heteroclitus (embryos and larvae)	residual chlorination + thermal stress	0.01 -1.0	7.5-60 min	decreased hatching larval death	Middaugh et al. (1977c)

chlorinated. Toxicity of the unchlorinated effluents was attributed to MBAS and ammonia.

Tests at Site II indicated that chlorinated effluents retained for 30 to 60 days were not toxic to sockeye fingerlings and alevins and pink salmon alevins after 26 days of exposure.

In tests at Site III, with fingerling sockeye salmon, chlorinated sewage effluents (measured residual 0.85 mg/l) resulted in 50% mortality after 48 min. Fifty percent mortality occurred after 13 hr of exposure to the unchlorinated effluents. Sublethal exposures of fingerling sockeye salmon to the effluents from Site III (1 to 3 hr of exposure to 0.22 mg/l TRC) resulted in gill damage including hyperplasia, swollen epithelial cells, and separation of epithelium from pillar cells.

The toxicity of chlorination and heat to pink salmon, *Oncorhynchus gorbuscha*, and chinook salmon, *O. tshawytscha*, has been determined by Stober and Hanson (1974). Juveniles of each species were tested in sea water at five residual chlorination concentrations ranging from 0.05 to 1.0 mg/l and four temperatures from Δt 0 to 10°C where the LT_{50} (lethal time for 50% mortality) ranged from approximately 10 min at 0.5 mg/l residual chlorination for chinooks to 80 min for pinks.

Meldrim *et al.* (1974) and Meldrim and Fava (1977) in flowing water bioassays studied the effect of chemical pollutants on estuarine organisms. They found that white perch, *Morone americana*, consistently avoided residual chlorination levels as low as 0.08 mg/l at temperatures from 7 to 17°C. Silversides, *Menidia menidia*, also avoided 0.08 mg/l residual chlorination at temperatures from 8 to 28°C but showed a preference for 0.08 mg/l residual chlorination when fish acclimated to 7°C were exposed at 12°C. Mummichogs, *Fundulus heteroclitus*, and hog chokers, *Trinectes maculatus*, avoided residual chlorination levels as low as 0.03 mg/l.

Alderson (1972, 1974) found that the 48- and 96-hr Tl_m of residual chlorination for plaice larvae, *Pleuronectes platessa*, was 0.062 and 0.34 mg/l, respectively. Eggs were not affected when exposed to 0.075 and 0.04 mg/l chlorination for 8 days indicating that the egg membrane gives considerable protection over long periods. The 72- and 192-hr Tl_m for eggs was 0.7 and 0.12 mg/l residual chlorination, respectively. Larval Dover sole, *Solea solea*, were most sensitive to chlorine immediately after hatching. The 48-hr LC_{50} was 0.03 mg/l for stage 1c to 1d larvae and 0.06 mg/l residual chlorination for stage 4b larvae.

Gehrs et al. (1974) tested the sensitivity of carp eggs, *Cyprinus carpio*, to two of the compounds identified by Jolley (1973) 4-chlororesorcinol and 5-chlorouracil. Significant reductions in the hatchability of nonwater-hardened carp eggs were observed in concentrations of each compound as low as 0.001 mg/l.

Massive fish kills occurred on the James River, Virginia, during May and June 1973 (Virginia State Water Control Board, 1974). Species affected by the kill included spot, *Leiostomus xanthurus*; white perch, *Morone americana*; bluefish, *Pomatomus saltatrix*; grey seatrout, *Cynoscion regalis*; and menhaden, *Brevoortia tyrannus*. Most of the fish kill in the James River occurred adjacent to sewage treatment plants. Total residual chlorination levels as high as 0.7 mg/l were observed in the James River while effluents from both plants showed more than 3.0 mg/l.

Distress symptoms of dying fish included spiral swimming patterns, broken vertebral columns, listless floating, inverted swimming, distension of the air bladder, loose body scales, mucous on the skin, and hemorrhaging along fins and body surface.

Live box tests conducted adjacent to the James River Municipal Treatment Works (MTW) demonstrated a correlation between rates of effluent chlorination and mortality of juvenile spot and croaker. At an average daily chlorine feed of 1200 lb (total flow of water was approximately 10 mgd during tests) and measured residual chlorination concentration of 3.0 mg/l, 100% of caged fish died within 20 hr. When the chlorine feed rate was reduced to approximately 400 lb/day, only 20% mortality was observed after 20 hr.

On-site aquaria tests complimented results of the cage tests. Water from an area adjacent to the outfall of the James River MTW was pumped through aquaria containing juvenile spot, *Leiostomus xanthurus*. Mortalities ranged from 91 to 100% after 40 to 85 min of exposure prior to reduced chlorination. After chlorination rates were reduced, mortalities were 0 to 26% after 120 min of exposure.

Continuous flow laboratory bioassays were also conducted. The estimated 96-hr LC_{50} for juvenile spot was 0.09 mg/l residual chlorination. The estimated 24-hr LC_{50} was 0.14 mg/l and the 6-hr LC_{50} was 0.28 mg/l residual chlorination.

Separate field studies on the spot, *Leiostomus xanthurus*, revealed that up to 40% of juveniles from the 1973 year class exhibited vertebral column deformities. The abnormal "bent back" forms have become identifiable as "tags" to distinguish the 1973 year class in 1975 population samples from the James River (Labbish Chao and John V. Merriner, Virginia Institute of Marine Science, personal communication). However, this phenomenon is now complicated by the revelation of Kepone levels in the James River and the situation cannot be clearly traced to chlorination (Couch *et al.*, 1977).

Additional laboratory studies of the toxicity of residual chlorination to marine and estuarine fishes have recently been completed. Morgan and Prince (1977) exposed eggs and larvae of white perch (*Morone americana*), striped bass (*Morone saxatilis*) and blueback herring (*Alosa aestivalis*), and eggs of Atlantic silversides (*Menidia menidia*) and tidewater silversides (*Menidia beryllina*), to residual chlorination for preestablished time intervals. It was learned that most of the developing eggs had LC_{50} values between 0.20 and 0.40 mg/l residual chlorination. For larvae the LC_{50} values were from 0.20 to 0.32 mg/l residual chlorination. Age-dependent responses were observed for both eggs and larvae. The general response for an egg was decreasing sensitivity to chlorine with increased age. In larvae, the trend was increased sensitivity to chlorine with larval age.

Middaugh *et al.* (1977a) investigated the toxicity of residual chlorination to early life history stages of the striped bass, *Morone saxatilis*. Beginning 8 to 9 hr after fertilization, developing embryos were exposed continuously to residual chlorination in flowing water tests conducted at 1 to 3 ppt salinity and 18°C. Fifty-six percent of the control group hatched. None of the embryos exposed to 0.21 mg/l residual chlorination hatched. Only 3.5% of the embryos exposed to 0.07 mg/l and 23% of those exposed to 0.01 mg/l residual chlorination hatched. Incipient LC_{50} bioassays were also conducted with larval and juvenile striped bass. The estimated incipient LC_{50} was 0.04 mg/l residual chlorination for 2-day-old prolarvae, 0.07 mg/l for 12-day-old larvae and 0.04 mg/l for 30-day-old juveniles in tests at 1 to 3 ppt salinity and 18°C.

Histological examination of 30-day-old juveniles which survived exposure in the incipient LC_{50} bioassay indicated gill and pseudobranch damage for fish exposed to 0.21 to 2.36 mg/l residual chlorination.

Twenty-four-day-old striped bass larvae showed significant and reproducible avoidance responses to measured residual chlorination concentrations of 0.79-0.82 mg/l and 0.29-0.32 mg/l. No avoidance occurred at 0.16-0.18 mg/l residual chlorination.

Middaugh et al. (1977b) also learned that juvenile spot, *Leiostomus xanthurus*, are sensitive to chlorination. In flowing sea water tests (20 to 24 ppt salinity) the estimated incipient LC_{50} was 0.12 mg/l residual chlorination at 10°C and 0.06 mg/l at 15°C. Histological examination of spot used in the incipient LC_{50} bioassay at 15°C indicated pseudobranch and gill damage in individuals exposed to 1.57 mg/l residual chlorination. Spot exposed to lower concentrations, 0.02 to 0.06 mg/l showed no consistent tissue damage.

Spot demonstrated temperature-dependent avoidance responses to residual chlorination. At 10°C a concentration of 0.18 mg/l was required for significant avoidance; at 15 and 20°C, spot showed significant avoidance of residual chlorination concentrations as low as 0.05 mg/l.

In tests to determine the effects of simultaneous thermal stress and residual chlorination on specific developmental stages of the mummichog, *Fundulus heteroclitus*, Middaugh et al. (1977c) learned that the larvae were more sensitive than developing embryos. Replicate groups of 30 individuals in four embryonic stages (1- to 2-cell, gastrula, onset of circulation and 10-day embryos) and two larval stages (1-day and 7-day old) were exposed to combinations of thermal stress and residual chlorination in a matrix design. Test temperatures were 24°C (the acclimation temperature), 29 and 34°C. Four exposure levels of residual chlorination; 0.00 (control), 0.05, 0.1, 0.5 and 1.0 mg/l were used at each temperature. Exposures were for 7.5, 15, 30 and 60 min at each temperature-residual chlorination and developmental stage.

For the embryonic stages the total number of successfully hatched larvae was used as the criterion to measure effect. For the larval stages, survival 24 hr after exposure was used.

Data were analyzed by a four-way analysis of variance (ANOVA). The Student-Neuman-Keuls' (SNK) procedure determined where differences in sample means occurred. In the embryonic stages, temperature was the most important main variable. Only one embryonic stage (gastrula) was confounded by second order interactions (temperature × duration of exposure × residual chlorination). Both of the larval stages showed significant higher order interactions for all combinations of test parameters, thus suggesting the presence of synergistic effects of the three main experimental variables.

In California, Young (1964) observed tumor-like sores around the mouths of white croaker, *Genyonemus lineatus*, collected near the Hyperion sewage outfall in Santa Monica Bay. Although no direct evidence was discovered to link occurrence of lesions with chlorinated sewage effluents, there was a general decline in fitness of croakers and other species found in close proximity to the outfall area.

Relevant to the decline of fitness of marine fishes exposed to chlorinated effluents is the report by Grothe and Eaton (1975) that fresh water fishes suffered anoxia due to chlorine oxidation of hemoglobin to methemoglobin. Fish exposed to fresh water containing chloramines exhibited striking elevations in blood concentrations of methemoglobin, up to 29 to 32%. This results in decreased tissue oxygen delivery of 30% in the test conditions.

IMPLICATIONS OF ECOSYSTEM IMPACT OF CHLORINATION IN MARINE WATERS

The foregoing information summarizes uncritically the available literature on effects of chlorination upon estuarine and marine organisms. It is appropriate to focus on the few available implications of published and unpublished research on effects of chlorine upon marine organisms.

As we have seen during this symposium, chlorine reacts rapidly, and in yet to be defined ways, in marine ecosystems. Differences in natural systems, including the presence of organic and inorganic compounds, interactions involving photolysis, specific reactions favored by specific temperatures, and suspended particles, all combine to give us a highly dynamic set of events. Couple these physical-chemical parameters with the uncertainty of organism responses and bioassays and we have a truly difficult task to produce relevant or meaningful standards.

Furthermore, much of the current available literature has accepted the amperometric titrator as actual representation of true residual chlorination concentrations. We know now from the discussions of marine chemistry during this symposium that these readings actually represent abstractions of chemical processes yet to be described in the case of marine waters. Therefore, without clear definitions of the physical-chemical parameters much of the reported data reviewed in this paper cannot be meaningfully compared.

Although this paper and others consistently refer to impacts of chlorination of marine waters, we must realize that many, if not the majority, of the effects studied are resulting from bromination processes. Typically sea water at 30-33 ppt salinity contains approximately 60 ppm of bromide salts. This bromide is oxidized to bromine during chlorination or by addition of other oxidants including ozone. Therefore, environmentally the results of chlorination would be analogous to a cascade of chemical transformations, favoring bromination in increasing salinities in an estuary (Dove, 1970; Carpenter and Macalady, this conference and 1977).

The by-products of chlorination addressed by Jolley (1973) impact estuarine environments via fresh water effluents. Few investigators have addressed the long-range aspects of chlorination, that is the holistic effects of chlorination although it is obviously of considerable concern to government, health and environmental officials (Anonymous, 1977). In one such experiment, Davis *et al.* (1977) and Domey *et al.* (in press), communities of benthic algae and invertebrates were clearly impacted by continuous low chlorination rates (0.125-0.50 mg/l nominal NaOCl). Measurements of ATP (Erickson and Foulks, in press; Gentile, 1976) in marine waters also were affected by chlorination.

Finally, very little is known of the fate of compounds induced by chlorination processes. Whether by-products are degraded, partitioned, trophically transferred or biologically accumulated has not yet been demonstrated. Clearly these are important issues, especially with regard to ocean outfalls, and criteria for water quality management. Projects which initiate the research for these needed data are summarized in Davis et al. (1977). The results of several of these investigations are found in these symposia.

REFERENCES

Alderson, R. 1972. "Effects of Low Concentrations of Free Chlorine on Eggs and Larvae of Plaice, *Pleuronectes platessa* L.," p. 312-315. In *Marine Pollution and Sea Life.* Fishing News, Ltd. London.

Alderson, R. 1974. "Seawater Chlorination and the Survival and Growth of the Early Developmental Stages of Plaice, *Pleuronectes platessa* L., and Dover sole, *Solea solea* (L.)," Aquaculture 4:41-53.

Anonymous. 1977. *Unnecessary and Harmful Levels of Domestic Sewage Chlorination Should be Stopped.* Comptroller General-Report to Congress. CED-77-108.

*Basch, R. E. and J. G. Truchan. 1973. *Calculated Residual Chlorine Concentrations Safe for Fish.* Interim report. Michigan Water Resources Commission, Bureau of Water Management. Lansing, Michigan.

*Block, R. M., G. R. Helz and W. P. Davis. 1977. "Fate and Effects of Chlorine in Marine Waters," *Chesapeake Sci.* 18(1):92-101.

*Bongers, L. H., T. P. O'Connor and D. T. Burton. 1977. *Bromine Chloride—an Alternate to Chlorine for Fouling Control in Condenser Cooling Systems.* EPA 600/7-77-053. U.S. Environmental Protection Agency. 60 pp. + appendix.

Brungs, W. A. 1973. "Effects of Residual Chlorine on Aquatic Life," *J. Water Pollut. Control Fed.* 45(10):2180-2192.

Brungs, W. A. 1976. *Effects of Wastewater and Cooling Water Chlorination on Aquatic Life.* EPA-600/3-76-098. U.S. Environmental Protection Agency. 52 pp.

Capuzzo, J. M., S. A. Lawrence and J. A. Davidson. 1976. "Combined Toxicity of Free Chlorine, Chloramine and Temperature to Stage I Larvae of the American Lobster, *Homarus americanus.*" *Water Research* 10:1093-1099.

Carpenter, J. H. and D. Macalady. 1977. "Problems in Measuring Residuals in Chlorinated Water," *Chesapeake Sci.* 18(1):112.

Carpenter, J. H. and C. A. Smith. In press. "Reactions in Chlorinated Sea Water." In R. L. Jolley, D. H. Hamilton and H. Gorchev (Eds.) *Water Chlorination: Environmental Impact and Health Effects* Vol. 2. Ann Arbor Science Publishers, Inc., Ann Arbor, Michigan.

Carpenter, E. J., B. B. Peck and S. J. Anderson. 1972. "Cooling Water Chlorination and Productivity of Entrained Phytoplankton," *Mar. Biol.* 16:37-40.

---

*Literature not cited.

Couch, J. A., J. T. Winstead and L. R. Goodman. 1977. "Kepone-Induced Scoliosis and its Histological Consequences in Fish," *Science* 197(4303):585-587.

*Davis, W. P. and D. P. Middaugh. 1977. "Impact of Chlorination Processes on Marine Ecosystems," *Proc. Conf. Estuarine Pollution Control.* EPA Off. Water Planning and Standards. 755 pp. + microfiche.

Davis, W. P., D. P. Middaugh, J. H. Carpenter, G. R. Helz and M. H. Roberts, Jr. 1977. "The Chemistry and Ecological Effects of Chlorination of Seawater—Summary of EPA Research Projects," *Program Review Proceedings of: Environmental Effects of Energy Related Activities on Marine/Estuarine Ecosystems.* EPA 600/7-77-111. U.S. Environmental Protection Agency. 135 pp.

Davis, W. P., B. S. Hester, R. L. Yoakum and R. G. Domey. 1977. "Marine Ecosystem Testing Units: Design for Assessment of Benthic Organism Responses to Low-Level Pollutants," *Helgoländer Wiss. Meersunters* 30:673-681.

*Doudoroff, P. and M. Katz. 1950. "Critical Review of Literature on the Toxicity of Industrial Wastes and Their Components to Fish," *Sewage Ind. Wastes* 22(11):1432-1458.

Dove, R. A. 1970. *Reaction of Small Dosages of Chlorine in Seawater.* Central Electricity Generating Board. Central Electric Research Laboratories. Surrey, G. B. (Service Research Rept. 42/70-microfiche) 189 pp.

*Fair, G. M., J. C. Morris, S. L. Chang, I. Weil and R. P. Burden. 1948. "Chlorine as a Water Disinfectant," *J. Am. Water Works Assoc.* 40: 1051-1061.

Gehrs, C. W., L. D. Eyman, R. L. Jolley and J. E. Thompson. 1974. "Effects of Stable Chlorine-Containing Organics on Aquatic Environments," *Nature* 249:675-676.

Gentile, J. H., S. Cheer and N. Lackie. 1973. "The Use of ATP in the Evaluation of Entrainment." Unpublished data. Environmental Research Laboratory, U.S. Environmental Protection Agency. Narrangansett, Rhode Island.

Gentile, J. H., J. Cardin, M. Johnson and S. Sosnowski. 1976. *Power Plants, Chlorine and Estuaries.* EPA-600/3-76-055. U.S. Environmental Agency.

Gibson, Charles I., Thomas O. Thatcher, Charles W. Apts. 1975. "Some Effects of Temperature, Chlorine and Copper on the Survival and Growth of the Coon Stripe Shrimp, *Pandalus danae*," *Proc. Thermal Ecol. Symp.* Augusta, Georgia. April 1975. In press.

Grothe, D. R. and J. W. Eaton. 1975. "Chlorine-Induced Mortality in Fish," *Trans. Am. Fish.* 104(4):800-802.

*Literature not cited.

Hamilton, D. H., D. A. Flemer, C. V. Keefe and J. A. Mihursky. 1970. "Power Plants: Effects of Chlorination on Estuarine Primary Productivity," *Science* 169:197-198.

Helz, G. R., R. Sugam and R. Y. Hsu. In press. "Chlorine Degradation and Volatile Halocarbon Generation in Estuarine Waters." In R. L. Jolley, D. H. Hamilton and H. Gorchev. (Eds.) *Water Chlorination: Environmental Impact and Health Effects* Vol. 2. Ann Arbor Science Publishers, Inc., Ann Arbor, Michigan.

Hirayama, K. and R. Hirano. 1970. "Influence of High Temperature and Residual Chlorine on Marine Phytoplankton," *Mar. Biol.* 7:205-213.

*Holland, G. A., J. E. Lasater, E. D. Neumann and W. E. Eldridge. 1964. *Toxic Effects of Organic and Inorganic Pollutants on Young Salmon and Trout.* Dep. Fish. Res. Bull. No. 5. State of Washington. 264 pp.

*Houghton, G. U. 1946. "The Bromide Content of Underground Waters. Part II: Observations on the Chlorination of Water Containing Free Ammonia and Naturally Occurring Bromide," *J. Soc. Chem. Ind.* 65: 324-328.

*Ingols, R. S., H. A. Wyckoff, T. W. Kethley, H. W. Hodgden, E. L. Fincher, J. C. Hildebrand and J. E. Mandel. 1953. "Bactericidal Studies of Chlorine," *Ind. Eng. Chem.* 45:995-1000.

James, W. G. 1967. "Mussel Fouling and Use of Exomotive Chlorination," *Chem. Ind. (Lond.)* 24:994-996.

*Johanneson, J. K. 1958. "The Determination of Monobromamine and Monochloramine in Water," *Analyst* 83:155-159.

*Johanneson, J. K. 1960. "Bromination of Swimming Pools," *Am. J. Public Health* 50:1731.

Jolley, R. L. 1973. *Chlorination Effects on Organic Constituents in Effluents from Domestic Sanitary Sewage Treatment Plants.* Ph.D. Dissertation. University of Tennessee. 339 pp.

Jolley, R. L., C. W. Gehrs and W. W. Pitt. "Chlorination of Cooling Water: A Source of Chlorine-Containing Organic Compounds with Possible Environmental Significance," p. 21-28. In C. E. Cushing (Ed.) *Radioecology and Energy Resources.* Dowden, Hutchinson and Ross, Inc. Stroudsburg, Pennsylvania.

Laubusch, E. J. 1971. "Chlorination and Other Disinfection Processes," In *Water Quality and Treatment.* Am. Water Works Assoc. 654 pp.

*Lewis, B. G. 1966. *Chlorination and Muscle Control. I. The Chemistry of Chlorinated Seawater. A Review of the Literature.* Lab. Note No. RD/L/N/106/66. Central Electric Research Laboratory.

Macalady, D. L., J. H. Carpenter and C. A. Moore. 1977. "Sunlight-Induced Bromate Formation in Chlorinated Seawater," *Science* 195: 1335-1337.

---

*Literature not cited.

Markowski, S. 1959. "The Cooling Water of Power Stations: A New Factor in the Environment of Marine and Freshwater Invertebrates," *J. Animal Ecol.* 28:243-258.

Markowski, S. 1960. "Observations on the Response of Some Benthonic Organisms to Power Station Cooling Water," *J. Animal Ecol.* 29:349-357.

*McKee, J. E. and H. W. Wolf. 1963. *Water Quality Criteria.* 2nd Ed. Publication 3A. California State Water Quality Control Board. Sacramento. 548 p.

McLean, R. I. 1972. "Chlorine Tolerance of the Colonial Hydroid, *Bimeria franciscana*," *Chesapeake Sci.* 13:229-230.

McLean, R. I. 1973. "Chlorine and Temperature Stress in Estuarine Invertebrates," *J. Water Pollut. Control Fed.* 45:837-841.

Meldrim, J. W., J. J. Gift and B. R. Petrosky. 1974. "The Effect of Temperature and Chemical Pollutants on the Behavior of Several Estuarine Organisms," Ichthyological Associates, Inc., Bulletin 11:1-129.

Meldrim, J. W. and J. A. Fova, Jr. 1977. "Behavior Avoidance Responses of Estuarine Fishes to Chlorine," *Chesapeake Sci.* 18(1):154-157.

*Merkens, J. C. 1958. "Studies on the Toxicity of Chlorine and Chloramines to the Rainbow Trout," *Water Waste Treat. J.* 7:150-151.

Middaugh, D. P., J. A. Couch and A. M. Crane. 1977a. "Responses of Early Life History Stages of the Striped Bass, *Morone saxatilis*, to Chlorination," *Chesapeake Sci.* 18:141-153.

Middaugh, D. P., A. M. Crane and J. A. Couch. 1977b. "Toxicity of Chlorine to Juvenile Spot, *Leiostomus xanthurus*," *Water Res.* In press.

Middaugh, D. P., J. M. Dean, R. G. Domey and G. Floyd. 1977c. "Effect of Thermal Stress and Chlorination on Early Life Stages of the Mummichog, *Fundulus heteroclitus*," *Marine Biol.* In press.

Moore, E. W. 1951. "Fundamentals of Chlorination of Sewage and Wastes," *Water Sewage Works* 98:130-136.

Morgan, R. P. and R. G. Stross. 1969. "Destruction of Phytoplankton in the Cooling Water Supply of a Stream Electric Station," *Chesapeake Sci.* 10:165-171.

Morgan, R. P. and R. D. Prince. 1977. "Chlorine Toxicity to Eggs and Larvae of Five Chesapeake Bay Fishes," *Trans. Am. Fish. Soc.* 106(4): 380-385.

Muchmore, D. and D. Epel. 1973. "The Effects of Chlorination of Wastewater on Fertilization in Some Marine Invertebrates," *Mar. Biol.* 19:93-95.

*Rosenberger, D. R. 1971. *The Calculation of Acute Toxicity of Free Chlorine and Chloramine to Coho Salmon by Multiple Regression Analysis.* Thesis. Michigan State University. East Lansing, Michigan.

*Literature not cited.

*Russell, P. P. and A. J. Horne. 1977. *The Relationship of Wastewater Chlorination Activity to Dungeness Crab Landings in the San Francisco Bay Area.* Univ. Calif. School of Public Health. Berkeley UCB/SERL Report 77-1. 37 pp.

*Sawyers, C. N. and P. L. McCarty. 1969. *Chemistry for Sanitary Engineers.* 2nd Ed. McGraw-Hill. New York.

Serviszi, J. A. and D. W. Martens. 1974. *Preliminary Survey of Toxicity of Chlorinated Sewage to Sockeye and Pink Salmon.* Int. Pac. Salmon Fish. Comm. Prog. Rep. 30:1-42.

Stober, Q. J. and C. H. Hanson. 1974. "Toxicity of Chlorine and Heat to Pink, *Oncorhychus gorbuscha,* and Chinook Salmon, *O. tshawytscha,*" *Trans. Am. Fish. Soc.* 103(3):569-576.

Sugam, R. and G. R. Helz. 1975. "Apparent Ionization Constant of Hypochlorous Acid in Seawater," *Environ. Sci. Technol.* 10:384-386.

Tsai, C. 1968. "Effects of Chlorinated Sewage Effluents on Fishes in Upper Patuxent River, Maryland," *Chesapeake Sci.* 9(2):83-93.

Tsai, C. 1970. "Changes in Fish Populations and Migrations in Relation to Increased Sewage Pollution in Little Patuxent River, Maryland," *Chesapeake Sci.* 11(1):34-41.

Tsai, C. 1973. "Water Quality and Fish Life Below Sewage Outfalls," *Trans. Am. Fish. Soc.* 102(2):281-292.

Tsai, C. 1975. *Effects of Sewage Treatment Plant Effluents on Fish: A Review of the Literature.* Chesapeake Research Consortium Publication 36:1-229.

Turner, H. J., D. M. Reynolds and A. C. Redfield. 1948. "Chlorine and Sodium Pentachlorophenate as Fouling Preventatives in Seawater Conduits," *Ind. Eng. Chem.* 40:450-453.

Virginia State Water Control Board. 1974. *James River Fish Kill.* Report 73-025. Bureau of Surveillance and Field Studies, Division of Ecological Studies. 61 pp.

Waugh, G. D. 1964. "Observations on the Effects of Chlorine on the Larvae of Oysters, *Ostrea edulis* L., and Barnacles, *Eliminius modestus* Darwin," *Ann. Appl. Biol.* 54:423-440.

*White, G. C. 1972. *Handbook of Chlorination.* Van Nostrand Rheinhold Co. New York. 744 pp.

*White, G. C. 1973. "Disinfection Practices in the San Francisco Bay Area," *J. Water Pollut. Control Fed.* 46:89-101.

*Whitehouse, J. W. 1975. *Chlorination of Cooling Water: A Review of Literature on the Effects of Chlorine on Aquatic Organisms.* Job VJ 440-RD/L/M496;1-22. Central Electricity Research Laboratory.

Young, P. H. 1964. "Some Effects of Sewage Effluents on Marine Life," *California Fish and Game* 50(1):33-41.

*Literature not cited.

DISCUSSION

James H. Carpenter, University of Miami. Are there any studies of behavioral response of marine fish?

Davis. Yes. The ones we are operating and the ones John Meldrim is operating are simplified types of avoidance responses in juvenile fish. In both cases the fish were successful and the spot were successful in avoiding lethal concentrations, that is, based on the LC_{50}'s at $10°C$ and above. But in the case of the spot, they were unsuccessful in avoiding lethal concentrations below $10°C$. There are, of course, the behavioral papers referred to in the previous presentation.

Arthur Brooks, University of Wisconsin-Milwaukee. In Bill Brungs' recent review article there are a number of estuarine studies made in the last ten months referring to chlorine and invertebrates.

Davis. The questions and statements refer basically to the review of Brungs and, I might add another review by Whitehouse in 1975, which both deal with estuarine invertebrates. I have avoided discussing these in great detail in the oral presentation. We did summarize them in the version that will be printed. We avoided this from the simple point of view that we can't tell from the studies in most cases just what technique they are using for residual chlorine measurements and, therefore, I don't find them comparable.

Donald J. Modell, Electrode Corporation. Due to papers discussing the measurement of HOCl and total chlorine residual by Dr. Johnson, how are you planning to separate the various components of total chlorine residual via measurement methods to ensure that your results on HOCl and chlorinated organics can be directly correlated to marine ecosystem effects? And what measurement methods can or are you planning to use to obtain meaningful results?

Davis. I would like to challenge that meaningful characteristic, right there.

Modell. Meaningful to Dr. Johnson.

Davis. Meaningful to Dr. Johnson may not necessarily be meaningful to the environment. What we are looking for are ecological effects. So in essence we are looking for summation effects. Now, just in a very general way, if you are going to run LC_{50} tests side by side you need something to measure for comparison. We do our calculations, our dilutions, and use an amperometric titrator to give us a number. Now if we were to compare at a given moment the reactions of spot, a small juvenile fish, to the reactions of a typical estuarine decapod *Paleomynetes*, we would get vastly different responses. Now at this point we ask ourselves your question, "What is meaningful?" We have a number to index ourselves to and we have references in terms of LC_{50} of what the organism response is. In general terms the response of the fish is much more sensitive than for this particular vertebrate. Therefore, we have to ask the question, "Why?" We can make a suggestion—epithelium. Maybe this is reinforced by the observation made from the swimming pool study where an algae with a siliceous shell and other particular adaptive mechanisms became relatively immune to chlorine effects compared to a fish which has a weak point at the gill surface. Now, did I approach your question or did I avoid it?

Modell. As I understand it you will continue to use present measurement methods.

Davis. We will use standard methods because what else do we have available. In the meantime, we are asking ourselves, "What else is going to become available?"

T. O. Thatcher, Batelle-Northwest Marine Research Laboratory. There is going to be quite a bit of information appearing in the literature concerning organisms exposed to chlorine in salt water. In reporting LC_{50} values, what physical and chemical parameters of the test water should also be reported with the total residual oxidant data to make the information more useful?

Davis. Well, I can't really comment too much on the bromine levels right now until we have the analytical technique. Other things? Of course it is standard to say temperature. In marine water I think you will have to define very carefully the site from which you get your data. If you wanted to transfer the data we are getting at Bears Bluff to the northwest —I would be anxious about that. For example, we have a very high sediment load in our estuaries. One part of our strategy for next year is to actually work more with the influence of entrained particles and the effects on what is called chlorine demand. Other factors are sunlight, oxygen and contained organics.

William A. Brungs, U.S. Environmental Protection Agency. I think it's a good question because I'm not a chemist, but if there is any possibility, additional information should be included with the published data on LC_{50}. Maybe a few years from now somebody can go back and make some back calculations and perhaps use those data. Whereas right now it would be very difficult because most people do not explain the test conditions adequately to make that transposition.

George C. White, Consulting Engineer. Because the residual chemistry in sea water is so complex you do not know what is reacting. Maybe a model could be developed with part of the model based on the chlorine consumption of the test water. You can measure dosage. That is, you know what you are putting in and you can measure the chlorine consumption. Thus you could make a model for each site and work back from that. But I would not want to trust the residuals.

Davis. No. I think the residual measurements have too much weight put on them.

Robert J. Huggett, Virginia Institute of Marine Science. Bioassay results are dependent on many things but one which should be noted for most anadromous fish in the Chesapeake Bay is that they appear to be more sensitive to chlorine derivatives in the spring soon after returning from the ocean.

Also, one point that must not be forgotten when we work on chlorine in the environment is that even though we may not know what we are measuring when we analyze for chlorine-chloramines, we do get a meter reading. This meter reading seems to be proportional to the chlorine-related toxicity of the water and can be used to predict acute biological effect after bioassays have been performed. It is essential, therefore, that we not lose sight of the fact that some management decisions can be made based on "meter readings" and that what is killing the animals (bromine, chlorine, hypochlorous acid, monochloramine or so forth) is proportional to the chlorine dose.

Davis. I think another point that you have just illustrated is the need for going ahead with field management and not waiting for bioassays.

. . .

According to T. O. Thatcher, the following points were made during the intermission which are appropriate to include in the context of the discussion because of their relevance to data which should be included in reports:

George R. Helz, University of Maryland, suggested that the pH value of the test water, the phenol content and total proteinaceous nitrogen should be included.

Edward G. Wolf, Batelle-Pacific Northwest Laboratories, and Ronald M. Block, University of Maryland, suggested that total organic carbon, iron, manganese, turbidity, total nitrogen, total organic nitrogen, nitrate, nitrite, ammonia and salinity should be included.

16

CHLORINATED COMPOUNDS FOUND IN WASTE TREATMENT EFFLUENTS AND THEIR CAPACITY TO BIOACCUMULATE

Herbert L. Kopperman

> Department of Chemistry
> University of Minnesota-Duluth
> Duluth, Minnesota 55812

Douglas W. Kuehl and Gary E. Glass

> Environmental Research Laboratory-Duluth
> U.S. Environmental Protection Agency
> Duluth, Minnesota 55804

ABSTRACT

As part of an ongoing research program to assess possible long-term environmental effects due to the formation of stable reaction products during disinfection processes, fish (fathead minnows, *Pimephales promelas*) and water from the 9-month chronic toxicity tests at two wastewater treatment plants in Michigan are being analyzed for chemical residues at this laboratory (formerly named the National Water Quality Laboratory).

Gel permeation chromatography was used for sample cleanup and gas chromatography/mass spectrometry was used for sample analysis. Di- and trichlorophenols, di- and trichlorobenzenes and trichloroanisoles were either not detected or detected at lower levels in the fish from nondisinfected effluent exposures compared to fish exposed to chlorinated effluent. Tetra- and pentachlorophenols, PCB's, DDT's, toxaphene components, chlordane and nonachlor were found in all fish raised in the sewage effluent. Tribromoanisole was tentatively identified in fish that lived in BrCl-treated wastewater.

Reports appear to be conclusive in support of the argument that even the most gentle chlorination conditions will cause chlorine to be incorporated into organic molecules. The incorporation of chlorine into an organic molecule increases its lipophilic character and at the same time normally causes an increase in the observed toxicity or bioaccumulation, or both.

The persistence of these compounds is now becoming a concern. Not all organochlorine compounds bioaccumulate to high levels. The data suggest that polar compounds are more easily biodegraded and that nonpolar (highly lipophilic) compounds accumulate. Some investigators have been able to demonstrate positive correlation between the n-octanol/water partition coefficients for given compounds and their ability to bioaccumulate in various species of fish.

CHLORINATION

Chlorination has been used extensively for disinfection of waste effluents by both industry and municipalities. In addition, it was assumed that chlorine was oxidizing the organic compounds to innocuous substances (CO_2, H_2O and easily degradable organics) and thereby lowering the BOD of the effluent. It has now been shown that although chlorine does oxidize many compounds, it also creates many organochlorine compounds which are more toxic. Carlson (1975) has shown that chlorinated phenols lower the BOD by destroying the bacteria. Carlson et al. (1975) have also shown that phenol incorporates chlorine very readily over a wide range of pH, while under conditions similar to disinfection (Table I).

Chlorination of water supplies has received considerable attention (Jolley, 1973, 1975; Jolley et al., 1975; Glaze and Henderson, 1975; Glaze et al., 1973; Bellar et al., 1974) in the last few years (Tables II, III and IV). People are becoming concerned not only with the production of toxic materials but also with the production of possible carcinogens.

Table I. Percentage of Available Chlorine Incorporated by Various Compounds Treated with 7 x 10^{-4} M Aqueous Chlorine at Three pH Levels (3, 7 and 10) for 20 min, at 25C[a]

Compound[b]	pH		
	3	7	10
Phenol	97.8	97.6	97.6
Anisole	80.7	11.4	2.8
Acetanilide	55.3	3.4	-
Toluene	11.1	2.9	-
Benzyl alcohol	2.3	-	-
Benzonitrile	2.1	-	-
Nitrobenzene	1.8	-	-
Chlorobenzene	1.8	-	-
Methylbenzoate	1.8	-	-
Benzene	1.5	-	-

[a] R. M. Carlson et al., 1975.
[b] Aqueous concentration = 9.5 ± 0.6 x 10^{-4} M.

Table II. Tentative Identification and Concentration of Chlorine-Containing Constituents in Chlorinated Effluents (Oak Ridge, Tennessee) by Liquid Chromatographic Techniques[a]

Organic Compound	Concentration (μg/l)
5-chlorouracil	4.30
5-chlorouridine	1.70
8-chlorocaffeine	1.70
6-chloroguanine	0.90
8-chloroxanthine	1.50
2-chlorobenzoic acid	0.26
5-chlorosalicylic acid	0.24
4-chloromandelic acid	1.10
2-chlorophenol	1.70
4-chlorophenylacetic acid	0.38
4-chlorobenzoic acid	1.10
4-chlorophenol	0.69
3-chlorobenzoic acid	0.62
3-chlorophenol	0.51
4-chlororesorcinol	1.20
3-chloro-4-hydroxybenzoic acid	1.30
4-chloro-3-methylphenol	1.50

[a] Jolley, 1975.

Table III. Chlorinated Organics in Wastewater Effluent (Denton, Texas) Identified by Gas Chromatography/Mass Spectrometry Methods (Glaze and Henderson, 1975)

Compound[a]	Concentration[b] (µg/l)	Compound[a]	Concentration[b] (µg/l)
Chloroform[c]	-	Chloro-α-methyl benzyl alcohol[e]	-
Dibromochloromethane[c]	-	Dichloromethoxytoluene[e]	32
Dichlorobutane (-)[d]	27	Trichloromethylstyrene (220)[d]	10
3-chloro-2-methylbut-1-ene[c]	285	Trichloroethyl benzene (208)[d]	12
Chlorocyclohexane (118)[d]	20	Dichloro-α-methyl benzyl alcohol (190)[d]	10
Chloroalkyl acetate (-)[d]	-	Dichloro-bis (ethoxy) benzene (220)[d]	30
o-dichlorobenzene[e]	10	Dichloro-α-methyl benzyl alcohol (190)[d]	-
Tetrachloroacetone[e]	11	Trichloro-α-methyl benzyl alcohol[e]	25
p-dichlorobenzene[c]	10	Tetrachlorophenol[c]	30
Chloroethylbenzene[e]	21	Trichloro-α-methyl benzyl alcohol[e]	50
Pentachloroacetone[c]	30	Trichlorocumene (222)[d]	-
Hexachloroacetone[c]	30	Tetrachloroethylstyrene (268)[d]	-
Trichlorobenzene[c]	-	Trichlorodimethoxybenzene (240)[d]	-
Dichloroethylbenzene[c]	20	Tetrachlorodimethoxytoluene (258)[d]	40
Chlorocumene (154)[d]	-	Dichloroaniline derivative (205)[f]	13
N-methyl-trichloroaniline (209)[d]	10	Dichloroaromatic derivative (249)[f]	15
Dichlorotoluene[e]	-	Dichloroacetate derivative (203)[f]	20
Trichlorophenol[e]	-	Trichlorophthalate derivative (296)[f]	-
		Tetrachlorophthalate derivative (340)[f]	-

[a] Compounds may be listed more than once if gas chromatographic retention times indicate distinct positional isomers.
[b] Quantitative values should only be considered as estimates because explicit recovery data are not available for the authors' extraction system.
[c] Completed identification based on mass spectrometric interpretation and confirmed by comparison with a reference spectrum.
[d] Fragmentation pattern tentatively suggests proposed compound; probable molecular weight indicated in parentheses.
[e] Probable identification based on mass spectral interpretation.
[f] Mass spectral information too incomplete to propose a structure; probable molecular weight indicated in parentheses.

Table IV. Concentration of Organochlorine Compounds (μg/l) in Water from Sewage Treatment Plants (Several Cities) Measured by Gas Chromatography/Mass Spectrometry Techniques[a]

Compound	Influent Before Treatment	Effluent Before Chlorination	Effluent After Chlorination
Methylene chloride	8.2	2.9	3.4
Chloroform	9.3	7.1	12.1
1,1,1-trichloroethane	16.5	9.0	8.5
1,1,2-trichloroethylene	40.4	8.6	9.8
1,1,2,2-tetrachloroethylene	6.2	3.9	4.2
Σ Dichlorobenzenes	10.6	5.6	6.3
Σ Trichlorobenzenes	66.9	56.7	56.9

[a]T. A. Bellar et al., 1974.

EFFECTS OF DISINFECTION ON CHEMISTRY OF ORGANIC MOLECULES

To answer some of the questions being raised, a program was initiated in Grandville and Wyoming, Michigan, to observe the effect of various disinfection processes upon the chemistry of organic molecules in municipal waste effluent. The four treatments examined were (a) chlorination, (b) chlorination followed by dechlorination with SO_2 treatment, (c) ozonation and (d) chlorobromination (Figure 1).

One of the main objectives of this study was to investigate the chlorinated effluent for newly formed chloro compounds. It has been shown that phenols will incorporate chlorine under the conditions normally encountered in water treatment plants (Carlson et al., 1975). The number of chlorines incorporated depends on the reactivity of the particular phenol toward electrophilic attack. Phenol itself was not identified in either the water or the fish; however, dichloro- through pentachlorophenols were identified. An attempt was made to identify as many organics as possible by using the GC/MS (gas chromatography/mass spectrometry) system. The usual chlorinated pesticides and PCBs were identified in the fish as well as many aromatic hydrocarbons (Table V). Di- and trichlorobenzene were only observed in the fish reared in the chlorinated effluents (Figure 2).

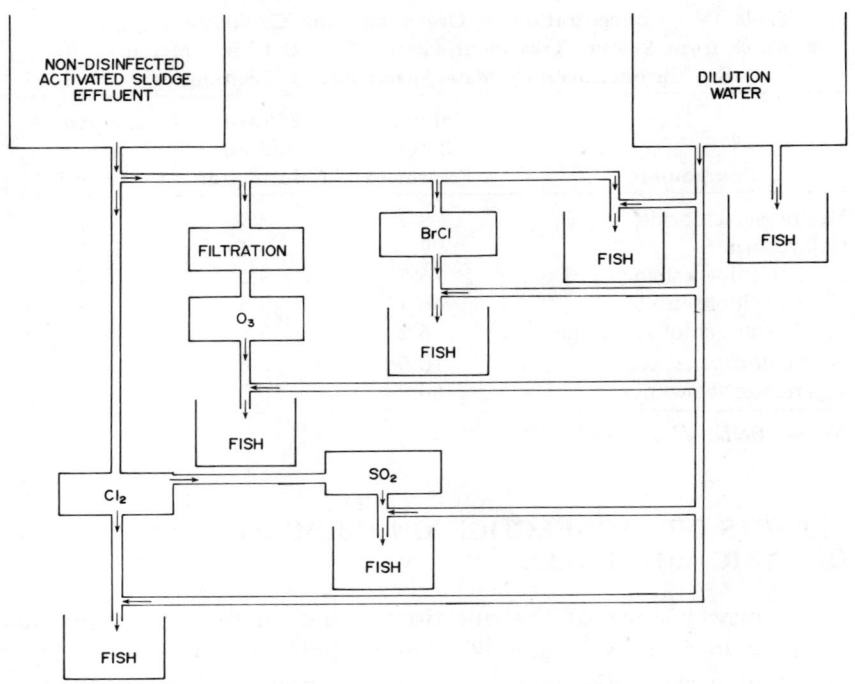

Figure 1. Schematic diagram indicating the experimental disinfection processes used in the chronic bioassay tests at the Environmental Protection Agency's Grandville, Michigan, project.

An unreported compound type, anisole, was identified in both the water and the fish. Analysis of the effluent indicated that the parent phenols were present in larger quantities than the corresponding anisoles, and the fish anlysis showed that the anisoles had become the more abundant of the two. Figure 3 shows a representative sample of anisole data illustrated by one of these compounds, pentachloroanisole. Experiments have not been carried out to determine whether or not activated sludge sewage treatment or fish possess the ability to methylate phenols; therefore, it is difficult to speculate on the bioconcentration factor of anisoles. Many of the PCBs and chlorinated pesticides were difficult to analyze using 36 liters of effluent; however, they were concentrated to such an extent that the analysis of 25 g of fish was not difficult.

Table V. Chlorinated Organic Compounds Identified by
Gas Chromatography/Mass Spectrometry in Fish Raised in
Disinfected Municipal Effluent (Grandville, Michigan)

Compound	Identification[a]
Dichlorobenzene	A
Trichlorobenzene	A
Trichlorophenol	B
Trichloroanisole	B
Tetrachlorophenol	B
Tetrachloroanisole (2)	A/B
Pentachlorophenol	B
Pentachloroanisole	B
Trichloropropylmethoxynaphthalene	C
Tetrachloropropylmethoxynaphthalene	C
cis-chlordane	B
trans-chlordane	B
cis-nonachlor	B
trans-nonachlor	B
p,p'-DDE	B
p,p'-DDT	B
Tribromoanisole	A
Dichlorobiphenyl (3)	A
Trichlorobiphenyl (5)	A
Tetrachlorobiphenyl (6)	A
Pentachlorobiphenyl (5)	A

[a] A, identification based on MS; B, identification supported with retention time; C, fragmentation pattern suggests a compound containing groups indicated.

BIOACCUMULATION

It is apparent that some compounds have a greater tendency to accumulate in the tissue than others. This phenomenon has been investigated by Macek *et al.* (1975) and Neely *et al.* (1974). They have shown that bioaccumulation is directly related to the partition coefficient (P) of the compound.

The use of the *n*-octanol/water partitioning system in correlating observed responses to compounds is not new. In 1899 Overton and Meyer correlated the solubility characteristics (*n*-octanol/water partition coefficient, P) of anesthetics to their

DICHLOROBENZENE

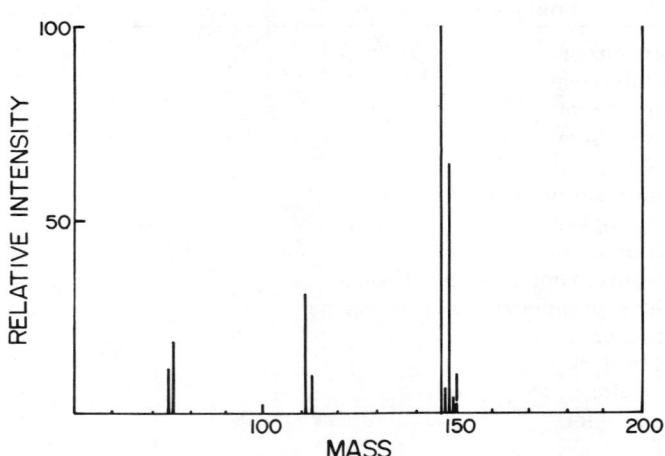

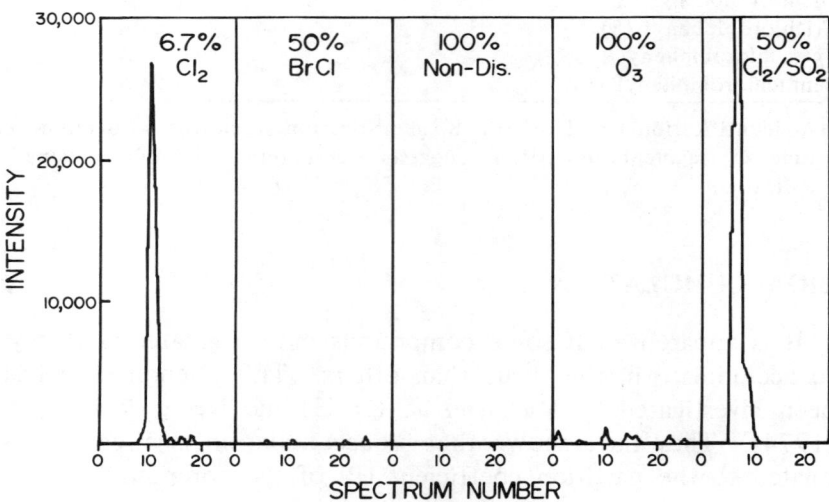

Figure 2. Mass spectrum and reconstructed mass chromatograms (M^+ 146) of dichlorobenzene found in fish raised in chlorinated sewage effluent. The disinfectant and percent effluent concentration are given in the lower figure. The spectrum number is a guide to the gas chromatographic retention time.

PENTACHLOROANISOLE

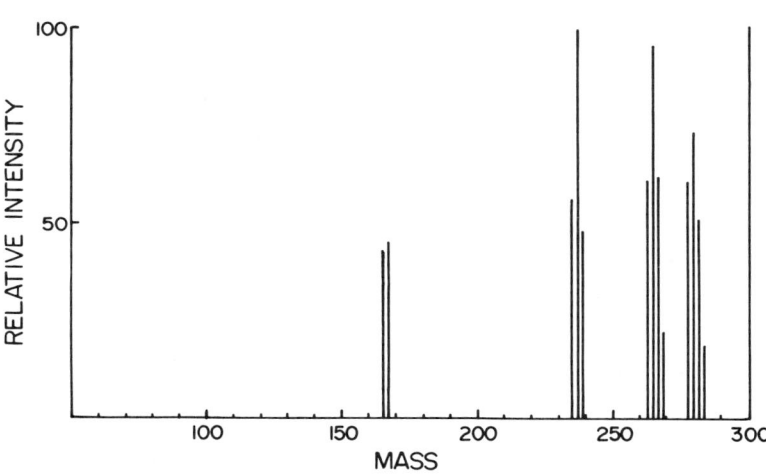

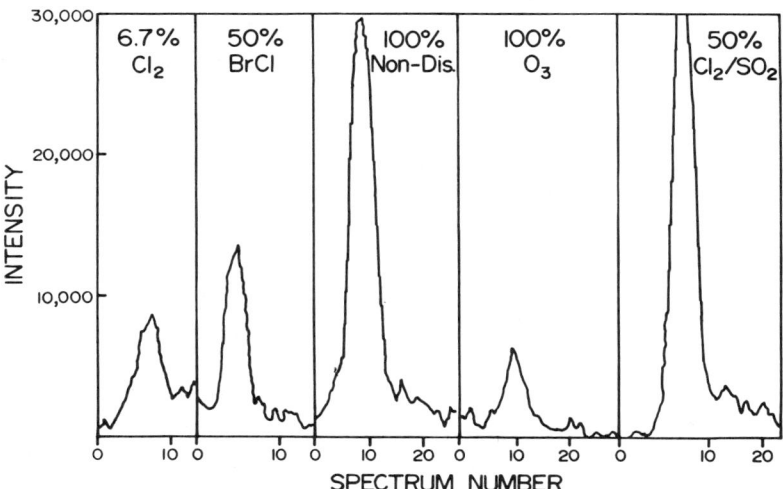

Figure 3. Mass spectrum and reconstructed mass chromatogram (M^+ 278) of pentachloroanisole found in fish raised in sewage effluent. The disinfectant and percent effluent concentration are given in the lower figure. The spectrum number is a guide to the gas chromatographic retention time.

observed biological response (log 1/C) (Overton, 1899; Meyer, 1899; Meyer and Hemmi, 1935; Meyer, 1937):

$$\log 1/C = a \log P + b \qquad (1)$$

In this equation, C is the molar concentration of a given compound required for a particular response (narcosis, LC50, LD50). This original work dealt with isonarcosis in tadpoles, and the chemicals studied were alcohols, ketones, esters and aromatic hydrocarbons. Ferguson (1939) placed the Overton-Meyer theory on a quantitative basis by defining the free energy change, ΔG, in terms of the chemical potential, μ.

In the 1960's Hansch and co-workers (1969) began applying these theories and were able to correlate many of the previously reported data by using the n-octanol/water system as the basic criterion. Electronic and steric factors have since been included to improve the correlation. Hansch introduced the need of a parabolic relationship in addition to a straight line relationship because as a compound became more lipophilic (higher partition coefficient, P) it tended to remain in the membranes and not be transferred to the active site and therefore caused a decrease in the observed activity. A parabola would be the best empirical representation for this type of observed biological response.

The onset of bioaccumulation is observed when the compounds start to remain in the cell membranes. It has been shown that the most likely place for the accumulation of lipophilic compounds is in the lipids of the animal, fish or bird. At this point these compounds are transferred from water to birds throughout the environment by the food-chain mechanism.

Kapoor et al. (1973) studied food-chain mechanisms extensively and have shown how various chemicals are transported by this means. Bemelmans and ten Noever de Brauw (1974) observed this process in chickens that had been fed contaminated food. The contamination involved chloroanisoles of the same type as those isolated from fish raised in the chlorine-treated effluent from Grandville, Michigan. Chlorinated anisoles have been reported (Engel et al., 1966; Curtis et al., 1972; Cserjesi and Johnson, 1972) to impart a distinct musty taint to both eggs (yolks) and flesh of chickens. One source of the contamination was the wood shavings in the poultry houses, which were produced from lumber treated with technical-grade pentachlorophenol (impurities constitute mainly isomeric tetrachlorophenols). The bacteria in the poultry litter were shown to be capable of methylating the chlorophenols on the surface of the wood shavings. It was also shown that some bacteria were capable of dechlorinating the higher chlorinated phenols. The problem persisted even when wood shavings were not used. This resulted in the discovery that the poultry feed was contaminated with trichloroanisole, which has a very low taste threshold (Table VI).

Table VI. Taste Detection Threshold of Compounds in Aqueous Solution[a]

Compound	Concentration (μg/g)
Pentachloroanisole	4×10^{-3}
2,3,4,6-tetrachloroanisole	4×10^{-6}
2,4,6-trichloroanisole	3×10^{-8}
2,3,6-trichloroanisole	3×10^{-10}

[a]Curtis et al., 1972.

Bemelmans and ten Noever de Brauw (1974) were unable to ascertain the cause of the tainted feed. Investigation has determined that fish products have been used in various formulations of chicken feed and in light of the contaminated Grandville fish it is highly possible that contaminated fish could have been involved. Thus aqueous chlorination may create a problem as large as the initial one that required its use as a disinfectant.

Anisoles are more persistent as compared to the parent phenols. The methylation of phenol increases the partition coefficient from 1.48 to 2.04. Chlorination also increases the partition coefficient (0.7 units for aromatics and 0.4 units for aliphatics). Both of these increase the bioconcentration factor.

Neely et al. (1974) have shown with trout (*Salmo gairdneri* Richardson) that the bioconcentration factors of organic compounds when correlated to their n-octanol/water partition coefficients (P) resulted in a straight line (Equation 2):

$$\log (\text{bioconc. factor}) = 0.542P + 0.124 \quad r = 0.948 \quad s = 0.342 \quad (2)$$

In their study a series was chosen so that the partition coefficients were spread over a wide range. The bioconcentration factor was determined by the ratio of the rate of chemical uptake (k_i) in the test solution to the rate of chemical clearance (k_c) in fresh water. Figure 4 gives a graphical representation of their results. Table VII illustrates the predictive application of Equation 2. Mosquitofish (*Gambusia affinis*) were used for these bioassays, which demonstrate the importance of the use of partition coefficients as a predictive tool.

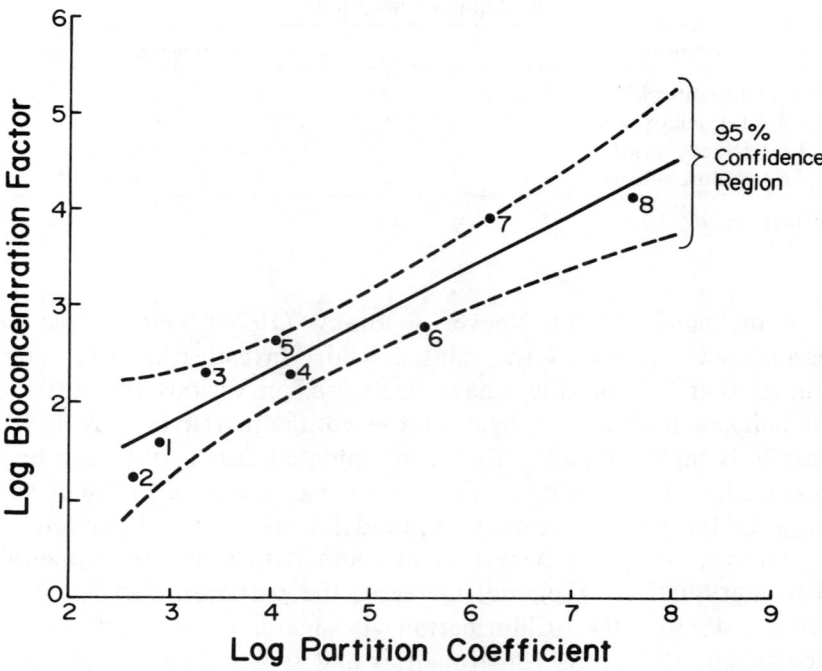

Figure 4. Linear regression between logarithms of *n*-octanol/water partition coefficient and bioconcentration of chemicals in trout muscle (Neely *et al.*, 1974). Compounds plotted are: (1) 1,1,2,2-tetrachloroethylene; (2) carbon tetrachloride; (3) *p*-dichlorobenzene; (4) diphenylether; (5) diphenyl; (6) 2-biphenylphenylether; (7) hexachlorobenzene; and (8) 2,2',4,4'-tetrachlorodiphenylether.

Table VII. The Use of Regression Equation 2 for Predicting the Bioconcentration Factor[a]

Chemical	Log Partition Coefficient	Log Bioconcentration Factor	
		Calculated	Experimental
Endrin	5.60	3.47 ± 0.989	3.17
Chlorpyrifos[b]	4.82	2.87 ± 0.963	2.67
3,5,6-trichloro pyridinol	1.35	0.88 ± 1.139	0.49

[a]Neely et al., 1974.
[b]0,0-diethyl-0-(3,5,6-trichloro-2-pyridyl) phosphorothioate.

A standard approach to determining the bioconcentration factor should be defined so that data can be reported in a standard way and, more importantly, can be more easily compared to previously reported data. A reasonable standard expression would be the ratio of the initial uptake rate (k_i) in the test solution to the clearance rate (k_c) in fresh water. With this in mind three basic patterns have been observed during a 28-day exposure: (a) $k_i \gg k_c$, continuous increase in residue concentration during exposure; (b) $k_i \cong k_c$, initial residue concentration increase, until an unchanging level is observed; and (c) $k_i \ll k_c$, initial residue concentration increase followed by a continuing decrease in concentration with exposure time. The initial rate constant will be dependent on the partition coefficient (P) and independent of k_c because the compound has not reached the site of action in large enough quantities to be of concern under the conditions of the bioassay. In order that k_c be independent of k_i in the second step, the fish have to be removed from the test solution and placed in fresh water. In this way one should be able to observe only the clearance mechanism. There may be several steps in each of these two general mechanisms, but the rate should be dependent only on the slowest step.

These examples illustrate some of the possible ramifications of the use of partition coefficients in an attempt to predict not only toxicity but bioaccumulation as well.

SUMMARY

A method is needed to rapidly screen the large number of organic compounds which, if allowed to be discharged, will contaminate our environment. The pharmaceutical industry has used water-solvent partition coefficients with great success in determining which compounds might be most biologically active. These coefficients can also be used as a predictive tool in the environmental world to study both toxicity and bioconcentration of organic contaminants. Unfortunately, the problems that arise with bioaccumulation of organic compounds which have been discharged into the environment are not immediately apparent. By the time a problem is identified it is usually too late and a major segment of the ecosystem has been contaminated. New products should be manufactured with this in mind, and a complete data base of bioaccumulation data needs to be developed to set reasonable guidelines which industry can use.

The incorporation of chlorine into compounds during the disinfection of waste effluents with chlorine is an undesirable end result of effluent treatment in that compounds become more persistent and bioaccumulate to a greater extent.

EXPERIMENTAL

Compound Library Generation

A large number of compounds which were predicted to be found in the chlorinated effluent were purchased or prepared by one of the following two procedures.

Procedure A

Five hundred milligrams of the starting organic compound was dissolved in 20 ml $CHCl_3$. Chlorine gas was bubbled into the solution at room temperature for 30 min. The excess chlorine was then stripped off with $CHCl_3$.

Procedure B

Five hundred milligrams of the starting organic compound was dissolved in 20 ml CCl_4. Chlorine gas was bubbled into the solution at room temperature for 30 min. The solution was then irradiated with UV light for 10 min at 254 nm. The excess chlorine was then stripped off with CCl_4.

GC retention time and mass spectral data were then obtained on each product. The mass spectra were loaded onto magnetic tape for library searches during sample investigations.

Water Analysis

The three water samples taken for examination were: (1) 8.0 gal of nondisinfected effluent, (2) 9.0 gal of chlorinated effluent, and (3) 4.5 gal of chlorinated effluent adjusted to pH 2.0. Samples were obtained by running the desired effluent through a glass column (2.5 x 20 cm) filled with a 50-ml bed volume of XAD-2 resin (Rohm and Haas). Each column was washed with 200 ml of 0.1 M NaOH, 200 ml 0.1 M HCl, and finally 200 ml of methanol. The acid and base washes were neutralized and extracted with three 50-ml portions of ether. The extracts were concentrated to 10 ml and the base extract methylated with diazomethane for GC and GC/MS analysis.

Fish Analysis

Whole fish (30 to 50 g) were blended with enough anhydrous sodium sulfate to make a dry powder and extracted with four 100-ml portions of hexane/ether (1:1). The extracts were concentrated, and the resulting oil residue was diluted with cyclohexane to 1 g/5 ml. The oil residue solution was placed on a 2.5-cm x 30-cm SX-2 (Bio-Rad Laboratory) gel permeation column and eluted with cyclohexane at 3.5 ml/min. The first 150 ml was vented, and the next two 60-ml fractions were collected. The collected fractions were combined and concentrated by using a Kuderna-Danish flask fitted with a three-ball Snyder column. Volumes were adjusted with hexane for GC and GC/MS analysis. Fish were obtained from the following bioassay tanks: (a) 6.7% Cl_2, (b) 50% Cl_2/SO_2, (c) 50% BrCl, (d) 100% O_3, and (3) 100% nondisinfected.

Instrumental Conditions

All gas chromatograms were run on a Varian Areograph 1700 equipped with a 6-ft x 1/8-in. glass column packed with 3% OV-101 on 80/100 mesh gas chrom Q. All GC effluent was monitored by electron capture-flame ionization detectors with a 1:50 splitter. Injection port and detector block temperatures were maintained at 230°C.

The mass spectrometer system was a Varian MAT CH-5 equipped with a Varian MAT mass spectrosystem 100 MS data system. All mass spectra were obtained at 70 eV with a 5-sec-per-mass decade scan.

ACKNOWLEDGMENTS

The authors acknowledge the help and assistance of M. DeGraeve, R. Ward and others, who conducted the bioassays, and to G. Veith for collecting samples and providing useful advice concerning the analysis.

REFERENCES

Bellar, T. A., J. T. Lichtenberg and R. C. Kroner. 1974. *The Occurrence of Organohalides in Chlorinated Drinking Water.* EPA-670/4-74-008. U.S. Environmental Protection Agency. November. 21 pp.

Bemelmans, J. M. H. and M. C. ten Noever de Brauw. 1974. "Chloroanisoles as Off-Flavor Components in Eggs and Broilers." *J. Agr. Food Chem.* 22:1137.

Carlson, R. M. 1975. Personal communication. See also R. M. Carlson and R. Caple, "Organo-Chemical Implications of Water Chlorination" (this conference).

Carlson, R. M., R. E. Carlson, H. L. Kopperman and R. Caple. 1975. "The Facile Incorporation of Chlorine into Aromatic Systems During Aqueous Chlorination Processes." *Environ. Sci. Technol.* 9:674. (And references cited therein.)

Cserjesi, A. J. and E. L. Johnson. 1972. "Methylation of Pentachlorophenol by *Trichoderma virgatum.*" *Can. J. Microbiology* 18:45.

Curtis, R. F., D. G. Land, N. M. Griffiths, M. Gee, D. Robinson, J. L. Peel, C. Dennis and J. M. Gee. 1972. "2,3,4,6-Tetrachloroanisole Association with Musty Taint in Chickens and Microbiological Formation." *Nature* 235:223.

Engel, C., A. P. deGroot and C. Weurman. 1966. "Tetrachloroanisole: A Source of Musty Taste in Eggs and Broilers." *Science* 154:270.

Ferguson, J. 1939. "The Use of Chemical Potentials as Indices of Toxicity." *Proc. Roy. Soc.* (London) Ser. B, 127:387.

Glaze, W. H. and J. E. Henderson, IV. 1975. "Formation of Organo-Chlorine Compounds from the Chlorination of a Municipal Secondary Effluent." *J. Water Poll. Control Fed.* 47:2511.

Glaze, W. H., J. E. Henderson, IV, J. E. Bell and V. A. Wheeler. 1973. "Analysis of Organic Materials in Wastewater Effluents after Chlorination." *J. Chrom. Sci.* 11:580.

Hansch, C. 1969. "A Quantitative Approach to Biochemical Structure–Activity Relationships." *Acc. Chem. Res.* 2:232. (And references cited therein.)

Jolley, R. L. 1973. *Chlorination Effects on Organic Constituents in Effluents from Domestic Sanitary Sewage Treatment Plants.* Ph.D. Thesis. University of Tennessee, Knoxville.

Jolley, R. L. 1975. "Chlorine-Containing Organic Constituents in Sewage Effluents." *J. Water Poll. Control Fed.* 47:601. (And references cited therein.)

Jolley, R. L., S. Katz, J. E. Mrochek, W. W. Pitt, Jr. and W. T. Rainey. 1975. "Analyzing Organics in Dilute Aqueous Solutions." *Chem. Technol.* 5:312. (And references cited therein.)

Kapoor, I. P., R. L. Metcalf, A. S. Hirwe, J. R. Coats and M. S. Khalsa. 1973. "Structure Activity Correlations of Biodegradability of DDT Analogs." *J. Agr. Food Chem.* 21:310.

Macek, K. J., M. E. Barrow and R. F. Frasny. 1975. "International Joint Commission Symposium on Structure-Activity Correlations in Studies of Toxicity and Bioconcentration with Aquatic Organisms," p. 119. (And references cited therein.)

Meyer, H. 1899. *Arch. Exptl. Pathol. Pharmakol.* 42:109. As cited, p. 134 in A. Burger (Ed.) *Medicinal Chemistry*, 3rd ed. 1970. Wiley-Interscience, New York.

Meyer, K. H. 1937. *Trans. Faraday Soc.* 33:1063. As cited, p. 134 in A. Burger (Ed.) *Medicinal Chemistry*, 3rd ed. 1970. Wiley-Interscience, New York.

Meyer, K. H. and H. Hemmi. 1935. *Biochem. Z.* 277:29; 282:444, 447. As cited, p. 134 in A. Burger (Ed.) *Medicinal Chemistry*, 3rd ed. 1970. Wiley-Interscience, New York.

Neely, W. B., D. R. Branson and G. E. Blau. 1974. "Partition Coefficient to Measure Bioconcentration Potential of Organic Chemicals in Fish." *Environ. Sci. Technol.* 8:1113.

Overton, E. 1899. *Viertljahrsschr. Naturforsch. Ges. Zurich* 44:88. As cited, p. 134 in A. Burger (Ed.) *Medicinal Chemistry*, 3rd ed. 1970. Wiley-Interscience, New York.

DISCUSSION

Alan H. Molof, Polytechnic Institute of New York. There are two parts to my question. First, in your testing protocol, did you use a clean tap water with the disinfectant you used for the activated sludge effluent? This would be valuable since it would relate to the effect of the disinfectants themselves, such as the effect of chlorine itself on fish. Second, have you or anyone else looked into the effects of disinfectants themselves in drinking water on man? Even though the conference title is "Environmental Impact of Water Chlorination," there does not seem to be any coverage of this subject. The importance of this area comes from the use of chlorine residuals in our tap water.

Kopperman. Well, again it comes to grant money I am assuming such studies are being done in other places such as at the EPA Laboratory in Cincinnati.

Herman F. Kraybill, National Cancer Institute. Bioaccumulation and partition coefficients cause a problem in bioassay for carcinogenicity of organochlorine compounds. In setting an MTD (maximum tolerated dose) one proceeds for 6 to 7 months in the test and then the rodents being tested are killed due to the concentration increase in the adiposity. Even at 0.5 MTD, lethalities occur several months later and one is not able to pursue the bioassay. One has to start over at much lower doses where the bioaccumulation will not overload the metabolism.

Kopperman. Actually all of us are a little overweight. We have probably stored more than our share of some of the chlorinated compounds. All sorts of things might happen if we were to go on crash diets. These compounds have to go somewhere when you start getting rid of fat tissue. They go right into your system.

Jerry J. Nelson, U.S. Energy Research and Development Administration. You have what appears to be classic linear first-order kinetics. Have you made any attempt to build a model for the uptake and movement such as a Van Dyne type descriptive model?

Kopperman. No. I have not because as it turns out I am no longer associated with EPA. I am not presently funded to do this type of research.

Nelson. In addition, the slide showing initial uptake rates which overshoot and decay may involve two processes such as sorption and intake with subsequent desorption. Or do you think this perhaps is decay in addition to clearance due to metabolic decomposition or some such mechanism?

Kopperman. Yes, those are possible mechanisms.

17

INVESTIGATING THE EFFECTS OF CHLORINATED ORGANICS

Carl W. Gehrs and George R. Southworth

Environmental Sciences Division
Oak Ridge National Laboratory
Oak Ridge, Tennessee 37830

ABSTRACT

The recent identification of stable chlorine-containing organics arising from the chlorination of natural waters has revealed a group of reaction products whose toxicities to aquatic organisms are unknown. In this paper we present information on the toxicity of two chlorinated compounds (5-chlorouracil and 4-chlororesorcinol) and a mixture of identified chlorinated organics to zooplankton and fish. We compare data emphasizing differences in relative toxicity depending on the response parameter used. Problems associated with studying individual compounds and complex mixtures are discussed and a systematic approach for overcoming the identified shortcomings is presented.

INTRODUCTION

Research designed for evaluating the environmental implications of chemical releases has as its goal the development of the data necessary for determining the probability of, and kinds of, adverse effects that may impinge on man or his environment. Critical to attaining this goal is an understanding of the kinetics, transformation products and ultimate fate in the environment of the chemical of concern.

Chlorine has been used as a disinfectant for many years, with an estimated 60,000 tons being added to effluents from sewage treatment plants in the United States during 1962 (Laubusch, 1962). Although the fate of reactive chlorine residuals in aquatic environments is very well documented (Palin, 1950; Draley, 1972), little is known about either the fate or effects of the recently identified stable chlorine-containing organics (Jolley, 1973). We have decided to emphasize the stable chlorine-containing organics in our effects research because they are more persistent and, consequently, have a greater potential for adversely impacting the aquatic environment. The purpose of this paper is to present information on the effects of chlorinated organics and to outline a method for organizing a research program aimed at evaluating the potential problem arising from chlorinated organics.

METHODS AND MATERIALS

In the initial investigations into the potential formation of stable chlorine-containing organics, Jolley (1973 and 1975) employed high-resolution liquid chromatography to analyze secondary sewage effluents treated to 1 mg/l chlorine residual with ^{36}Cl-tagged chlorine gas or hypochlorite solution. He found more than 60 specific peaks, of which 17 individual compounds have been tentatively identified at microgram-per-liter concentrations. The compounds identified are of three classes: phenols, purines and pyrimidines, and aromatic acids. Two compounds were chosen for initial investigation: the pyrimidine, 5-chlorouracil (because of its potential for incorporation into DNA as a base analog) and a phenol, 4-chlororesorcinol (because of the toxicity of phenols and the relatively high concentration—1.2 µg/l—found by Jolley). Later it was decided to produce and test a synthetic mixture made up into a stock solution of 1 g/l total chlorinated organics in the proportions in which they were found in the treated secondary effluent (Table I).

Toxicity studies were conducted using two species of aquatic organisms, the zooplankter *Daphnia magna* and the fish *Cyprinus*

Table I. Composition[a] of Synthetic Chlorinated Effluent

Compound	Percent
5-chlorouracil[b]	20.8
5-chlorouridine	8.2
8-chlorocaffeine	8.2
6-chloroguanine	4.3
8-chloroxanthine	7.2
5-chlorosalicylic acid	1.2
4-chloromandelic acid	5.3
2-chlorophenol	8.2
3-chlorophenol	2.5
4-chlorophenol	3.5
4-chloro-3-methylphenol	8.2
4-chlorophenylacetic acid	1.8
4-chlorobenzoic acid	5.3
2-chlorobenzoic acid	1.3
3-chloro-4-hydroxybenzoic acid	6.3
3-chlorobenzoic acid	3.0
4-chlororesorcinol[c]	5.9

[a] Composition was determined using the data of Jolley (1973).
[b] 4.3 $\mu g/l$.
[c] 1.2 $\mu g/l$.

carpio (carp). The parameters used were mortality and maturation in the zooplankter and hatching success (mortality) and malformation of fry for the fish.

Zooplankton Studies

Zooplankton studies were conducted in 100-ml beakers containing 80 ml of appropriate concentration of the test solutions in spring water. Ten replicates were used at each concentration with two individuals (12 ± 12 hr old) added to each beaker to initiate a test. Five concentrations of each test material plus a control (0.00, 10^{-2}, 10^{-1}, 10^0, 10^1 and 10^2 mg/l) were used. The initial water temperature was 21°C. Beakers were kept at 21 ± 1°C in a constant-temperature chamber having a 12-hr light/12-hr dark cycle during the testing period. Animals were censused daily (Gehrs and Jolley, 1975) for mortality and production of young, with dead animals immediately removed.

Animals were fed 0.1 ml of a trout chow and water mixture twice weekly (Gehrs, 1972). At seven-day intervals, animals were transferred to fresh solution, with young counted and removed. An LC50 compilation was used to evaluate mortality effects. Maturation was determined by counting the number of beakers at each concentration of each chemical which contained young after the initial 7-day period.

Fish Egg Hatching Studies

The effects of the chemicals on hatching success of fish eggs were determined following modified procedures of Blaylock and Griffith (1971). Gravid carp were collected and spawned artificially in the laboratory. Eggs were fertilized immediately in the test solutions as would be expected in the natural environment so that they were subjected to the toxicant during the hardening procedure. After 30 min, the test solutions were decanted and fresh solutions of the appropriate concentration added. Two replicates of 200-400 eggs were used for each concentration. For the duration of the test (about 3 days) eggs were held in an environmental chamber at 26°C ± 1°C with a 12-hr light/12-hr dark photoperiod. Twice daily during this period solutions were changed and dead eggs recorded and removed. Six concentrations plus controls were used for the tests (0.00, 10^{-3}, 10^{-2}, 10^{-1}, 10^0, 10^1 and 10^2 mg/l). Concentrations were based on serial dilutions prepared in the laboratory rather than on *in vivo* measurements. While the latter is preferable, the capabilities for such measurements have not yet been realized.

RESULTS

Zooplankton Studies

When mortality was used as the response variable (Figure 1), the three test solutions (synthetic mixture, 5-chlorouracil and 4-chlororesorcinol) were dissimilar in their level of activity. 5-Chlorouracil caused no change in median survival times (when compared to controls) at concentrations of up to 1 mg/l (three

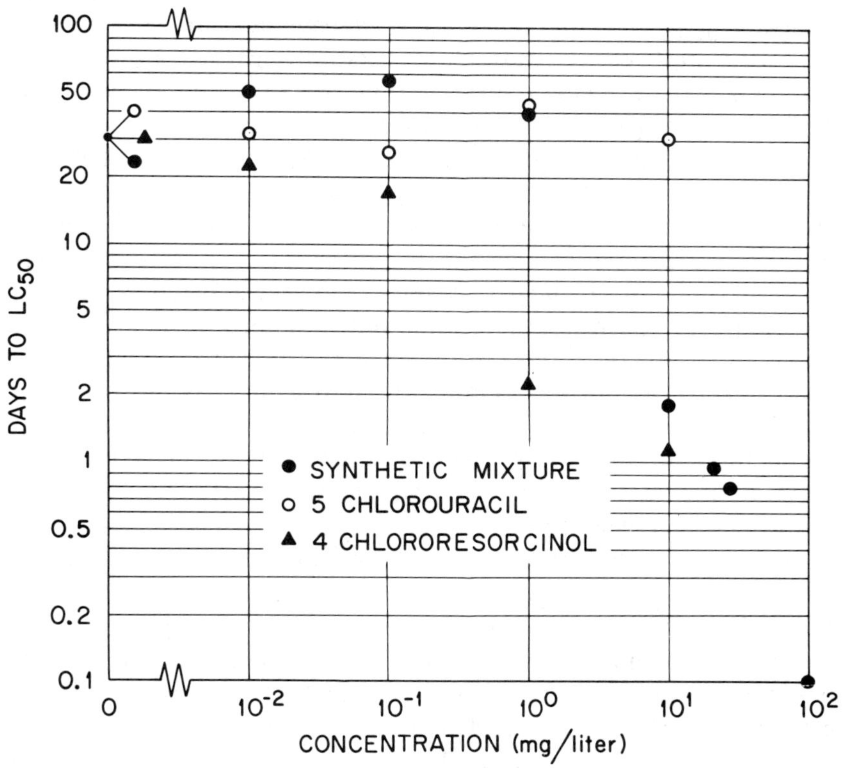

Figure 1. Days to LC50 for *Daphnia magna* in various concentrations of a synthetic mixture of stable chlorine-containing organics, 5-chlorouracil and 4-chlororesorcinol. Data on 5-chlorouracil and 4-chlororesorcinol from Gehrs and Jolley, 1975.

orders of magnitude above effluent concentrations; 4.3 µg/l, Jolley, 1973). Both 4-chlororesorcinol and the synthetic mixture, however, produced discernible changes in median survival times over the ranges tested. 4-Chlororesorcinol reduced median survival time approximately 20% (from 30 to 24 days) at the lowest concentration (10^{-2} mg/l) used, with higher concentrations causing further reductions until the 10^1 mg/l level resulted in a median survival time of slightly more than one day. The synthetic mixture originated a different response with all concentrations $\leq 10^0$ mg/l increasing the median survival time. The 10^{-1} mg/l concentration, for example, caused an almost doubling of the median survival time as compared to controls (~58 days in 10^{-1} mg/l concentration, 30 days in controls). A major decrease in survival time at concentrations $\geq 10^{-1}$ mg/l was seen for the synthetic mixture, suggesting a threshold toxicity level at concentrations somewhere between 10^{-1} and 10^1 mg/l. Comparison of the responses of this variable (median survival time) in *D. magna* to the three test substances reveals an interesting point. One compound, 5-chlorouracil, caused no apparent effect over the entire range of concentrations tested. The second, 4-chlororesorcinol, initiated a decrease in median survival times at even the lowest concentration (10^{-2} mg/l) while the synthetic mixture effected an increase in median survival time until $\geq 10^1$ mg/l was reached, after which a negative effect was noted.

While mortality data are useful for comparing relative toxicities of substances, quite often it is the more subtle effects that might impinge upon large segments of the population that are more ecologically important. For zooplankton this includes parameters such as production of young. In a study on the effects of 5-chlorouracil and 4-chlororesorcinol on *D. magna* using this response variable, all concentrations (10^{-3} mg/l to 10^1 mg/l) of 4-chlororesorcinol and all but the lowest concentration (10^{-2} mg/l) of 5-chlorouracil caused at least a 50% decrease in the number of young produced during the first seven days of free life of the zooplankton. When these data were manipulated to provide specific birth rates (number of young/adult/day), all but the 10^{-2} mg/l concentration of 5-chlorouracil revealed a statistically significant ($P \leq 0.05$) decrease (Gehrs and Jolley, 1975).

No young were produced in any concentrations (controls to 10^0 mg/l) of the synthetic mixture during the first 7 days of a study designed to investigate the response of this parameter. Specific birth rates in this investigation were consequently determined after a 2-week period, and the research was not an exact replication (experimentally the time frame was different) of that conducted on the individual compounds. The data from the two studies were therefore normalized to show percent of specific birth rate of controls (Table II). Whereas both 4-chlororesorcinol and 5-chlorouracil caused decreases in specific birth rates, the synthetic mixture (containing similar concentrations of each of the two compounds, as well as 15 other compounds) caused specific birth rates greater than controls. Apparently, some antagonistic interaction took place in the synthetic effluent, resulting in a specific birth rate response dissimilar to that produced by either individual compound.

Fish Hatching Studies

Both 5-chlorouracil and 4-chlororesorcinol have been found to significantly lower ($P \leq 0.05$) hatching success of carp eggs at concentrations as low as 10^{-3} mg/l (approximately that measured in the effluent) (Gehrs et al., 1974). The synthetic mixture gave drastically different results (Figure 2). Not only was there no lowering of hatching success at levels similar to the 10^{-3} mg/l, but there was no apparent effect over the next three higher orders of magnitude of concentrations. The breakpoint in toxicity (hatching success < 10%) of the synthetic mixture was approximately 30 mg/l (geometric mean). Concentrations this high were not investigated in the 5-chlorouracil and 4-chlororesorcinol studies and, consequently, no real comparison of data at this level can be made. If, however, one

Table II. Specific Birth Rate (Number of Young/Adult/Day)[a] for Zooplankton in Different Concentrations of 5-Chlorouracil, 4-Chlororesorcinol and the Synthetic Mixture

Concentration (mg/l)	5-Chlorouracil	4-Chlororesorcinol	Synthetic Mixture
10^{-4}		0.57	1.89
10^{-3}		0.37	1.55
10^{-2}	1.50	0.36	1.61
10^{-1}	0.81	0.43	0.86
10^{0}	0.64	0.00	1.30
10^{1}	0.73	0.00	

[a]The data have been normalized as percent of specific birth rate of controls (1.00).

336 WATER CHLORINATION

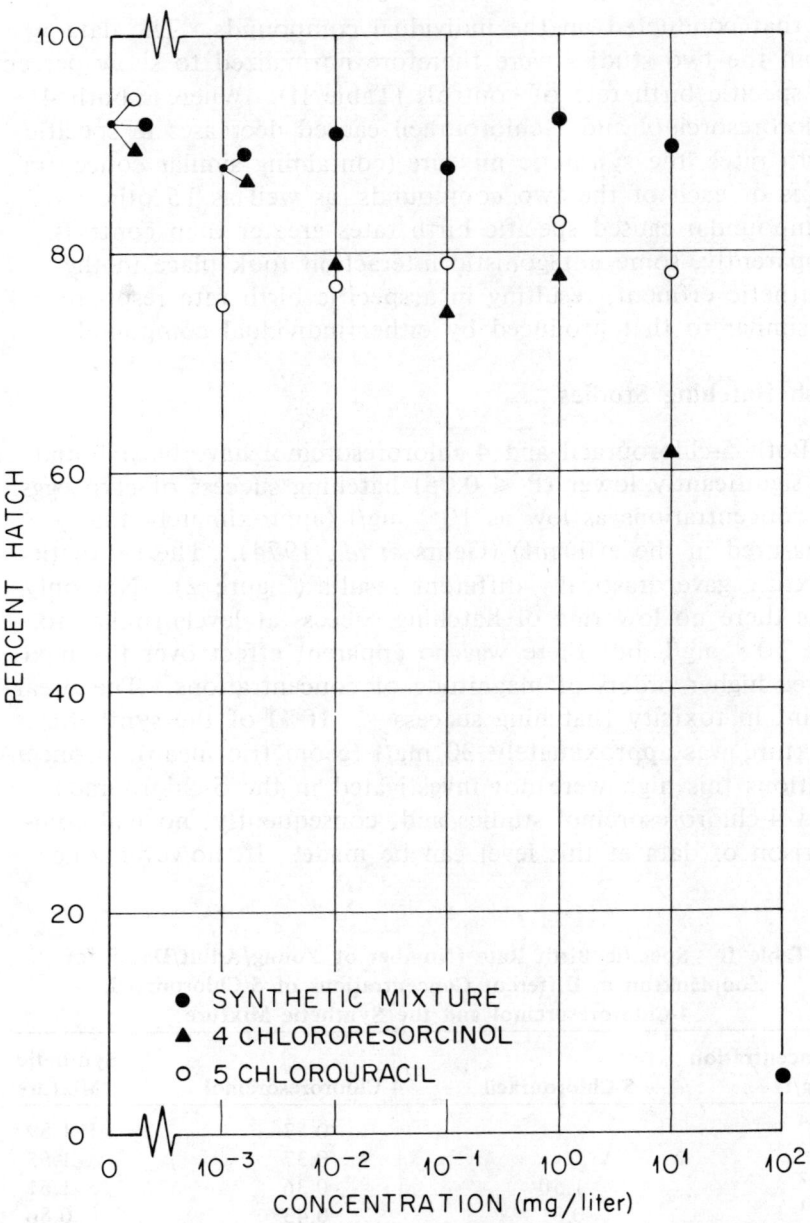

Figure 2. Percent hatch of *Cyprinus carpio* in various concentrations of a synthetic mixture of stable chlorine-containing organics, 5-chlorouracil, and 4-chlororesorcinol. Data on 5-chlorouracil and 4-chlororesorcinol from Gehrs *et al.*, 1974.

extrapolates the higher concentration (10^0 and 10^1 mg/l) data from 4-chlororesorcinol, a response would be seen similar to that generated by the synthetic mixture.

The percent of hatched embryos that were malformed is an important parameter that should be included in analyses designed to assess effects of contaminants on exposed fish populations, because malformed individuals would not be expected to survive and reproduce. Malformation augments the negative response of nonhatching by effectively removing a discrete subset of the hatched embryos from the population. Eyman et al. (1975) observed a positive correlation between concentration of 5-chlorouracil and percent malformation in carp with a linear dose-response relationship originating at 0.5 mg/l. No similar response was found for either 4-chlororesorcinol or the synthetic mixture. Unfortunately, at the acute toxicity threshold concentration for the synthetic mixture, the concentration of 5-chlorouracil was less than 0.5 mg/l. Consequently, nothing can be said concerning possible interactions of the chemicals in the mixture relative to this parameter.

DISCUSSION

The research results presented above call attention to two types of problems faced in assessing the potential environmental impact associated with any group of chemicals, in this case chlorinated organics. The first problem is the differential response seen both within a species (when different parameters are used as variables) and between species. The second problem is that of interpreting and quantifying the potential toxic interactions between chemicals in a complex mixture.

The data on zooplankton exemplify the differences in response, depending on the parameter studied, and show the necessity of utilizing more than one parameter when attempting to evaluate the potential ecological effects of a chemical on a population. Although no effect on median survival times results from 5-chlorouracil, decreases in specific birth rate were observed. If only the former parameter (median survival times) were used there would be no potential identified for 5-chlorouracil adversely affecting a zooplankton population. Zooplankton populations are cropped very heavily, however, with as much as 20% of the population removed each day (Hall, 1964). While the levels of 5-chlorouracil causing a decrease in specific birth rate are above the concentrations identified in effluents, and consequently not expected to adversely affect zooplankton, a decrease in the production of young could have detrimental effects on zooplankton populations.

If the purpose of toxicity research is to evaluate the potential effects on an ecosystem or the environment, then it is necessary to determine the response of several components of the ecosystem. It is only necessary to look at the compilation tables of Becker and Thatcher (1973) to verify this statement. For example, the 96-hr TL_m is shown as 3.4 and 90.0 mg/l, respectively, for *Lepomis* sp. and *Physa* sp.

The difficulty involved in evaluating the potential effects of a mixture of chemicals (*i.e.*, an effluent) is exemplified by the differing responses observed for the individual compounds as compared to those of the synthetic mixture. An obvious question is how does one design research to evaluate the potential effects of mixtures of chemicals (whether they are chlorinated organics or anything else)? The approach taken is dependent on the type of information desired. If the concern is simply the immediate area into which an effluent is released, then testing the composite mixture by using several different parameters for several components of the ecosystem (as discussed above) is sufficient. If the concern is broader, and includes the potential effect on the environment downstream from the release, then the necessary research is more difficult to design. The environment (both abiotic and biotic components) will modify the effluent (chemical mixture), altering the composition of the materials that will be confronted at different distances from the release. All compounds will not behave similarly with simple dilution of the effluent being the only variable necessary to evaluate potential effects. Factors such as sedimentation, physical degradation and microbial degradation are very real activities functioning to alter the effluent. As an example, we recently found differences in photolysis and microbial degradation of two of the components of chlorinated effluents, 5-chlorouracil and 4-chlororesorcinol (Southworth and Gehrs, 1976). Furthermore, if research were conducted solely on the composite mixture, then effects of individual compounds may be overlooked, as would be the case for effects of both 5-chlorouracil and 4-chlororesorcinol on fish eggs.

While the above statements suggest problems with the total effluent, evaluation of the toxic effects of each of the individual compounds also has limited value for predicting potential effects of complex mixtures on natural systems. This is the result of the numerous types of interactions that can occur between components of a complex mixture that substantially alter the relationship between predicted toxicity (based on data from individual compounds) and actual response resulting from the mixture.

Interactions can range from antagonism, where the response to the mixture is less than would be predicted from the individual compounds, to synergism, where the response is greater than anticipated from the toxicity levels of the individual compounds. Only in the case of an additive interaction, where the observed effect to the mixture resembles that predicted from the relative toxicities of the individual compounds, would predictions from individual compounds approximate the response caused by the mixture. None of these interactions occurring between compounds would be discovered if only individual compounds were tested. The number of mixtures of compounds necessary for evaluation would be extremely large, being given by the general expression:

$$_nC_r = \sum_{r=1}^{n} \frac{n!}{(n-r)!\, r!}$$

where $_nC_r$ = total number of possible interactions
  n = number of compounds
  r = number of compounds to be examined for interactions in a specific subset

Even in a simple situation of five individual compounds, the number of possible interactions for evaluation would be thirty-one. Such an approach would be both extremely time-consuming and expensive.

The approach we believe appropriate combines both the previously mentioned systems (composite mixtures and individual compounds). It is designed to determine the potential for interactions between classes or groups of compounds in the mixture (effluent) rather than attempt to look at all possible interactions initially (which we have shown above to be an astronomical undertaking). In this system, representative compounds from the various classes present in the mixture would be used to provide an early estimate of potential interaction. Selection of the model compounds for use can be based on several criteria (*e.g.*, amount of data currently existing, concentration in effluent, structure, etc.). For the stable chlorine-containing organics, initial evaluation would be concerned with three classes of compounds (phenols, pyrimidines and purines, and aromatic acids) rather than the seventeen (and there are certainly more) individual compounds identified. Depending on data derived from these investigations, evaluation of other inter- and intraclass interactions would be undertaken. For example, if a pyrimidine and phenol showed no interaction while a pyrimidine and aromatic acid produced an additive response, we would next pursue the interactions that might arise from the pyrimidine and aromatic acid rather than the pyrimidine and phenol.

We do not mean that this approach should be taken with the exclusion of investigating either composite effluents or individual compounds. On the contrary, all three approaches should be used. What we suggest, however, is a method that will aid in providing the most timely data (including future research needs) for the effort and money expended.

To conclude, what is necessary for evaluating the environmental impact of water chlorination is a systematic approach. This paper has been designed to address only the effects segment of this research. To attain a full understanding of, and consequently the capability of predicting effects of chlorination requires research on chlorine kinetics, transformation, metabolism and bioconcentration as well.

ACKNOWLEDGMENTS

Research sponsored by Energy Research and Development Administration under contract with Union Carbide Corporation. Publication No. 864, Environmental Sciences Division, Oak Ridge National Laboratory.

REFERENCES

Becker, C. D. and T. O. Thatcher. 1973. *Toxicity of Power Plant Chemicals to Aquatic Life.* WASH-1249. Battelle Pacific-Northwest Laboratories. Richland, Washington. June.

Blaylock, B. G. and N. A. Griffith. 1971. "A Laboratory Technique for Spawning Carp." *Prog. Fish. Cult.* 33:48-50.

Draley, J. E. 1972. *The Treatment of Cooling Water with Chlorine.* ANL/ES-12. Argonne National Lab. Chicago, Illinois.

Eyman, L. D., C. W. Gehrs and J. J. Beauchamp. 1975. "Sublethal Effects of 5-Chlorouracil on Carp (*Cyprinus carpio*) Larvae." *J. Fish. Res. Bd. Can.* 32:2227-2229.

Gehrs, C. W. 1972. *Aspects of the Population Dynamics of the Calanoid Copepod,* Diaptomus clavipes. Ph.D. Thesis, Univ. of Oklahoma, Norman.

Gehrs, C. W., L. D. Eyman, R. L. Jolley and J. E. Thompson. 1974. "Effects of Stable Chlorine-Containing Organics on Aquatic Environments." *Nature* 249:675-676.

Gehrs, C. W. and R. L. Jolley. 1975. "Chlorine-Containing Stable Organics: New Compounds of Environmental Concern." *Verh. Internat. Verein. Limnol.* 19:2185-2188.

Hall, D. J. 1964. "An Experimental Approach to the Dynamics of a Natural Population of *Daphnia galeata mendotae.*" *Ecology* 45:94-112.

Jolley, R. L. 1973. *Chlorination Effects on Organic Constituents in Effluents from Domestic Sanitary Sewage Treatment Plants.* ORNL/TM-4290. Oak Ridge National Laboratory. Oak Ridge, Tennessee.

Jolley, R. L. 1975. "Chlorine-Containing Organic Constituents in Chlorinated Effluents from Sewage Treatment Plants." *J. Water Poll. Control Fed.* 47:601-618.

Laubusch, E. J. 1962. In J. S. Sconce (Ed.) *Chlorine, Its Manufacture and Use.* Am. Chem. Soc. Monogr. Series No. 154. Reinhold Publishing Corp., New York.

Palin, A. T. 1950. "A Study of Chloro Derivatives of Ammonia and Related Compounds, with Special Reference to their Formation in the Chlorination of Natural and Polluted Waters." *Water Water Eng.* 54:248-256.

Southworth, G. R. and C. W. Gehrs. 1976. "Photolysis of 5-Chlorouracil in Natural Waters." *Water Res.* 10:967-971.

DISCUSSION

Mathilde J. Kland, Lawrence Berkeley Laboratory. What was the dark reaction control in the 5-chlorouracil reaction?

Gehrs. I am going to call on George Southworth to answer that question.

George Southworth, Oak Ridge National Laboratory. The controls were maintained under the same conditions, that is, temperature and pH, but with no light.

Kland. Did you identify the product or products of the light reaction?

Southworth. No. We don't know specifically, but it is degraded to the point where it has little UV absorption.

Kland. Under the influence of UV radiation, uracil and methyluracil both add water across the double bond.

Southworth. We have postulated that it is a hydrolysis mechanism as has been demonstrated for halouracils by Garrett and co-workers at the University of Florida. The reaction involves dehalogenation and loss of chromophoric properties.

Kland. Did you examine the behavior of the solution obtained after light exposure in the presence of acid and base? The hydrate undergoes ring opening to the corresponding aldehyde, dehydration or reconstitution, for example to the original uracil, depending on pH.

Southworth. Afterwards? Yes. You are talking about reversibility as in the hydration of uracil. There is no evidence of that in our experiments.

Greg L. Seegert, University of Wisconsin at Milwaukee. Your graphs on the effects of 5-chlorouracil and 4-chlororesorcinol on the hatching success of carp eggs indicated a significant depression in hatching success over a range of concentrations from 0.001 to 10 mg/l. However, this effect was not seen at comparable levels when these chemicals were tested in combination with a series of chloro-organic compounds. Do you have an explanation for this?

Gehrs. I am glad you raised the point. You are fortifying the concern we are raising with respect to mixtures and complex effluents. We can't postulate what the interactive activity might be in the synthetic effluent. I believe our results to be quite valuable, however, in showing the complexity of the problem. While we can look at form, or structure, and at least calculate or predict effects from these data, what happens when we move into the natural system and find all of the materials together?

SECTION IV.
MODELING AND PREDICTION

Carl W. Gehrs, Session Chairman
>Environmental Sciences Division
>Oak Ridge National Laboratory
>Oak Ridge, Tennessee 37830

The first two days of our discussions have been aimed at presenting information raising questions with respect to what types of research are still needed. After our initial planning for this conference, we came to the realization that one of the things that we did not have included was a time when individuals who are involved in assimilating data and coming up with generalizations from these data could present the information. And that is the purpose of this section this morning. It's to allow individuals who have taken a look at the data which are available on chlorine to present to us their concepts and their ideas with respect to the question, "What can we do with the data so that we can use it?"

18

MODELING RESIDUAL CHLORINE LEVELS: CLOSED-CYCLE COOLING SYSTEMS

Guy R. Nelson

    Industrial Environmental Research Laboratory
    U.S. Environmental Protection Agency
    Cincinnati, Ohio 45268

ABSTRACT

    A mathematical model which predicts residual chlorine levels in cooling tower blowdown streams at any time during the chlorination cycle is discussed in this paper. To quantify the absence or presence of residual chlorine in the blowdown, the model interprets residual chlorine as negative chlorine demand.

    The general model has eight variations applying to specific chlorination program characteristics. The program characteristics affecting the general model are:

    1. Split stream vs no split stream chlorination (the fraction of the recirculating water chlorinated).
    2. Residual data feedback vs no residual data feedback (the type of chlorine feed equipment used).
    3. Positive vs negative demand at the end of the chlorine feed period (the time length of the chlorine feed period).

The variations to the model are useful not only in predicting residual chlorine levels in the blowdown, but also in making alterations in existing chlorination programs which minimize chlorine waste, provide more disinfecting efficiency, and reduce residual chlorine levels in the blowdown.

INTRODUCTION

Many cooling tower systems use chlorine or hypochlorites to control bacteria and slime growth in the condenser tubes carrying cooling water in the plant process. The biological growth, if left uncontrolled, causes excessive tube blockages, poor heat transfer and accelerated system corrosion—all of which reduce plant efficiency. For any cooling tower system the length of time of the chlorine feed period and the number of chlorine feed periods per day, week or month change as the biological growth problem changes. In most cooling tower systems, the chlorine is added at or near the condenser inlet in enough quantity to produce a free residual chlorine level of 0.1 to 0.6 mg/l in the water leaving the condenser. The amount of chlorine added to maintain the free residual chlorine depends upon the amount of chlorine demand agents and ammonia in the water.

Chlorine and ammonia react to form chloramines. These chloramines constitute the combined residual chlorine of the water. This combined residual chlorine is less efficient and slower in providing biological control than free residual chlorine (AIChE, 1972; Draley, 1972a; White, 1972; Puckorius, 1975).

Although chlorination is effective for slime control in the condenser tubes of cooling tower systems, its application may result in residual chlorine in the blowdown discharged to the receiving water. The effects of residual chlorine on aquatic life are of great concern (Hamilton *et al.*, 1970; Brooks and Baker, 1972; Draley, 1972a; Brungs, 1973). Data on the toxicity of residual chlorine to aquatic organisms is available. W. A. Brungs' publication, "Effects of Residual Chlorine on Aquatic Life: Literature Review," recommends criteria for maximum residual chlorine concentrations in receiving waters (Brungs, 1973). For the intermittent presence of residual chlorine, not to exceed 2 hr/day, the criteria indicate a maximum tolerable concentration of 0.2 mg/l. For the continuous presence of residual chlorine, the maximum tolerable concentration is 0.01 mg/l. These concentrations would not protect trout, salmon and some important food organisms and are potentially lethal to sensitive life stages of sensitive fish species. Brungs recommends lower criteria to protect trout and salmon. These lower criteria are 0.04 mg/l for the intermittent presence (2 hr/day) of residual chlorine and 0.002 mg/l for the continuous presence of residual chlorine.

The potential effects of residual chlorine on aquatic organisms require the measurement, prediction and control of residual chlorine in effluent discharges to the aquatic environment. The purpose of this paper is to discuss a model which predicts levels of residual chlorine in the recirculation systems and blowdown streams of cooling towers. The model can be used to improve chlorination programs resulting in less chlorine waste, more disinfecting efficiency and reduced environmental impact. Although the paper is directed toward the application of chlorine in condenser cooling systems of thermal-electric power plants, the model presented can be used in other industrial chlorination programs where the conditions are similar.

MODEL ANALYSIS

In an Environmental Protection Series report, U.S. Environmental Protection Agency (EPA), I develop and analyze a model which calculates the residual chlorine level in cooling tower blowdown (Nelson, 1973). The vocabulary, concepts and notations that apply to the model are defined in the appendix. The mathematical expression of the model itself is contained in the appendix also. To illustrate the model's utility, I apply it to studies of three hypothetical chlorination programs on two cooling tower systems. One system is a mechanical draft cooling tower; the other is a natural draft cooling tower.

Six cooling tower characteristics affect the model's results in both cooling tower systems: (a) the cooling system volume, V; (b) the recirculation rate, Q_R; (c) the initial chlorine demand in the blowdown, C_{BO}; (d) the flashing rate of residual chlorine in the atmosphere, F; (e) the initial residual chlorine level in the condenser effluent, C_{RO}; and (f) the blowdown flow rate, Q_B. These characteristics manifest themselves in the predictive model through the terms discussed below.

The V/Q_R term (expressed in minutes) is the ratio of the water volume of the entire cooling system to the recirculating water flow rate. This term indicates the length of time for a given water parcel to make one pass through the system. With conventional system design, this value is 10 to 15 min for mechanical draft towers and 20 to 22 min for natural draft towers.

The dimensionless term Q_B/Q_R is the ratio of the blowdown water flow rate to the recirculating water flow rate. Its value, along with the V/Q_R term, expresses the length of time that a conservative chemical remains in the system.

The dimensionless term C_{RO}/C_{BO} is the ratio of the initial residual chlorine level in the recirculating water (C_{RO}) to the initial chlorine demand in the blowdown (C_{BO}).

The dimensionless term F is the fraction of the residual chlorine which flashes or decomposes as it passes through the cooling tower fill. The theoretical range of the value of F can be 0.0 to 1.0. Probably F is related in some yet unquantified way to tower inlet temperature and to cooling range and/or the water-to-air ratio; the chemical form of the chlorine residual may also be a factor. The concept of F is documented (Draley, 1972a, 1972b; White, 1972). In the draft Environmental Impact Statement for the Davis Besse steam-electric generating plant, Draley suggests an F value of 0.5 for combined residual chlorine in the natural draft tower system (Draley, 1972b).

Table I lists the values of cooling tower and chlorination program characteristics found in typical cooling tower systems. The table also lists the values which were applied to my hypothetical cooling tower systems. Table II and Figures 1, 2 and 3 summarize the results of the model's application on the cooling systems.

Program Study One

In this first study, the chlorination program characteristics do not include residual data feedback or split stream chlorination. The expression for TSA of models NNN and NNP is the following (see appendix):

Table I. Cooling Tower and Chlorination Program Characteristics (Nelson, 1973)

Characteristic	Typical Values		Program Study Values
	Minimum	Maximum	
A. Chlorination program:			
Chlorination cycle	8 hr	7 days	24 hr
Chlorine feed period	10 min	30 min	15 min
Split stream Cl_2	Optional		Program study 3
Residual feedback	Optional		Program study 2
B. Cooling tower:			
Q_B/Q_R	0.008	0.015	0.010
F	0.300	0.600	0.400
C_{RO}	-0.100	-1.000	0.400
C_{BO}	0.500	3.000	0.667
C_{RO}/C_{BO}	-0.100	-1.200	-0.600
V/Q_R[a]	10.000	15.000	10.000
V/Q_R[b]	20.000	22.000	20.000

[a] Mechanical draft.
[b] Natural draft.

Table II. Program Study Results

Program Study	Time Chlorine in Blowdown (min)	Maximum Concentration (mg/l)
Number 1		
a. Natural draft	0	0.000
b. Mechanical draft	83	0.349
Number 2	15	0.033
Number 3	0	0.000

350 WATER CHLORINATION

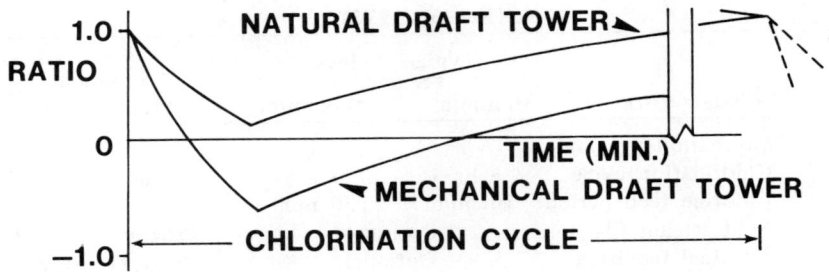

Figure 1. Program study one.

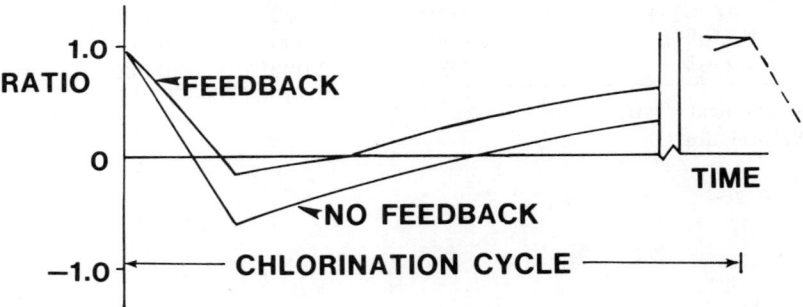

Figure 2. Program study two.

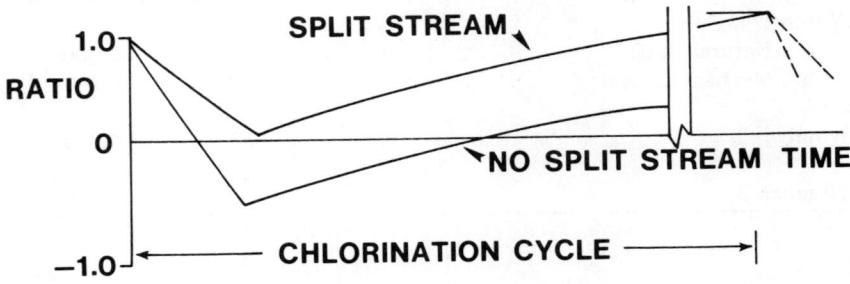

Figure 3. Program study three.

$$\text{RATIO} = X_1 (1-e^{-Y_1 t}) + e^{-Y_1 t}$$

In the natural draft program, RATIO remains positive throughout the chlorine feed period. This indicates that there is no residual chlorine in the blowdown. In the mechanical draft program, RATIO becomes negative during TSA. The boundary condition, t = 15 min, determines its value at the end of TSA. The first appearance of residual chlorine in the blowdown occurs when there is zero chlorine demand in the sump, or RATIO = 0. The time during the chlorine feed period when this occurs is 8.75 min. Therefore, residual chlorine is present in the blowdown for 6.25 min during TSA.

A residual first appears at 8.75 min after the start of the chlorine feed period. This residual increases in value until the end of the feed period, at which time it is at its maximum value. Since the model defines residual chlorine as negative chlorine demand, RATIO is at its minimum value at this time.

The TSB expression does not apply to the natural draft program, because its terminal demand value is not negative. The TSB expression for the mechanical draft program is the following:

$$\text{RATIO} = X_2 (1-e^{-Y_2 t}) + (C_B/C_{BO})_B\, e^{-Y_2 t}$$

The value of RATIO at the start of TSB is -0.523. RATIO increases in value through TSB until it reaches the value of zero. The boundary condition, RATIO = 0, quantifies the time of TSB. For this case it is 76 min. Figure 1 illustrates the value of RATIO through the complete chlorination cycle for both programs under program study one.

To minimize repetition, the other two program studies concentrate on alterations to the mechanical draft program. The concepts in program studies two and three are applicable to natural draft programs.

Program Study Two

This program study analyzes the effect of residual feedback on the value of RATIO for the mechanical cooling tower draft.

Model NRN applies in this case—because the value of RATIO is negative at the end of TSA. The comparison of the two studies shows a marked reduction in the concentration of residual chlorine in the blowdown of the cooling tower in study two. The use of residual data feedback in the mechanical draft program not only reduces the concentration of residual chlorine in the blowdown, but also reduces the length of time during the chlorination cycle in which the blowdown contains residual chlorine.

Program Study Three

In this program study, the chlorination program for the mechanical draft tower includes split stream chlorination and excludes residual data feedback. In this case, Model SNP applies—because the value of RATIO at the end of TSA is positive.

There are two substeps to TSA in model SNP. The first substep applies until there is residual chlorine in the recirculating water after the split stream is remixed with the remaining streams. The flashing term (F) does not appear in the substep 1 expression because the chlorine demand is positive (no residual chlorine). Negative demand appears in the recirculating water when RATIO = $S(1 - C_{TO}/C_{BO})$. Once the condition is reached, then the substep 2 expression applies for the rest of TSA.

The positive demand value at the end of TSA indicates that there is no residual chlorine in the blowdown. Therefore, TSB does not apply. RATIO regenerates back to the value of positive unity before the start of the next chlorine feed period.

This study shows the value of split stream chlorination in minimizing or eliminating residual chlorine in the blowdown.

CONCLUSIONS

The model discussed in this paper predicts the value of RATIO during the chlorination cycle of a cooling tower system. RATIO is defined as the ratio of the blowdown chlorine demand value at any time during the chlorination cycle to its value at the start of the chlorination cycle. A positive value of RATIO indicates that there is no residual chlorine present in the blowdown. A negative value of RATIO indicates that residual chlorine is in the blowdown. RATIO's rate of decrease and increase during the chlorination cycle, and its minimum value at the end of the chlorine feed period, are all functions of the cooling tower and chlorination program characteristics of specific systems.

The analysis of the models shows potential methods of reducing or eliminating residual chlorine levels in the blowdown by taking advantage of some of the optional program characteristics. These potential methods include one or more of the following alternatives:

1. Installing residual data feedback equipment into the chlorine feed system.
2. Practicing split stream chlorination.
3. Reducing the chlorine feed period, if possible.
4. Reducing the initial residual chlorine level in the condenser effluent.
5. Increasing the water volume of the cooling tower. This alternative may not apply to existing cooling towers because it involves the system design. The alternative can apply to systems on the engineering drawing boards. This alternative may have other advantages—such as an extra supply of water for fire protection.
6. Cutting off the blowdown when residual chlorine appears in the sump. The blowdown flow can resume after the residual is dissipated by the flashing effect and the make-up water chlorine demand. The length of time during which the blowdown can be eliminated is a function of the system's upper limit on dissolved solids.
7. Mixing the blowdown with another stream which has a high chlorine demand.

In cases where limited field analyses have been performed, trends are identified which conform to model predictions (Baker, 1973; Draley, 1973). For field verification it is necessary to use the amperometric procedure to measure low levels of residual chlorine (Brungs, 1973). The *Standard Methods* amperometric procedure (American Public Health Association, 1965) for determining residual chlorine in aqueous solutions is not applicable to all cooling tower waters. Some cooling tower waters contain copper, turbidity, natural buffering and water treatment chemicals. These constituents may produce interferences in the analytical procedure which result in erroneous residual chlorine readings. A report by Manabe (1972), which describes a modification of the procedure, is available. The modification increases the efficiency of both the sampling and the titrating procedures.

Even without substantial field verification, the models have utility. They can be used to modify existing and proposed chlorination programs in cooling tower systems. The modifications can provide increased chlorination efficiency and reduced residual chlorine levels in the blowdown.

REFERENCES

American Institute of Chemical Engineers (AIChE). 1972. "Cooling Towers." pp. 79-80.

American Public Health Association. 1965. *Standard Methods for the Examination of Water and Wastewater*, 12th ed. New York. p. 103.

Baker, R. J. 1973. *Fate and Disposal Characteristics of Residual Chlorine Discharged to Receiving Streams.* Wallace and Tiernan.

Brook, A. J. and A. L. Baker. 1972. "Chlorination at Power Plants: Impact on Phytoplankton Productivity." *Science* 176:1414-1415.

Brungs, W. 1973. "Effects of Residual Chlorine on Aquatic Life: Literature Review." *J. Water Pollut. Control Fed.* 45:2180-2193.

Draley, J. E. 1972a. *The Treatment of Cooling Waters with Chlorine.* ANL/ES-12. Argonne National Laboratories, Argonne, Illinois. 11 pp.

Draley, J. E. 1972b. "Davis Besse Nuclear Power Station Draft Environmental Impact Statement. Appendix B." U.S. Atomic Energy Commission. November. (Final statement issued March 1973).

Draley, J. E. 1973. *Cl_2 Experiments at the John E. Amos Plant.* ANL/ES-23. Argonne National Laboratory, Argonne, Illinois.

Hamilton, D. H., Jr., D. A. Flemer, C. W. Keefe and J. A. Mihursky. 1970. "Effects of Chlorination on an Estuarine Primary Production." *Science* 169:197-198.

Manabe, R. 1972. *Measuring Residual Chlorine Levels in Cooling Water/ Amperometric Method.* EPA 660/2-73-039. U.S. Environmental Protection Agency.

Nelson, G. R. 1973. *Predicting and Controlling Residual Chlorine in Cooling Tower Blowdown.* EPA-R2-73-273. U.S. Environmental Protection Agency.

Puckorius, P. 1975. Personal communication. Zimmite Corp., West Lake, Ohio.

White, G. C. 1972. *Handbook of Chlorination.* Van Nostrand Reinhold Company, New York.

APPENDIX A
ALTERNATE MODEL EXPRESSIONS
FOR SPECIFIC CHLORINATION PROGRAMS

The models for the specific chlorination programs are given in Tables IIIa and IIIb.

Table IIIa. Programs Without Split Stream Chlorination

Model	Equations		
	TSA	TSB	TSC
NNN	1A	2A	4A
NNP	1A		3A
NRN	5A	2A	4A
NRP	5A		3A

Table IIIb. Programs With Split Stream Chlorination

Model	Equations			
	TSA		TSB	TSC
	Sub 1	Sub 2		
SNN	6A	7A	2A	4A
SNP	6A	7A		3A
SRN	8A	9A	2A	4A
SRP	8A	9A		3A

Equations referred to in Tables IIIa and IIIb:

$$\frac{C_B}{C_{BO}} = X_1 (1-e^{-Y_1 t}) + e^{-Y_1 t} \qquad (1A)$$

where
$$X_1 = \frac{\frac{Q_B}{Q_R} + (1-F)\left(\frac{C_{RO}}{C_{BO}} - 1\right)}{\frac{Q_B}{Q_R} + F}$$

356 WATER CHLORINATION

and $\quad Y_1 = \dfrac{\dfrac{Q_B}{Q_R} + F}{\dfrac{V}{Q_R}}$

. . .

$$\dfrac{C_B}{C_{BO}} = X_2 (1-e^{-Y_2 t}) + \left(\dfrac{C_B}{C_{BO}}\right)_B e^{-Y_2 t} \qquad (2A)$$

where $\quad X_2 = \dfrac{\dfrac{Q_B}{Q_R}}{\dfrac{Q_B}{Q_R} + F}$

$Y_2 = \dfrac{\dfrac{Q_B}{Q_R} + F}{\dfrac{V}{Q_R}}$

and $\left(\dfrac{C_B}{C_{BO}}\right)_B$ = RATIO's value at the start of TSB

. . .

$$\dfrac{C_B}{C_{BO}} = 1 - \left(1 - \left(\dfrac{C_B}{C_{BO}}\right)_C\right) e^{-Y_3 t} \qquad (3A)$$

where $\quad Y_3 = \dfrac{Q_B}{V}$

and $\left(\dfrac{C_B}{C_{BO}}\right)_C$ = RATIO's value at the start of TSC

MODELING AND PREDICTION 357

$$\frac{C_B}{C_{BO}} = 1-e^{-Y_4 t} \qquad (4A)$$

where $Y_4 = \dfrac{Q_B}{V}$

. . .

$$\frac{C_B}{C_{BO}} = X_5 (1-e^{-Y_5 t}) + e^{-Y_5 t} \qquad (5A)$$

where $X_5 = \dfrac{\dfrac{Q_B}{Q_R} + (1-F)\dfrac{C_{RO}}{C_{BO}}}{\dfrac{Q_B}{Q_R} + 1}$

and $Y_5 = \dfrac{\dfrac{Q_B}{Q_R} + 1}{\dfrac{V}{Q_R}}$

. . .

$$\frac{C_B}{C_{BO}} = X_6 (1-e^{-Y_6 t}) + e^{-Y_6 t} \qquad (6A)$$

where $X_6 = \dfrac{\dfrac{Q_B}{Q_R} + S\left(\dfrac{C_{TO}}{C_{BO}} - 1\right)}{\dfrac{Q_B}{Q_R}}$

$Y_6 = \dfrac{Q_B}{V}$

C_{TO} = chlorine demand in chlorinated condenser effluent at the start of TSA

S = chlorinated fraction of total recirculating water flow

$$\frac{C_B}{C_{BO}} = X_7 (1-e^{-Y_7 t}) + S \left(1 - \frac{C_{TO}}{C_{BO}}\right) e^{-Y_7 t} \qquad (7A)$$

where
$$X_7 = \frac{\frac{Q_B}{Q_R} + (1-F) S \left(\frac{C_{TO}}{C_{BO}} - 1\right)}{\frac{Q_B}{Q_R} + F}$$

$$Y_7 = \frac{\frac{Q_B}{Q_R} + F}{\frac{V}{Q_R}}$$

C_{TO} = chlorine demand in chlorinated condenser effluent at the start of TSA

S = chlorinated fraction of total recirculating water flow

. . .

$$\frac{C_B}{C_{BO}} = X_8 (1-e^{-Y_8 t}) + e^{-Y_8 t} \qquad (8A)$$

where
$$X_8 = \frac{\frac{Q_B}{Q_R} + S \frac{C_{TO}}{C_{BO}}}{\frac{Q_B}{Q_R} + S}$$

$$Y_8 = \frac{\frac{Q_B}{Q_R} + S}{\frac{V}{Q_R}}$$

C_{TO} = chlorine demand in chlorinated condenser effluent at the start of TSA

S = chlorinated fraction of total recirculating water flow

$$\frac{C_B}{C_{BO}} = X_9 \left(1-e^{-Y_9 t}\right) + \left(\frac{S}{S-1}\right)\left(\frac{C_{TO}}{C_{BO}}\right)e^{-Y_9 t} \quad (9A)$$

where
$$X_9 = \frac{\dfrac{Q_B}{Q_R} + (1-F)\,S\,\dfrac{C_{TO}}{C_{BO}}}{\dfrac{Q_B}{Q_R} + S + F - FS}$$

$$Y_9 = \frac{\dfrac{Q_B}{Q_R} + S + F - FS}{\dfrac{V}{Q_R}}$$

C_{TO} = chlorine demand in chlorinated condenser effluent at the start of TSA
S = chlorinated fraction of total recirculating water flow

APPENDIX B
GLOSSARY

Vocabulary

Free residual chlorine is that portion of the total residual chlorine which will react chemically and biologically as hypochlorous acid or hypochlorite ion.

Combined residual chlorine is that portion of the total residual chlorine which will react chemically and biologically as chloramines.

Total residual chlorine is the sum of the free and combined residuals. Unless otherwise specified, throughout this paper *residual chlorine* refers to the *total residual chlorine.*

Chlorine demand is the amount of chlorine (mg/l) required to be added to a water (sample) before any stable residual chlorine is formed. Organics and reducing agents in the water cause this demand. These materials have varying reaction rates with chlorine. The reaction rates cause the chlorine demand value to be time-dependent. For the purpose of this paper, the chlorine demand is that demand which reacts with chlorine within 5 min of exposure.

The *chlorination program* describes the manner in which chlorine is fed and controlled in the cooling tower system.

The *chlorination cycle* is the length of time between the start of two sequential chlorine feed periods.

Split stream chlorination is an alternate method of chlorine addition. It is the practice of splitting the total recirculation flow through the condenser into a number of separate streams. One of these streams is chlorinated at a time. The chlorinated stream is then mixed with the remaining streams. The presence or absence of split stream chlorination is a chlorination program characteristic.

Residual feedback describes a function performed by chlorine feed equipment. If the control system in the equipment is capable of adjusting the flow of chlorine to produce a constant residual in the recirculation water out of the condenser, the system has residual feedback. The presence or absence of residual feedback is a chlorination program characteristic.

Concepts

Residual Chlorine vs Chlorine Demand

The model expresses the change in the chlorine demand of the blowdown during the chlorination cycle. To quantify the absence or presence of residual chlorine in the blowdown, the model interprets residual chlorine as negative chlorine demand. By conceptual definition, residual chlorine is not present in the blowdown unless the chlorine demand is satisfied.

C_B

In the model development, the term C_B represents the chlorine demand in the blowdown at any time during the chlorination cycle.

C_{BO}

The term C_{BO} represents the chlorine demand in the blowdown at the beginning of the chlorination cycle.

RATIO

The term RATIO represents the ratio of the chlorine demand in the blowdown at any time during the chlorination cycle to its initial value at the beginning of the cycle (*i.e.*, RATIO = C_B/C_{BO}).

Time Steps

In order to predict the chlorination cycle, the model breaks down the cycle into time steps. Figure 4 illustrates the time step concept.

Time Step A (TSA)

TSA is the length of time during which chlorine is added to the system (the chlorine feed period).

Time Step B (TSB)

TSB is the length of time after TSA in which the value of RATIO is negative.

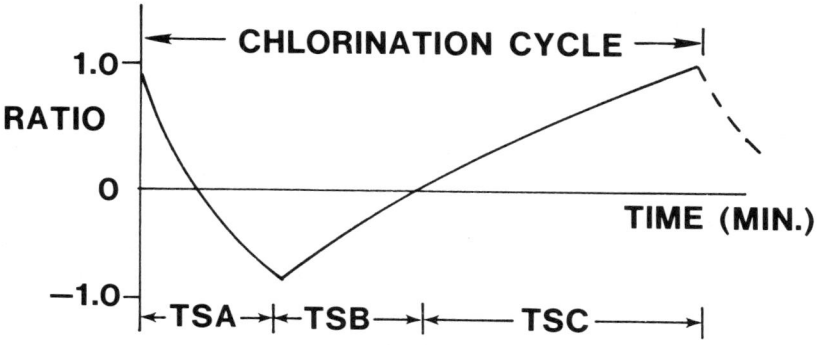

Figure 4. Time steps of the chlorination cycle.

Time Step C (TSC)

TSC is the length of time after TSA of a chlorination cycle in which the value of RATIO is zero or positive. TSC ends at the start of the next chlorination cycle.

Although equations can be (and have been) written for TSC, the numerical output is rather spurious and of no practical value in either environmental protection or chlorination program design. This is true primarly because the length of TSC is dictated by the biocidal requirements of the cooling system rather than any level or function of chlorine demand or residual obtainable from the equations. Also, extraneous factors such as dust washout or changes in make-up water characteristics are not accounted for. Finally, optimal use of the models for environmental protection and chlorine conservation require chlorine demand data at start of the cycle (C_{BO}).

Terminal Demand Value (TDV)

TDV is the value of RATIO at the end of TSA. It is the lowest value of RATIO during the chlorination cycle.

Positive vs Negative TDV

TDV can be either positive or negative at the end of TSA. If TDV is negative as shown in Figure 4, then the cycle contains three time steps (TSA, TSB and TSC). If TDV is positive or zero, then the cycle contains only two time steps (TSA and TSC). The positive vs negative TDV is a chlorination program characteristic.

Model Notation

There are eight specific models based upon the general model discussed in this paper. Each specific model applies to a set of chlorination program characteristics. Table IV is a matrix which defines the shorthand notation used to describe each model. For easy reference, the first letter in the shorthand notation refers to sidestream filtration or no sidestream filtration (S or N). The second letter refers to residual feedback (R or N). The third letter refers to positive TDV or negative TDV (P or N).

Table IV. Model Notation

	Residual Feedback		No Residual Feedback	
	(-) TDV	(+) TDV	(-) TDV	(+) TDV
Split stream chlorination	SRN	SRP	SNN	SNP
No split stream chlorination	NRN	NRP	NNN	NNP

For example, model NNN applies to a chlorination program which (a) does not have split stream chlorination, (b) does not have residual feedback, and (c) has a negative chlorine demand at the end of the chlorine feed period. Model SRP applies to a program which (a) has split stream chlorination, (b) has residual feedback, and (c) has a positive chlorine demand at the end of the chlorine feed period.

Cooling System Notation

Figure 5 is a schematic flow diagram of a typical cooling tower system;

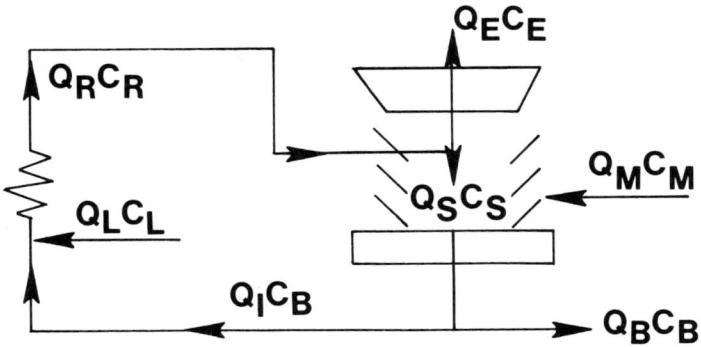

Figure 5. Cooling tower system.

where:

Q_B = blowdown flow rate in m^3/min (gpm);
Q_I = water flow rate into the condenser in m^3/min (gpm);

Q_L = chlorine flow rate into the condenser in m³/min (gpm);
Q_R = water flow rate returning to tower from the condenser in m³/min (gpm);
Q_E = water evaporation rate leaving tower stack in m³/min (gpm);
Q_S = water flow rate to the sump in m³/min (gpm);
Q_M = make-up water flow rate into the tower sump in m³/min (gpm);
V = cooling tower system volume in m³ (gal);
C_B = chlorine demand in the blowdown, (mg/l);
C_{BO} = chlorine demand in the blowdown at the start of the chlorine feed period (mg/l);
C_L = chlorine demand in the chlorine feed stream (mg/l);
C_R = chlorine demand in water returning to the tower from the condenser (mg/l);
C_{RO} = chlorine demand in the water returning to the tower from the condenser at the start of the chlorine feed period (mg/l);
C_E =- chlorine demand in the evaporated water leaving the tower (mg/l);
C_S = chlorine demand in the recirculating water entering the sump (mg/l);
C_M = chlorine demand in the make-up water entering the sump.

DISCUSSION

John E. Butts, Envirosphere Company. You mentioned that cooling tower blowdown could be held up to allow residual chlorine decay. In your experience what are the time ranges required to minimize the discharge of residual chlorine and what would be the approximate chlorine concentrations?

Nelson. It is highly specific to the type of cooling system that you are operating and the type of chlorination program you are applying to the system. I would say, in general, about two hr. And, in that time, you would probably notice about a 5% increase in total solids.

George C. White, Consulting Engineer. Plants are getting so large some blowdown lines are as large as 72 in. in diameter. How is it practical to shut down such a large pipe two or three times every 24 hr?

Nelson. In other words, you can't shut it off. You bring out an interesting point. For some of these already on-board systems, you cannot apply all the options I have listed. Hopefully, we can consider these kinds of options on the drawing board, so that we can try to solve some of the problems.

Robert J. Baker, Wallace and Tiernan Division of Pennwalt Corporation. An additional method of reducing or eliminating residual chlorine in tower blowdown is to chemically dechlorinate.

Nelson. Very good. Thank you, Bob. That should be added to the methods. An interesting thing about the model is that you can use the model to determine how much feed rate of sulfite or bisulfite to add to reduce the residual chlorine to the desired level.

William A. Brungs, U.S. Environmental Protection Agency. Cooling tower blowdown contains a much higher concentration of dissolved solids due to evaporation and other dissolved materials introduced intentionally. Did the presence of these materials affect the accuracy of the amperometric method?

Nelson. Yes. We have found that copper and iron in the cuprous and ferrous state did interfere with amperometric measurements in the field to the point where we were actually measuring a chlorine residual when none was present. We have since then solved the problem and can now get accurate amperometric measurements by adding a chelating agent to the water sample before we titrate. The reference is Manabe (1972). He describes not only the procedure for eliminating interferences, but also a good method for getting field data on cooling system blowdown. I don't know how many of you have tried but it is really an experience to get out in the field. At a desk we can draw all these curves that we want, but out in the field, cold and wet, with cooling tower drift all over you, it is different. Manabe takes the practical experience that he has gained and has developed a good procedure for field verification.

Thomas A. Miskimen, American Electric Power Service. The following represents a written comment submitted after the preceding discussion. The management of chlorine residual in cooling tower blowdown should not be isolated from the management of all other wastewater discharges from a power plant. For coal burning plants, there are advantages to using tower blowdown to transport fly ash. For any plant, the TSS of tower blowdown may prove unacceptable to the state water pollution control agency unless (a) either make-up water is treated for TSS removal or (b) blowdown is treated for TSS removal, such as being used for ash transport. The second choice is widely used.

19

A KINETIC MODEL FOR PREDICTING THE COMPOSITION OF CHLORINATED WATER DISCHARGED FROM POWER PLANT COOLING SYSTEMS

Milton H. Lietzke

    Chemistry Division
    Oak Ridge National Laboratory
    Oak Ridge, Tennessee 37830

ABSTRACT

    We are in the process of developing a kinetic model for predicting the composition of chlorinated water discharged from power plant cooling systems. As a start this model will contain three rate equations: the reaction of hypochlorous acid with ammonia, the reaction of hypochlorous acid with an organic amine, and the further reaction of hypochlorous acid with monochloramine. The simultaneous differential equations will be solved numerically to give the composition of the water as a function of time. Other rate equations will later be added to the model to account for other reactions that are known to take place. Eventually, the model will be incorporated into a large unified transport program.
    Kinetic and thermodynamic data for the most important chemical reactions that may occur during the chlorination of cooling waters are presented.

    Natural waters used for cooling purposes at power stations vary widely in composition. Depending on the source, such waters contain both organic and inorganic impurities. In addition to bacteria, algae, spores and viruses there will usually be traces of organic amines, other organic compounds, ammonia (or ammonium ion), traces of heavy metals and various anions. The analyses of two typical natural waters are shown in Table I.

Table I. Analyses of Two Typical Natural Waters

	Watts Bar Lake	Mississippi River
Sample data	July 1973	July 1974
pH	7.50	7.30
Chloride, mg/l	0.50	7.70
Organic carbon, mg/l	2.70	9.00
Inorganic carbon, mg/l	0.04	16.30
Organic nitrogen, mg/l	2.70	<0.05
Ammonia (as N), mg/l	0.50	0.15

In order to prevent slime formation in the cooling towers, chlorine is commonly added to the water. The chlorine is rapidly hydrolyzed to yield equimolar quantities of hypochlorous acid and hydrochloric acid. The hypochlorous acid dissociates into hydrogen ions and hypochlorite ions, the extent of dissociation being a function of both pH and temperature. Both the hypochlorous acid and the hypochlorite ion are powerful chlorinating agents. They will react rapidly, for example, with ammonia or ammonium ions to produce chloramines and with organic amines to produce N-chlorinated amines. The rates of these reactions are also functions of both pH and temperature. The treated cooling water containing the chloramines and unreacted chlorine is eventually returned to the biosphere.

Although the toxic nature of chloramines has long been recognized, it has only been within the last few years that national attention has been focused on the problem. For example, a massive fish kill in the cooling waters of a large generating station on the shore of Saginaw Bay has been attributed to a lethal concentration of residual chlorine in the water. Whether the chloramines or other chlorinated compounds will have a long-term toxic effect on man is at present unknown.

In any event it is now necessary to take cognizance of this problem in assessing the impact of power-generating stations on the environment.

In developing a model for predicting the composition of chlorinated water discharged from power plant cooling systems it is first necessary to have kinetic and thermodynamic data on the various chemical reactions that may occur. Unfortunately these data are scattered in the literature and in many cases difficult to find. Hence a literature search has been made and the available data on the most important reactions to be expected have been compiled and are presented in the appendix of this paper. In each case reference is made to the source of the data.

Examination of the appendix reveals that in many cases two rate constants are given for a particular reaction: a theoretical rate constant and an observed rate constant. The theoretical rate constant refers to the chemical reaction exactly as written. In practical calculations the theoretical rate constant is inconvenient to use since the reacting species may themselves participate in equilibria which are functions of pH. These equilibria are taken into account in the case of the observed rate constants. For example, consider the reactions between ammonia and hypochlorous acid (Equation 5, appendix). When the observed rate constant for this reaction is used it is only necessary to specify the total concentration of nitrogen as ammonia and ammonium ion and the total concentration of chlorine as hypochlorous acid and hypochlorite ion. Similar considerations apply to many of the other chemical reactions tabulated.

In order to predict the composition of chlorinated natural water returned to the environment from a cooling tower we are developing a kinetic model under the unified transport approach. At present the model considers three chemical reactions: the reaction of hypochlorous acid with ammonia, the reaction of hypochlorous acid with a composite organic amine, and the further reaction of hypochlorous acid with the monochloramine formed in the first reaction. These reactions are the following:

$$NH_3 + HOCl \rightleftharpoons NH_2Cl + H_2O$$
$$RNH_2 + HOCl \rightleftharpoons RNHCl + H_2O$$
$$NH_2Cl + HOCl \rightleftharpoons NHCl_2 + H_2O$$

The three differential equations representing the rates of these reactions in terms of the appropriate observed rate constants and the concentrations of the several species involved are solved simultaneously using the Runge-Kutta technique. Additional chemical reactions will be added to the model as seem warranted.

In addition to the kinetic model an equilibrium model has also been developed for the same chemical reactions. Since under some conditions of pH and temperature prevailing in natural waters the chlorination reactions are extremely rapid (a few seconds to a few minutes for 99% reaction), it is only necessary in these cases to perform an equilibrium calculation to predict the composition of the chlorinated water returned to the environment.

At the present time both of the programs are designed for calculations at a fixed temperature only. In the immediate future temperature-dependent expressions for the various rate and equilibrium constants will be incorporated into the models.

ACKNOWLEDGMENT

The author would like to express his sincere thanks to Dr. R. M. Rush for presenting this paper at the conference while the author was out of the country.

APPENDIX

Thermodynamic and kinetic data on the following reactions are summarized:

1. $NH_3 + H_2O \rightleftharpoons NH_4^+ + OH^-$
2. $NH_4 + (H_2O) \rightleftharpoons NH_3 + H^+$
3. $Cl_2 + H_2O \rightleftharpoons HOCl + H^+ + Cl^-$
4. $HOCl \rightleftharpoons H^+ + OCl^-$

5. $NH_3 + HOCl \rightleftharpoons NH_2Cl + H_2O$

6. $CH_3NH_2 + HOCl \rightleftharpoons CH_3NHCl + H_2O$

7. $CH_3NH_2 + H_2O \rightleftharpoons CH_3NH_3^+ + OH^-$

8. $(CH_3)_2NH + H_2O \rightleftharpoons (CH_3)_2NH_2^+ + OH^-$

9. $(CH_3)_3N + H_2O \rightleftharpoons (CH_3)_3NH^+ + OH^-$

10. $NH_2Cl + HOCl \rightleftharpoons NHCl_2 + H_2O$

11. $2NH_2Cl + H^+ \rightleftharpoons NH_4^+ + NHCl_2$

12. $2NH_2Cl \rightarrow NHCl_2 + NH_3$

13. $NH_2Cl + H_2O \rightleftharpoons NH_3 + HOCl$

14. $(CH_3)_2NH + HOCl \rightleftharpoons (CH_3)_2NCl + H_2O$

15. $CH_3NHCl + HOCl \rightleftharpoons CH_3NCl_2 + H_2O$

16. $2CH_3NHCl \rightleftharpoons CH_3NCl_2 + CH_3NH_2$

17. $CH_3CONHCH_3 + HOCl \rightleftharpoons CH_3CONClCH_3 + H_2O$

18. $CH_3CONHCH_2COO^- + HOCl \rightleftharpoons CH_3CONClCH_2COO^- + H_2O$

19. $NH_2CONH_2 + HOCl \rightleftharpoons NH_2CONHCl + H_2O$

20. Rate constants for N-chlorination of a number of additional compounds relative to ammonia.

1. $NH_3 + H_2O \rightleftharpoons NH_4^+ + OH^-$

$$K_1 = \frac{[NH_4^+][OH^-]}{[NH_3]}$$

$t°C$	$K_1 \times 10^5$	$t°C$	$K_1 \times 10^5$
0	1.51	25	1.81
10	1.62	40	2.00
15	1.70	50	1.95

Reference: H. Lunden, *J. Chim. Phys.* 5: 574 (1907).

2. $NH_4^+(H_2O) \rightleftharpoons NH_3 + H^+$

$$K_2 = \frac{[NH_3][H^+]}{[NH_4^+]}$$

t°C	$K_2 \times 10^{10}$	t°C	$K_2 \times 10^{10}$
0	(0.83)	30	8.06
5	1.25	35	11.30
10	1.86	40	15.70
15	2.73	45	21.40
20	3.98	50	28.90
25	5.68		

Reference: R. G. Bates and G. D. Pinching, *J. Res. Nat. Bur. Stand.* 42: 419 (1949).

3. $Cl_2 + H_2O \rightleftharpoons HOCl + H^+ + Cl^-$

$$K_3 = \frac{[Cl^-][H^+][HOCl]}{[Cl_2]}$$

t°C	$K_3 \times 10^4$
0	1.46
15	2.81
25	3.94
35	5.10
45	6.05

Reference: R. E. Connick and Yuan-tsan Chia, *J. Am. Chem. Soc.* 81: 1280 (1959).

Rate Constants:
$k_3 = 5.60$ sec^{-1} ± 0.45 if the reaction is monomolecular, and
$k_3 = 0.1$ 1/mole-sec if the reaction is bimolecular (the reaction mechanism has not been resolved).

Reference: A. Lifshitz and B. Perlmutter-Hayman, *J. Phys. Chem.* 64: 1663 (1960).

4. $HOCl \rightleftharpoons H^+ + OCl^-$

$$K_4 = \frac{[H^+][OCl^-]}{[HOCl]}$$

t°C	$K_4 \times 10^8$
0	2.0
5	2.3
10	2.6
15	3.0
20	3.3
25	3.7

Reference: G. M. Fair, J. C. Morris, S. L. Chang, I. Weil and R. P. Burden, *J. Am. Water Works Assoc.* 40: 1051 (1948).

5. $NH_3 + HOCl \rightleftharpoons NH_2Cl + H_2O$

$$K_5 = \frac{[NH_2Cl]}{[NH_3][HOCl]}$$

$K_5 = 3.6 \times 10^9$ at 25°C

Reference: J. E. Draley, ANL/ES-12 (1972), Argonne National Laboratory, Argonne, Illinois.

Rate constant: Over the temperature range 5-35°C the activation energy for the reaction is 3 kcal.
The theoretical rate constant is given by

$$k_5 = 9.7 \times 10^8 \exp(-3000/RT) \text{ 1/mole-sec}$$

Reference: J. C. Morris in *Principles and Applications of Water Chemistry*, S. D. Faust and J. V. Hunter (Ed.), John Wiley and Sons, Inc., New York (1967), p. 27.

k_5 (theoretical) is related to k_5 (obs.) by

$$k_5 \text{ (obs)} = \frac{k_5 \text{ (theoret)}}{1 + \frac{K_a K_b}{K_w} + \frac{K_a}{[H^+]} + \frac{K_b}{K_w}[H^+]}$$

where $K_a = K_4$; $K_b = K_1$; and K_w is given by $K_w = [H^+][OH^-]$ for the dissociation of water: $H_2O \rightleftharpoons H^+ + OH^-$.

Reference: I. Weil and J. C. Morris, *J. Am. Chem. Soc. 71*: 1664 (1949).

The temperature dependance of K_w (at saturation vapor pressure) is given by

$$\log K_w = \frac{P_1}{T} + P_2 \ln T + P_3 T + \frac{P_4}{T^2} + P_5,$$

where
$P_1 = 3.12860 \times 10^4$
$P_2 = 94.9734$
$P_3 = -0.097611$
$P_4 = -2.17087 \times 10^6$
$P_5 = -6.06522 \times 10^2$
$T = T°K$

Reference: F. H. Sweeton, R. E. Mesmer and C. F. Baes, Jr., *J. Solution Chem. 3*: 191 (1974).

6. $CH_3NH_2 + HOCl \rightleftharpoons CH_3NHCl + H_2O$

Rate constant: Over the temperature range 5-25°C the activation energy for the reaction is 1.9 kcal.
The theoretical rate constant is given by

$$k_6 = 7.8 \times 10^9 \exp(-1900/RT) \text{ 1/mole-sec}$$

Reference: J. C. Morris (1967), loc. cit., p. 32.

k_6 (theoretical) is related to k_6 (obs.) by

$$k_6 \text{ (obs)} = \frac{K_6 \text{ (theoret)}}{1 + \frac{K_a K_b}{K_w} + \frac{K_a}{[H^+]} + \frac{K_b}{K_w}[H^+]}$$

where $K_a = K_4$; $K_b = K_7$; and K_w is rep. as in 5 (above).

Reference: I. Weil and J. C. Morris, loc. cit.

7. $CH_3NH_2 + H_2O \rightleftharpoons CH_3NH_3^+ + OH^-$

$$K_7 = \frac{[CH_3NH_3^+][OH^-]}{[CH_3NH_2]}$$

t°C	-log K_7
0	3.449
10	3.405
20	3.380
30	3.367
40	3.375
50	3.386

Reference: D. H. Everett and W. F. K. Wynne-Jones, *Proc. Roy. Soc. (London), 177A*: 499 (1941).

8. $(CH_3)_2NH + H_2O \rightleftharpoons (CH_3)_2NH_2^+ + OH^-$

$$K_8 = \frac{[(CH_3)_2NH_2^+][OH^-]}{[(CH_3)_2NH]}$$

t°C	-log K_8
0	3.392
10	3.307
20	3.245
30	3.203
40	3.183
50	3.175

Reference: D. H. Everett and W. F. K. Wynne-Jones, loc. cit.

9. $(CH_3)_3N + H_2O \rightleftharpoons (CH_3)_3NH^+ + OH^-$

$$K_9 = \frac{[(CH_3)_3NH^+][OH^-]}{[(CH_3)_3N]}$$

t°C	-log K_9
0	4.591
10	4.407
20	4.261
30	4.141
40	4.059
50	3.992

Reference: D. H. Everett and W. F. K. Wynne-Jones, loc. cit.

10. $NH_2Cl + HOCl \rightleftharpoons NHCl_2 + H_2O$

$$K_{10} = \frac{[NHCl_2]}{[NH_2Cl][HOCl]}$$

$K_{10} = 1.35 \times 10^6$ at 25°C

Reference: Draley, loc. cit.

Rate constant: the activation energy for this reaction is 7.3 kcal. The theoretical rate constant is given by
$$k_{10} = 7.6 \times 10^7 \exp(-7300/RT) \text{ 1/mole-sec.}$$
In mildly acid solutions $k_{10} = k_{10}K_a/[H^+]$

where $K_a = K_2$.

Reference: J. C. Morris (1967) loc. cit, p. 28.

11. $2NH_2Cl + H^+ \rightleftharpoons NH_4^+ + NHCl_2$

$$K_{11} = \frac{[NH_4^+][NHCl_2]}{[H^+][NH_2Cl]^2}$$

$K_{11} = 6.7 \times 10^5$ at 25°C

Reference: G. M. Fair, J. C. Morris, S. L. Chang, I. Weil and R. P. Burden (1948), loc. cit.

12. $2NH_2Cl \rightarrow NHCl_2 + NH_3$

The activation energy for this reaction in the temperature range 7-49°C is 4.3 kcal.

Rate constant:
$$k_{12} = 80 \exp(-4300/RT) \text{ 1/mole-sec.}$$

Reference: M. L. Granstram, PhD Thesis, Harvard University, (1954).

13. $NH_2Cl + H_2O \rightleftharpoons HOCl + NH_3$

$$K_{13} = \frac{[HOCl][NH_3]}{[NH_2Cl]}$$

$K_{13} = 9.0 \times 10^{-2} \exp(-14{,}000/RT)$ moles/l.

Rate constant: as the first-order process (in the absence of competing NH_3) the specific rate is given by
$$k_{13} = 8.7 \times 10^7 \exp(-17{,}000/RT) \text{ sec}^{-1}$$

Reference: M. L. Granstrom, loc. cit.

14. $(CH_3)_2NH + HOCl \rightleftharpoons (CH_3)_2NCl + H_2O$

    Rate constant: over the temperature range 5-38°C (no change with temperature) the theoretical rate constant is given by $k_{14} = 3.3 \times 10^8$ 1/mole-sec.

    Reference: J. C. Morris (1967), loc. cit., p. 34.

    k_{14} (theoretical) is related to k_{14} (obs) by

    $$k_{14} \text{ (obs)} = \frac{k_{14} \text{ (theoret)}}{1 + \dfrac{K_a K_b}{K_w} + \dfrac{K_a}{[H^+]} + \dfrac{K_b}{K_w}[H^+]}$$

    where $K_a = K_4$; $K_b = K_8$; and K_w is represented as in 5 (above).

    Reference: I. Weil and J. C. Morris, loc. cit.

15. $CH_3NHCl + HOCl \rightleftharpoons CH_3NCl_2 + H_2O$

    Rate constant: $k_{15} = 1.1 \times 10^3$ 1/mole-sec at 25°C.
    The temperature coefficient has not been determined.

    Reference: J. C. Morris (1967), loc. cit., p. 32.

16. $2CH_3NHCl \rightleftharpoons CH_3NCl_2 + CH_3NH_2$

    Rate constant: $k_{16} = 4.7 \times 10^{-2}$ 1/mole-sec at 25°C.
    The temperature coefficient has not been determined.

    Reference: J. C. Morris (1967), loc. cit.

17. $CH_3CONHCH_3 + HOCl \rightleftharpoons CH_3CONClCH_3 + H_2O$

    Rate constant: $k_{17} = 1.4 \times 10^{-3}$ 1/mole-sec at 25°C.

    Reference: J. C. Morris (1967), loc. cit.

18. $CH_3CONHCH_2COO^- + HOCl \rightleftharpoons CH_3CONClCH_2COO^- + H_2O$

    Rate constant: $k_{18} = 5 \times 10^{-2}$ 1/mole-sec at 25°C.

    Reference: J. C. Morris (1967), loc. cit.

19. $NH_2CONH_2 + HOCl \rightleftharpoons NH_2CONHCl + H_2O$

   Rate constant: $k_{19} = 7.5 \times 10^{-2}$ 1/mole-sec at 25°C.

   Reference: J. C. Morris (1967), loc. cit.

20. Summary of rate constants for N-chlorination of a number of nitrogenous compounds relative to NH_3. F_o is the ratio of the rate constant for the particular reaction relative to that for the reaction with NH_3.

Table II. Rate Constants for N-Chlorination Relative to NH_3[a]

Nitrogenous Compound	$pK_o(25°C)$	$F_o = k_u/k_r$	log F_o
Methylamine	3.376	60[b]	1.78
Dimethylamine	3.226	54[b]	1.73
Diethylamine	3.067	23.0	1.36
Morpholine	5.300	9.0	0.95
Diethanolamine	5.120	9.3	0.97
Ethlaminoacetate	6.270	2.0	0.30
Glycine	4.221	22.0	1.34
Alanine	4.133	19.0	1.28
Leucine	4.256	14.0	1.15
(-) Alanine	3.765	33.0	1.52
Serine	4.795	6.7	0.83
Glycylglycylgylcine	6.090	2.3	0.36
Chloramide	ca. 15	5.5×10^{-5}	-4.26
N-chlormethylamine	ca. 13.8	1.8×10^{-4}	-3.74

[a]Based on reaction between HOCl and basic form of N compound.
[b]Computed from individual rate constants.
Reference: J. C. Morris (1967), loc. cit., p. 37.

DISCUSSION

Arthur S. Brooks, University of Wisconsin at Milwaukee. Have you used light as a factor in your model?

Richard M. Rush, Oak Ridge National Laboratory. I do not think it is planned although I gather from what I have heard in the last day or two that is might be useful to do so.

Brooks. Our work in Lake Michigan indiciates that light is a very important factor in determining the rate of chlorine dissipation in natural waters. Perhaps a light extinction coefficient could be incorporated in the model to account for various light levels in an effluent plume.

Rush. Very good.

J. Donald Johnson, University of North Carolina at Chapel Hill. Have you found any of the methylamine compounds in cooling tower systems? If so, how much of these compounds did you find?

Rush. We are not doing experimental work for this program. We will simply utilize kinetic and thermodynamic data on reactions for which that information is available. Obviously, we are looking at the first two very simple reactions. We do not have any results from this initial computation. We just got the program operational about the time that Dr. Lietzke left for Europe and so things are in a bit of a standstill right at the moment.

Robert L. Jolley, Oak Ridge National Laboratory. The type of program we are developing at this time is really a first approximation, and it will be more fully developed to represent the real situation as new program components are added. Because the kinetics of methylamine and HOCl reactions are known it was assumed, in order to start developing a working model, that the organic nitrogen which we have measured in cooling water samples was indeed methylamine. The program will be refined later.

20

ASSESSING TOXIC EFFECTS OF CHLORINATED EFFLUENTS ON AQUATIC ORGANISMS: A PREDICTIVE TOOL

Jack S. Mattice

Environmental Sciences Division
Oak Ridge National Laboratory
Oak Ridge, Tennessee 37830

ABSTRACT

Increased chlorination of surface waters has necessitated development of an analytical tool for assessing the toxicity of a chlorinated effluent so that environmental damage may be minimized. The tool proposed includes length of exposure as well as concentration as factors of importance and allows for protective limitations to be applied to releases on a site-specific basis. Cornerstones of the assessment tool are acute and chronic mortality thresholds. The thresholds are derived by (a) summarizing extant chlorine toxicity data (mostly as median tolerance limits) in log concentration-log duration plots, (b) bounding these data points from below with straight lines to estimated acute and chronic median lethal thresholds and (c) shifting the acute "median lethal threshold" to estimate a true mortality threshold (zero mortality) using an empirically derived relationship between exposure time necessary to yield fifty and zero percent mortality. Dose-time exposures for organisms entrained through the plant and into the discharge plume are compared with these thresholds to derive yes or no decisions regarding mortality. Limits may then be set to minimize environmental harm. Accurate plume dilution data are necessary for optimal application of this procedure, but parameters which increase discharge dilution rate are taken into consideration.

INTRODUCTION

Prediction of the magnitude of environmental perturbation resulting from release of chlorinated surface waters depends upon generic information dealing with (a) fate of the chlorine as regards both chemical form and concentration, (b) effect of these resultant forms on the individual organisms exposed and (c) translation of the effects on individuals to those on populations, communities and the ecosystem. Reaction products derived from chlorinated natural waters are determined by initial chlorine concentration, local water quality (especially relative concentrations of potential reactants), time of reaction and such physical conditions as pH, temperature, etc. Toxicity of chlorine compounds varies over a wide range; for example, some of the chloro-organic compounds are effective toxicants at parts per billion (ppb) concentrations (Allen et al., 1946 and 1948; Parker, 1935; Hopkins and Bean, 1966; Gehrs et al., 1974). In some instances, even the less chemically complex chloramines differ in toxicity from free chlorine (Butterfield, 1948; Douderoff and Katz, 1950; Moore, 1951; Merkens, 1958; Rosenberger, 1971; Holland et al., 1960). In addition, toxicity has been found to vary with temperature (Scheuring and Stetter, 1950; Dressel, 1971; Hoss et al., 1974; Stober and Hanson, 1974; Gregg, 1974), heavy metal exposure (Hoss et al., 1974; Wolf et al., 1975), species of organism, concentration and time of exposure, even for the same form of chlorine. Preliminary evidence also indicates that intermittent exposure to chlorine may be more efficacious than continuous exposure at the same mean concentration. Problems encountered in extrapolating individual effects to responses at higher levels of organization are as complex as those involved at the other generic information categories. The above complexity is confounded by the limited data base available from which to make valid generalizations for predictive purposes.

Despite the problems involved in predicting the impacts of chlorination, pragmatic considerations require the development of this capability. Water use by the power industry has been rapidly increasing (Krenkel, undated; Science and Public Policy Program, University of Oklahoma, 1975; Federal Power Commission, 1971). Since chlorine is the biocide specified for most power plants (Motley and Hoppe, 1970; Lee and Stratton, 1972; Becker and Thatcher, 1973), use of chlorine has paralleled this rapid increase. Expansion of other industries which discharge chlorinated effluents has also increased the potential for environmental damage. The potential problems revolving around use of chlorine would appear to indicate an obvious concomitant solution—elimination of chlorine from the list of acceptable biocides—except for the many advantages over currently available alternatives (Baker, 1959). The only rational policy would be to maintain functional capability of chlorine within the facility (*e.g.*, power plant) while controlling discharge levels to eliminate (or limit to acceptable levels) environmental effects. Reasonable limits on effluent releases must be based on the ability to predict the impact of various discharge levels.

In this paper, I will present in summary form, a method or tool [presented in full in Mattice and Zittel (1976) and summarized for another purpose in Mattice (1976)] which can be used to predict the mortality resulting from a specified chlorination schedule. This tool, which is aimed at filling the second of the generic information gaps described above, is based on comparison of chlorine dose-times (a) within an electric generating facility and (b) within its discharge area, with toxicity thresholds derived from the limited data base available. The tool is applicable in both marine and freshwater environs, but this presentation is confined to freshwater analyses.

Development of this tool was stimulated by the need for the Nuclear Regulatory Commission (formerly the U.S. Atomic Energy Commission) to respond to the requirements of the National Environmental Policy Act (NEPA, 1969). Early development of this tool by the Impact Assessments staff at Oak Ridge National Laboratory proceeded through the work of Coutant and co-workers (U.S.A.E.C., 1972; Coutant, 1974; Goodyear *et al.*, 1974). Subsequent modifications have been primarily due to the work of Mattice (U.S.A.E.C., 1973; Goodyear and Mattice, 1973; U.S.A.E.C., 1974; Mattice and Zittel, 1976; Mattice, 1976), although other members of the staff have also contributed significantly.

MORTALITY THRESHOLDS

In order to evaluate the effect of a proposed chlorination, it is necessary to know the vulnerability of freshwater organisms to chlorine exposures. Existing data regarding chlorine toxicity to freshwater organisms are presented as a plot of log chlorine concentration vs log exposure time (Figure 1), a form suggested by Coutant (1974). Concentrations are presented as total residual chlorine without regard to the type (free or combined) which predominated during the experiment. This appears reasonable as a first approximation (Mattice and Zittel, 1976). Data shown were derived using appropriate biological methodology and either accurate or conservative (*i.e.*, chemical analytical procedures which would tend to underestimate the concentrations of chlorine tolerated by the experimental organisms) chemical methods. Experimental conditions other than chlorine concentration were largely ignored in compiling and interpreting the data. Further information concerning species involved, investigators and methodology may be found in Mattice and Zittel (1976). A bibliography of abstracts of the papers from which data have been obtained is also available (Mattice and Pfuderer, 1976). Most of the data points (Figure 1) are dose-times yielding median (50%) mortality. The data shown represent 64 species, of which 8 are algae, 27 are invertebrates, 28 are fish and 1 is an amphibian.

Environmentally protective mortality thresholds were based on the most sensitive organisms for which data are available. This type of treatment is appropriate in cases in which it is necessary to make decisions based on limited data such as is true for chlorine toxicity. The form of the threshold lines (acute and chronic) was based on general toxicological principles. At concentrations of toxicant above some low value (the chronic threshold), increase in exposure time increases mortality at a given concentration. Below the chronic threshold, exposure time has no effect on mortality. On a plot of log concentration vs log exposure time the chronic threshold line would thus be parallel to the abscissa, and the acute threshold line would have a negative slope and end at the chronic threshold line. Preliminary threshold lines were drawn by enclosing the median

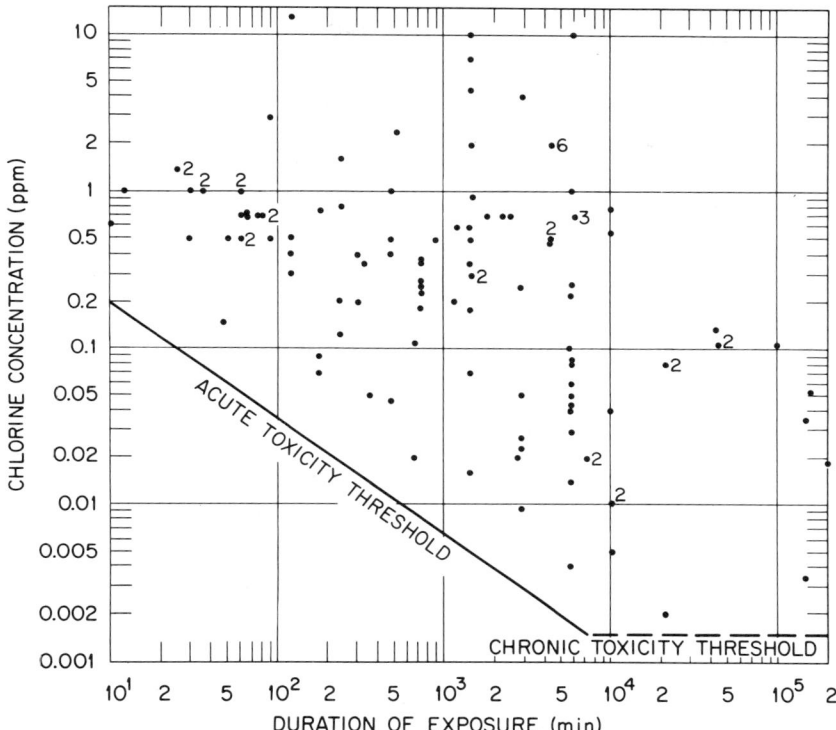

Figure 1. A summary of data on the toxicity of chlorine to aquatic life. Most points are for dose-times yielding median mortality. Acute and chronic toxicity thresholds (estimates of the 0% effect levels) are also shown (modified from Mattice and Zittel, 1976).

mortality data with two straight lines—the chronic and acute median mortality thresholds. Once these straight lines were drawn, it was necessary to shift the preliminary acute mortality threshold to estimate the true threshold (zero mortality). Mattice and Zittel (1976) found for fourteen species of organisms that for a given concentration of chlorine the maximum exposure time which would not result in any mortality was 37% of that which would cause death of 50% of the exposed organisms. The acute mortality threshold line shown in Figure 1 is the result of shifting the preliminary line to the left to exposure times only 37% as long. The threshold lines drawn indicate that for concentrations greater than 0.0015 mg/l, an increase in time of exposure generally increases the toxicity of chlorine. For concentrations less than 0.0015 mg/l, however, toxicity is no longer time-dependent and exposure for infinite time would not result in death. This concentration was thus designated the chronic mortality threshold.

Evaluation of a given exposure to chlorine can be made by comparing the dose-time combination with the mortality thresholds (Figure 1). If the dose-time combination is above the chronic mortality threshold and to the right of the acute mortality threshold, organisms exposed are killed; if not, they are not affected at all (*i.e.*, there are no sublethal effects). The latter results from the fact that a preponderance of the data involved mortality. Based on this procedure, exposure to 0.1 mg/l total residual chlorine for 10 min would not be lethal, while exposure to the same concentration for about 30 min would be lethal. In some cases (for example, if an area were populated entirely by highly tolerant organisms), this method could result in prediction of mortality when none would actually occur. However, because chlorine mortality thresholds are known for so few species, this type of conservative approach is necessary to ensure environmental protection. On the other hand, the procedure may underestimate sublethal effects. Chance of this latter, however, is reduced by inclusion of some sublethal effects as data points (see Mattice and Zittel, 1976).

Setting the acute and chronic toxicity thresholds as shown above (Figure 1) satisfies the criteria most important to preoperational evaluation of a chlorination. Any method for analysis must be (a) predictive, which implies a basis in principles of general applicability and (b) reliable, which implies a certain degree of conservatism. Although based on data from a limited number of species, the relationships between time and concentration and chlorine toxicity are similar to those found for other toxicants (Warren, 1971), indicating that the form used here is probably realistic. The predictions of mortality are conservative because the thresholds are set based on the most sensitive species at levels expected to be below lethal levels. Demonstration that a relationship exists between general species groups and relative toxicity, such as that postulated by Brungs (1973) for freshwater fish, may allow less conservative thresholds for some environments.

DEMONSTRATION OF THE ANALYTICAL TOOL

Three hypothetical examples will be used to demonstrate the procedures involved in using the proposed analytical tool as well as to indicate the efficient way in which the important time variable is treated (Figure 2). In each of the cases, plant specifications regarding chlorination are the same: chlorination proceeds at a rate sufficient to yield 2 mg/l initially; residence time in the facility following chlorination is 20 min; and discharge concentration is 1 mg/l total residual chlorine. Dilution rates following discharge differ in each case, with dilution being most rapid in Case III and least rapid in Case I. A given concentration isopleth (*e.g.*, 0.62) is shown at the same spatial location for each plume to indicate the equal volumes of water needed for the dilution, but the times differ in each case. None of the examples necessarily represent actual cases, but were chosen for purposes of demonstration.

Toxicity of both plant and plume entrainment may be assessed using plots of weighted concentration vs cumulative time since discharge. In each of the cases given, organisms entrained through the facility would be killed because the dose-time combination (see Plant Entrainment, Figure 3) lies to the right of the acute toxicity threshold and above the chronic toxicity threshold. Therefore, in this case, because further analysis need only concern organisms not passing through the plant, the times shown (Figure 3) refer to the interval since discharge. Such would also be the case for many industrial or waste treatment discharges. An organism entrained into a certain area of the discharge plume during dilution is exposed to a concentration bounded by those of the two surrounding isopleths for a time bounded by the total time for water movement between isopleths. In order to determine mortality, it is necessary to approximate both the dose-times of exposure over

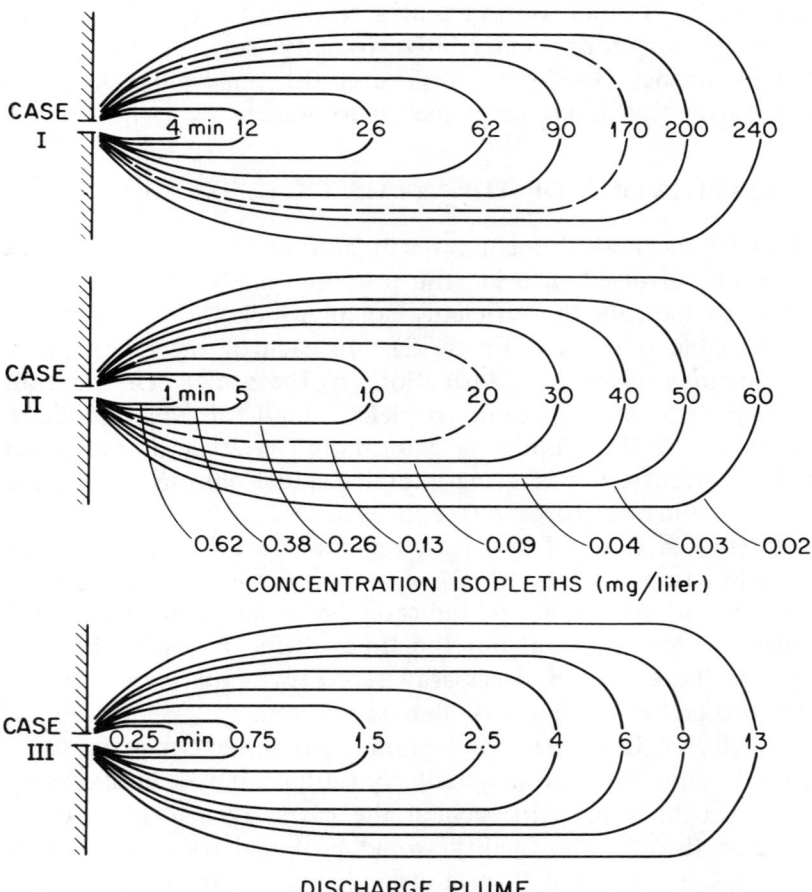

Figure 2. Comparison of the chlorine concentration in the discharge plumes of three hypothetical power plants showing three different rates of dilution. Concentration isopleths are identical for all the plumes, but the times to reach a given isopleth differ. Chlorine concentration at the point of discharge is given as 1 mg/l.

the periods between each concentration isopleth and the total dose-time of exposure in the plume. The concentration equal to the mean of two consecutive isopleths and applied over the total time between them was chosen as a reasonable estimate

MODELING AND PREDICTION 387

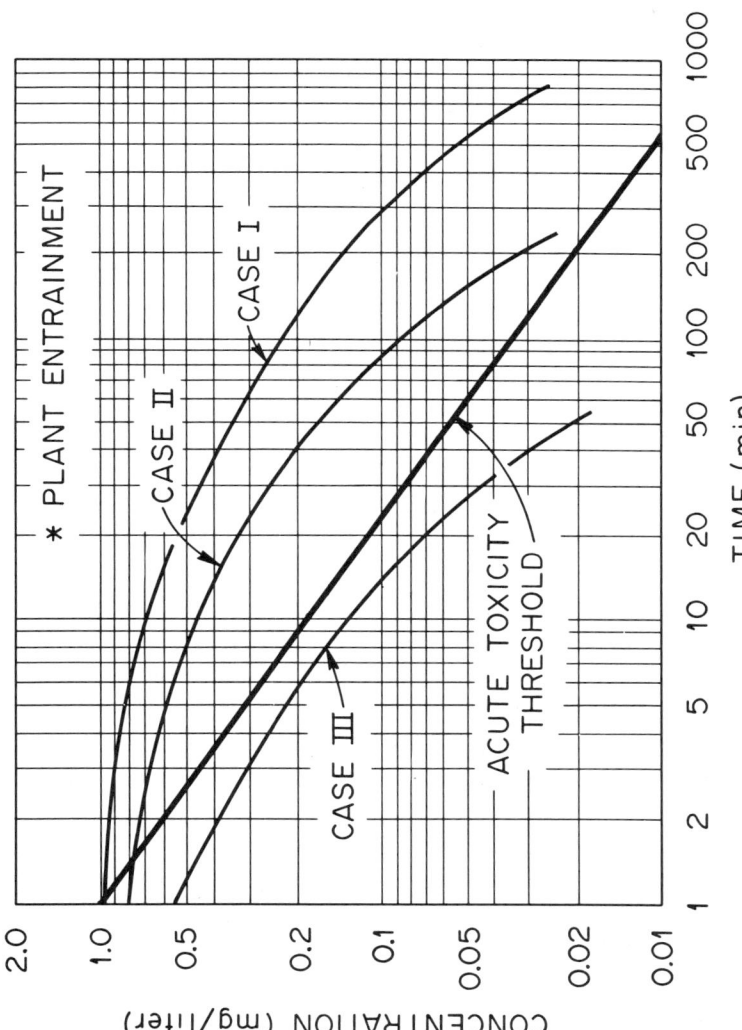

Figure 3. Chlorine dose-time combinations for organisms entrained into each of the plumes (Cases I, II and III) (Figure 2) prior to the 0.62 mg/l isopleth shown in relation to the acute toxicity threshold (Figure 1). Time shown is that from point of discharge. The dose-time for organisms entrained through the plant is also shown.

of the true exposure for that interval. Total exposure concentration for organisms entering the plume in a given area was estimated by computing the weighted average concentration for the total exposure period. Weighting was based on the exposure time at a given concentration relative to the total exposure duration. This method of computation assumes that chlorine doses are additive. Effects of entrainment in the plume differ for each of the "Cases." This is indicated clearly by the curves relating weighted concentration and cumulative time since discharge (Figure 3). For comparative purposes, only the concentration path for the organisms entrained into the plume prior to the first isopleth is shown for each case. In reality, a series of such plots would need to be made for organisms entrained into each area of the plume for each "Case." Mortality would result for organisms in Cases I and II, but not for Case III, because in the former cases one or more of the dose-times is enclosed by the acute and chronic toxicity thresholds; in the latter case none of the points is to the right of the acute threshold and above the chronic threshold. Because organisms entrained into the initial sector of the plume receive maximum exposures, no plume mortality would result at all in Case III, and it has not been considered further.

Further analyses of Cases I and II indicate that a larger area of the plume is impacted when dilution rate is less rapid (Tables I and II). In order to determine if organisms entrained into an area of the plume will be killed, the dose-time of exposure must be estimated. For example, organisms entrained prior to the sixth isopleth (0.04 mg/l, Figure 2 and Table I) will be exposed to 0.065 mg/l for 144 min in Case I or to the same concentration for 30 min in Case II. Comparison of these dose-times with the acute toxicity threshold (Figure 1) indicates that mortality will result from the former but not the latter case. In a full analysis, the full series of weighted mean dose-times might need to be compared with the thresholds. If any dose-time combination is above the thresholds, death is predicted for any organisms initially entrained in the area under consideration. By following a horizontal row, it is possible to find the area of entry into the plume which does not yield any lethal dose-times.

Table I. Concentration and Time of Exposure for Organisms Entrained into that Area of the Plume Enclosed by a Given Isopleth and Continuing in the Plume During Subsequent Dilution for Case I.

By following down a given column it is possible to observe the cumulative time course of exposure for organisms entrained into various areas of the plume as they reach each subsequent isopleth. Times shown are cumulative using discharge as time zero and concentrations are weighted according to percent of the total time since discharge. Lethal dose times for each case are enclosed.

Concentration Isopleth	Dose-Time Experienced for Each Initial Isopleth Encountered Upon Entrainment[a]								
	0.62 Dose Time	0.38 Dose Time	0.26 Dose Time	0.13 Dose Time	0.09 Dose Time	0.04 Dose Time	0.03 Dose Time	0.02 Dose Time	0.01 Dose Time
0.62	0.81 4	- -	- -	- -	- -	- -	- -	- -	- -
0.38	0.60 12	0.50 8	- -	- -	- -	- -	- -	- -	- -
0.26	0.45 26	0.39 22	0.32 14	- -	- -	- -	- -	- -	- -
0.13	0.30 62	0.23 58	0.23 50	0.20 36	- -	- -	- -	- -	- -
0.09	0.24 90	0.19 86	0.19 78	0.16 64	0.11 28	- -	- -	- -	- -
0.04	0.16 170	0.13 166	0.13 158	0.11 144	0.08 108	0.07 80	- -	- -	- -
0.03	0.14 200	0.11 196	0.11 188	0.09 174	0.07 138	0.06 110	0.04 30	- -	- -
0.02	0.12 240	0.10 236	0.10 228	0.08 214	0.06 178	0.05 150	0.03 70	0.03 40	- -
0.01	0.09 320	0.08 316	0.07 308	0.06 294	0.04 258	0.04 230	0.02 150	0.02 120	0.02 80

[a] Organisms entrained into the area between isopleths 0.62 and 0.38 are listed in column 0.38. Refer to Figure 2.

Table II. Concentration and Time of Exposure for Organisms Entrained into that Area of the Plume Enclosed by a Given Isopleth and Continuing in the Plume During Subsequent Dilution for Case II.

By following down a given column it is possible to observe the cumulative time course of exposure for organisms entrained into various areas of the plume as they reach each subsequent isopleth. Times shown are cumulative using discharge as time zero and concentrations are weighted according to percent of the total time since discharge. Lethal dose times for each case are enclosed.

Concentration Isopleth	0.62		0.38		0.26		0.13		0.09		0.04		0.03		0.02		0.01	
	Dose	Time	Dose	Time	Dose	Time	Dose	Time	Dose	Time	Dose	Time	Dose	Time	Dose	Time	Dose	Time
0.62	0.81	1	-	-	-	-	-	-	-	-	-	-	-	-	-	-	-	-
0.38	0.56	5	0.50	4	-	-	-	-	-	-	-	-	-	-	-	-	-	-
0.26	0.44	10	0.40	9	0.32	5	-	-	-	-	-	-	-	-	-	-	-	-
0.13	0.32	20	0.29	19	0.24	15	0.20	10	-	-	-	-	-	-	-	-	-	-
0.09	0.25	30	0.23	29	0.19	25	0.15	20	0.11	10	-	-	-	-	-	-	-	-
0.04	0.20	40	0.19	39	0.15	35	0.12	30	0.09	20	0.07	10	-	-	-	-	-	-
0.03	0.17	50	0.16	49	0.13	45	0.10	40	0.07	30	0.05	20	0.04	10	-	-	-	-
0.02	0.15	60	0.13	59	0.11	55	0.09	50	0.06	40	0.04	30	0.03	20	0.03	10	-	-
0.01	0.08	120	0.07	119	0.06	115	0.05	110	0.03	100	0.02	90	0.02	80	0.02	70	0.02	70

[a]Organisms entrained into the area between isopleths 0.62 and 0.38 are listed in column 0.38. Refer to Figure 2.

The areas enclosed in the boxes (Tables I and II) define the plume areas in which organisms would be killed for Cases I and II. The mean initial concentrations encountered by organisms entrained near the periphery of these areas are 0.07 for Case I and 0.20 for Case II. Thus, organisms in about 14 and 5 times the volume of the plant flow would be killed in Cases I and II, respectively, because dose-time combinations of organisms exposed in those volumes would be enclosed by the threshold lines (Figure 1). Of course, the ecosystem impact on the areas involved would have to be assessed on the basis of the relative importance of that volume and of the organisms in it to that of the rest of the ecosystem which is not directly impinged upon by the chlorine-caused mortality. Assessment at this level would involve population or ecosystem dynamics, including compensatory mechanisms and population interactions. Pertinent information at this level is sparse and limits real statements regarding impacts of any chlorine-caused mortality.

DISCUSSION

The proposed analytical tool for evaluating impacts associated with chlorination has a number of advantages: (a) application is relatively simple, (b) bias incorporated into the procedure favors environmental protection, (c) site-specific information regarding such factors as hydrology and facility design is incorporated and (d) flexibility exists, allowing inclusion of multiplant considerations and modification following appropriation of new data. The flexibility of this analytical tool is one of its greatest assets, because incorporation of topical information will probably not result in major modification of the tool itself, $i.e.$, the concepts and procedures involved in the analysis will remain constant. This is important since major efforts are being made in the areas of modeling the fate of chlorine following discharge, of determination of the toxicity of free and combined chlorine to aquatic organisms, and of population modeling under conditions of increased mortality. Other advantages are discussed further in Mattice and Zittel (1976).

As the tool evolves, the general trend of analyses will be from the more to the less conservative. It is obvious from the preceding discussion that the proposed method of estimating mortality (and chlorine limits based on them) is conservative at each step: setting the thresholds, estimating plume chlorine concentrations, and estimating the toxicity of each dose-time combination. Unless compensatory population mechanisms are taken into account, it is also conservative in extrapolating the resultant mortality to population effects. If environmental protection is the goal, each of these steps must be treated conservatively because of deficiencies in knowledge. As the information base grows in both depth and breadth, many of the conservative assumptions can be replaced by environmental data.

As with all summarizations of data, disadvantages of this method are based in the generalizations or assumptions made to facilitate collection. The most important of these is the assumption of equal toxicity of the reaction products (other than chlorine demand) resulting from chlorination. Some of the chloro-organics, such as the pyrimidine analog 5-chlorouracil, (Gehrs et al., 1974) may by their function in translation of the genetic material have relatively large effects at low concentrations. On the other hand, toxicity levels found by Zillich (1969), Basch et al. (1971) and Arthur et al. (1974) using chlorinated sewage effuents were in the same general range as those found following chlorination of less organically enriched waters (Mattice and Zittel, 1976). Most of the other assumptions of this method would be found in any method of toxicity assessment.

Although this method has advantages over other possible analytical methodologies, as discussed here and by Mattice and Zittel (1976), perhaps its greatest strength lies in its potential for use in setting realistic limits on discharges of chlorine. Toxicity of chlorine is influenced most strongly by concentration and time of exposure (Coutant, 1974; Goodyear et al., 1974). The importance of the time factor has played a role in development of complex discharge structures designed to dilute the effluent as rapidly as possible with the receiving water, but the time factor has largely been ignored in limits proposed by previous authors (Brungs, 1973; Collins and Deaner, 1973; Baker and Cole, 1974; Dawson, 1974; EPA, 1974; Gentile et al., 1974). The examples in Figure 2 and Table I indicate that, in using the proposed tool, time of exposure is a critical factor in determining toxicity of chlorine above the chronic mortality threshold. Limits may thus be adjusted to include consideration of site-specific dilution patterns and rates following release from the facility. These limits would thus have a realistic basis in the data available.

The tool proposed is not intended to represent the ultimate procedure for analyzing for impact of chlorine or for setting safe limits on releases. The information required for this does not exist. However, in being relatively conservative with respect to environmental protection and in having the broadest data base available, the tool probably represents the optimum given present knowledge and does allow interim decisions to be made regarding proposed chlorine utilization.

ACKNOWLEDGMENTS

I wish to thank Dr. Charles C. Coutant for presenting this paper to the Conference.

This research was supported by the Energy Research and Development Administration under contract with the Union Carbide Corporation. Publication No. 845, Environmental Sciences Division, ORNL.

REFERENCES

Allen, L. A., N. Blezard and A. B. Wheatland. 1946. "Toxicity to Fish of Chlorinated Sewage Effluents." *Surveyor* (England) 105:298.

Allen, L. A., N. Blezard and A. B. Wheatland. 1948. "Formation of Cyanogen Chloride During Chlorination of Certain Liquids: Toxicity of Such Liquids to Fish." *J. Hyg.* 46:184-195.

Arthur, J. W., R. W. Andrew, V. R. Mattson, D. T. Olson, G. E. Glass, B. J. Halligan and C. T. Walbridge. 1974. "Comparative Toxicity of Sewage-Effluent Disinfection to Freshwater Aquatic Life." National Water Quality Laboratory, U. S. Environmental Protection Agency, Duluth, Minnesota. 62 pp. (Unpublished manuscript).

Baker, R. J. 1959. "Types and Significance of Chlorine Residuals." *J. Am. Water Works Assoc.* 51:1185-1190.

Baker, R. and S. Cole. 1974. "Residual Chlorine: Something New to Worry About." *Ind. Water Eng.* (March/April) 10-18.

Basch, R. E., M. B. Newton, J. G. Truchan and C. M. Fetterolf. 1971. *Chlorinated Municipal Waste Toxicities to Rainbow Trout and Fathead Minnows.* Water Pollution Control Research Series No. 18050 GZZ 10/71, 49 pp.

Becker, C. D. and T. O. Thatcher. 1973. *Toxicity of Power Plant Chemicals to Aquatic Life.* WASH-1249, UC-11. Battelle Pacific Northwest Laboratories, Richland, Washington.

Brungs, W. S. 1973. "Effects of Residual Chlorine on Aquatic Life." *J. Water Pollut. Control Fed.* 45:2180-2193.

Butterfield, C. T. 1948. "Bactericidal Properties of Free and Combined Available Chlorine." *J. Am. Water Works Assoc.* 40:1305-1312.

Collins, H. F. and D. G. Deaner. 1973. "Sewage Chlorination Versus Toxicity—A Dilemma?" *J. Environ. Eng. Div. ASCE,* 99(EE): 761-772, Proc. Paper No. 19190.

Coutant, C. C. 1974. "Evaluation of Entrainment Effect," pp. 1-11. In L. D. Jensen (Ed.). *Proceedings of the Second Entrainment and Intake Screening Workshop.* The Johns Hopkins University Cooling Water Research Project, Report No. 15.

Dawson, G. W. 1974. *The Chemical Toxicity of Elements.* BNWL-1815, UC-70. Battelle Northwest Laboratory, Richland, Washington. 23 pp.

Doudoroff, P. and M. Katz. 1950. "Critical Review of Literature on the Toxicity of Industrial Wastes and Their Components to Fish." *Sew. Ind. Wastes* 22:1432-1458.

Dressel, D. M. 1971. *The Effects of Thermal Shock and Chlorine on the Estuarine Copepod (Acartia tonsa)."* M.S. Thesis. University of Virginia. 58 pp.

Environmental Protection Agency. 1974. "Steam Electric Power Generating Point Source Category—Proposed Effluent Limitations Guidelines and Standards." *Federal Register* 39:8294-8307.

Federal Power Commission. 1971. "The 1970 National Power Survey: Guidelines for Growth of the Electric Power Industry." Part I. Chapter 10. Disposal of waste heat from steam-electric plants. 20 pp.

Gehrs, C. W., L. D. Eyman, R. L. Jolley and J. E. Thompson. 1974. "Effects of Stable Chlorine-Containing Organics on Aquatic Environments." *Nature* 249:675-676.

Gentile, J. H. 1974. "Toxicity to Marine Organisms of Free Chlorine and Chlorinated Compounds in Seawater." Environmental Protection Agency, Research Development Report. 27 pp. (Unpublished manuscript).

Goodyear, C. P. and J. S. Mattice. 1973. "Testimony on Effect of Indian Point Unit 2 Chlorination on the Aquatic Biology of the Hudson

River." Indian Point Unit No. 2, ASLB Hearings, Docket No. 50-247, March 1. 13 pp.

Goodyear, C. P., C. C. Coutant and J. R. Trabalka. 1974. *Sources of Potential Biological Damage from Once-Through Cooling Systems of Nuclear Power Plants.* ORNL/TM-4180. Oak Ridge National Laboratory, Oak Ridge, Tennessee. 41 pp.

Gregg, B. C. 1974. *The Effects of Chlorine and Heat on Selected Stream Invertebrates.* Ph.D. Thesis, Virginia Polytechnic Institute and State University. 311 pp.

Holland, E. A., J. E. Laster, E. D. Neumann and W. E. Eldridge. 1960. *Chlorine and Chloramine Experiments. Toxic Effects of Organic and Inorganic Pollutants on Young Salmon.* Washington Department of Fisheries Research Bulletin No. 5, 188, 264 pp.

Hopkins, E. S. and E. L. Bean. 1966. *Water Purification Control.* The Williams and Winkins Co., Baltimore, Maryland.

Hoss, D. E., L. C. Coston, J. P. Baptist and D. W. Engel. 1975. "Effects of Temperature, Copper and Chlorine on Fish During Simulated Entrainment in Power-Plant Condenser Cooling Systems," pp. 519-527. In *Proceedings of Symposium on Physical and Biological Effects on the Environment of Cooling Systems at Thermal Discharges at Nuclear Power Stations,* International Atomic Energy Agency.

Lee, G. F. and C. L. Stratton. 1972. "Effect of Cooling Tower Blowdown Water on Receiving Water Quality—A Literature Review." Presented to the Lake Michigan Enforcement Conference, Chicago, Illinois, September. 53 pp.

Mattice, J. S. 1976. "A Method for Estimating the Toxicity of Chlorinated Discharges." pp. 142-155. In Sigma Research Inc. (Ed.) *Report of a Workshop on the Impact of Thermal Power Plant Cooling Systems on Aquatic Environment.* Special Report No. 38, Volume 2. Electric Power Research Institute, Palo Alto, California.

Mattice, J. S. and H. A. Pfuderer. 1976. *Chemistry and Effects of Chlorine on Aquatic Systems: An Annotated Bibliography.* ORNL/EIS-82. Oak Ridge National Laboratory, Oak Ridge, Tennessee. 69 pp.

Mattice, J. S. and H. E. Zittel. 1976. "Site Specific Evaluation of Power Plant Chlorination: A Proposal." *J. Water Poll. Control Fed.* 48:2284-2308.

Merkens, J. C. 1958. "Studies on the Toxicity of Chlorine and Chloramines to the Rainbow Trout." *Water Waste Treat. J.* 7:150-151.

Moore, E. W. 1951. "Fundamentals of Chlorination of Sewage and Waste," *Water Sewage Works.* 98:130-136.

Motley, F. W. and T. C. Hoppe. 1970. "Cooling Tower Design Criteria and Water Treatment." Presented at Cooling Tower Institute Meeting, Aspen, Colorado. June.

National Environmental Policy Act of 1969 (NEPA). 1970. PL 91-190, 83 Stat. 852. January 1.
Parker, A. 1935. "Some Problems of Water Supply." *J. Soc. Chem. Ind.* 34:49.
Rosenberger, D. R. 1971. *The Calculation of Acute Toxicity of Free Chlorine and Chloramines to Coho Salmon by Multiple Regression Analysis.* Thesis. Michigan State University, East Lansing. 72 pp.
Scheuring, L. and H. Stetter. 1950. "Experiments on the Effects of Chlorine on Fish and Other Aquatic Organisms." *Vom Wasser* 18:101-140.
Stober, Q. J. and C. H. Hanson. 1974. "Toxicity of Chlorine and Heat to Pink (*Oncorhynchus gorbuscha*) and Chinook Salmon (*O. tshawytscha*)." *Trans. Am. Fish. Soc.* 103:569-576.
U.S. Atomic Energy Commission. 1972. "Final Environmental Statement Related to Operation of Shoreham Atomic Power Station." Docket No. 50-322, September.
U.S. Atomic Energy Commission. 1973. "Draft Environmental Statement Related to the Proposed Newbold Island Nuclear Generation Station Units 1 and 2." Docket Nos. 50-354 and 50-355. July.
U.S. Atomic Energy Commission. 1974. "Final Environmental Statement Related to the Proposed Seabrook Stations Units 1 and 2." Docket Nos. 50-443 and 50-444. December.
University of Oklahoma, The Science and Public Policy Program. 1975. Chapter 13, Energy Consumption. In *Energy Alternatives: A Comparative Analysis.* University of Oklahoma, Norman.
Warren, C. E. 1971. *Biology and Water Pollution Control.* W. B. Saunders, Philadelphia. 434 pp.
Water Resources Council. 1968. *The Nation's Water Resources.* U.S. Government Printing Office, Washington, D. C.
Wolf, E. G., M. J. Schneider and T. O. Thatcher. "Bioassays on the Combined Effects of Chlorine, Heavy Metals, and Temeprature on Fish and Fish Food Organisms." Battelle-Northwest Research Laboratory. Unpublished manuscript.
Zillich, J. S. 1969. "The Toxic Effects of the Grandville Wastewater Treatment Plant Effluent to the Fathead Minnow, *Pimephales promelas.*" Michigan Water Resources Commission, Department of Water Management Report. 9 pp.

DISCUSSION

Joseph E. Draley, Argonne National Laboratory. For large bodies of water, dilution is the most rapid method of reducing chlorine concentrations, except when there is fast-acting chlorine demand in the admixed water (especially for the case in which chlorine is discharged). Thus the isopleth method shown, corrected for currents, provides a reasonable approximation for some cases only, and it is important to identify those cases to which it can be applied. The other cases should be addressed too, such as (a) flowing rivers, where slugs of chlorine and derivatives flow downstream and (b) reactions between substances in the receiving water and the chlorine and chlorine derivatives that remove the toxicant.

Charles C. Coutant, Oak Ridge National Laboratory. I agree that the examples shown are simplistic. However, if we assume that dilution is the only cause for decline in toxic chlorine concentration, the same method can be applied to any known series of isopleths. The actual case will probably never be a nice simple series of ellipses like I showed in the slides. Time-dependent chlorine demand could be factored into such a model if it were known.

Guy R. Nelson, U.S. Environmental Protection Agency. Can the model predict residual chlorine levels in the plume based on an' increasing and/or decreasing amount of residual chlorine in the discharge?

Coutant. This method is based on given residual chlorine levels and cannot predict them. The method is probably not applicable in situations in which organisms are exposed to an increasing residual chlorine concentration because the time-dose would probably be underestimated.

A. D. Parsons, Wisconsin Electric Power Company. Is the model developed by Dr. Mattice adaptable to fixed biological populations such as the benthos and fish positioned within the discharge, as well as moving planktonic populations?

Coutant. Yes. That's a very good question and one that we just didn't get into because of time. It bears on Joe's comment on the plume. In most situations the plume is going to move back and forth. The wind is going to change. It's going to run this direction one time and another direction another time due to tides, winds and other situations. And if you picture a stationary organism at one point out there in the overall potentially affected area, the chances are he's never going to be exposed to a constant concentration. He's going to have ups and downs of concentration. He'll be exposed at some times and not others. For the same type of analysis, you just sort of turn the tables and say we've got a fixed spot out here with a plume waving back and forth and the organism is exposed to a time series of concentrations and you do the same sort of analysis. An important question comes up there though, that may come up in entrainment perhaps to a lesser extent, and that is repeated exposure. Are repeated exposures additive? Or, if an organism survives one cycle, is it sort of back to normal, and you start from zero again on the next cycle? Those questions need to be resolved.

Robert A. Goldstein, Electric Power Research Institute. Written question submitted after the discussion ended: How do you sum exposures to varying concentrations? For instance, if an organism is exposed to 10 ppm for 10 min and then immediately following to 5 ppm for 10 min, how do you relate the total exposure to your threshold curve which is based on exposure to a constant concentration for a given period of time?

Jack S. Mattice, Oak Ridge National Laboratory. The dose-time estimated by this method is the weighted mean of the two exposures for the total time, that is, 7.5 ppm for 20 min. In the case specified, the method might not be particularly accurate, but organisms in a plume would not be likely to encounter this step-function exposure.

SECTION V.
ROUNDTABLE DISCUSSION

21
ROUNDTABLE DISCUSSION

ROUNDTABLE PARTICIPANTS

Carl W. Gehrs, Moderator
Environmental Sciences Division
Oak Ridge National Laboratory
Oak Ridge, Tennessee 37830

William A. Brungs
National Water Quality Laboratory
U.S. Environmental Protection Agency
Duluth, Minnesota 55804

Robert B. Cumming
Biology Division
Oak Ridge National Laboratory
Oak Ridge, Tennessee 37830

Joseph E. Draley
Assistant Director
Argonne National Laboratory
Argonne, Illinois 60439

D. Heyward Hamilton
Division of Biomedical and
Environmental Research
Energy Research and
Development Administration
Washington, D. C. 20545

J. Carrell Morris
Division of Engineering
and Applied Physics
Harvard University
Cambridge, Massachusetts 02138

George Clifford White
Consulting Engineer
San Francisco, California 94118

DISCUSSION

Carl W. Gehrs, Oak Ridge National Laboratory. One of the major purposes of this conference was to bring together scientists of diverse background in order to interact and develop, if you will, some hybrid vitality or viability. In this last session, it is our goal to present and to allow a forum for a semiformal interaction. Somewhat more formal than we had at the dinner on Wednesday night and last night at the social hour; more formal in the sense that the interacton will be recorded and will consequently be included in the proceedings. As Bob mentioned at the beginning of the session today, those of you who will be presenting a question or comment, kindly identify yourself and write it out, and give them either to Bob or myself or send them to us so that we can include them in this discussion. We will give each of our roundtable participants approximately three to five minutes to summarize and to present their feelings now with respect to where we are and where we need to go. Following this we will then open it up to other participants for questions and comments.

William A. Brungs, U.S. Environmental Protection Agency. As you all learned yesterday, quite thoroughly I think, and I hope we all benefit from it, the marine toxicity data and the marine chemistry of chlorination information is extremely difficult to interpret at this time. I hope we will benefit from this in the sense that maybe by including as much information as possible in reports on toxicity tests, it will be useful later on in interpreting really what levels of residual chlorine you might have had. Somebody brought the point up yesterday, and I think it might be quite useful.

There was a little discussion yesterday on the other types of chloro-organics that are formed when you chlorinate various types of wastewaters. I think the fact that we had only two papers or so that discussed this is an indication that it's just a very new concept. It's something that is so new, we really have no idea or way of interpreting its potential impact. We know they are formed. We know now they bioaccumulate in fish. Nobody has even an inkling of what that means. So I would hope over the next few years a few people who have unlimited funding will be able to answer all these questions for us and maybe even the utilities could help out.

I think, also, as you heard people talk, especially if you were somewhat familiar with what they were going to say, you could read between their lines and sort of keep picking out what might be considered to be a series of research needs. I'm only going to mention a few because sometimes we have people in attendance who are just going to begin to do things. I just spoke to some gentleman who is going to do chronic bioassays with copepods in marine water. Maybe between some comments I can make and other ideas he may have gained this week, he'll be able to do something that can be even more fruitful than he had originally anticipated.

I think it's apparent that, from Art Brooks, Will Davis and others, that various factors can influence a toxicity residual chlorine, that is, things such as temperature and water quality. We keep getting inconsistent results. I think one reason we get inconsistent results is that there may be slight differences small enough that some people, because of experimental design, cannot see it. For example, using another type of method for sequential temperatures exposure might show differences in toxicity. So a little more effort in that area would be well worthwhile.

Also the idea of side-by-side comparison of free versus combined chlorine in fresh water would be extremely useful. It's difficult to compare somebody's study with free chlorine in the Columbia River with somebody else's combined chlorine data in the Ohio River. Much of my comments, I must emphasize, are for fresh water. I am not about to get into the same problem that Will Davis has right now.

I think also, as we heard yesterday and has been discussed and presented for the last several years, there's a concept of relative sensitivity of fresh water fish that needs a little more clarification. We had in the past talked about some monads being more sensitive than warm water fish. And, as indicated yesterday again by Art, there are some warm water forage fish of great importance that are extremely sensitive, the same as some monads. So I think we need to also look at some other important fish families that have not been looked at in the past. I don't know of any really good data on the catfish family in fresh water. They could be extremely sensitive or they might be resistant. Without checking a few more of these I think we may be making some erroneous estimates.

A key point I want to discuss is in Jack Mattice's paper which was very well given by Chuck Coutant. I hope he's around to answer questions if it is necessary. I want to bring up a critical point here. The line on the slide, as you noticed, was not a line that would estimate a 50% lethal concentration. He mentioned it was an adjustment down by a factor of 0.37. In other words, if you multiply the LC50 by 0.37, you obtain an estimated concentration which allows 100% survival. In playing around with criteria in this review paper I'm finishing up now, I started looking at the old concept of application factors, whereby you compare acute and chronic toxicity results to get a ratio which is called an application factor. Using that concept in this criteria evaluation, you come up with an application factor of 1/8. Which means that you take 1/8 of the LC50 and you get a chronically safe concentration in fresh water. If you take 1/3 of that same LC50 value, you're getting something that would just keep them from dying. So the difference between 1/3 and 1/8, 1/3 being no mortality and 1/8 being no chronic effects at all, is extremely narrow. So criteria for short-term exposure and criteria for continuous chronic exposure are not going to be tremendously different. I think I'll stop with that point. Maybe I've planted a few seeds that will carry us through the rest of the morning, along with all the other interesting points to come up.

Robert B. Cumming, Oak Ridge National Laboratory. The job of the biomedical scientist in dealing with the kind of problems that we have been discussing for the last several days is to provide information on compounds which might have an adverse effect on human health. At the present time that's about where we stand. I would like to suggest that many of the biomedical scientists aspire to be able to do a little more than that. That is, to provide information, quantitative information, that may be useful in making risk-benefit calculations.

I'd like to talk a little about risk and make a few statements which are largely my own opinion. Carl has said we want to stimulate discussion, and I think that if I say a few things that a substantial number of people disagree with that will tend to stimulate discussion. The first is that all human activity, regardless of what it is, carries with it risk. Risk cannot be completely avoided. It can only be minimized or controlled—understood. It's not realistic to say that no risk is acceptable. And that includes cancer risk. We can't realistically expect to live in a society with zero risk. One or two comments creep in every now and then about risk relative to whether or not it is a voluntary risk or an involuntary risk. I would like to suggest that most of the risks that we accept in this society are really involuntary. To say that it's fine to accept a large risk from cigarette smoking, because that is a voluntary risk, but we cannot accept a small risk from something else, because it is involuntary, is not an altogether realistic approach to the problem. Most so-called voluntary risks in this society are really not voluntary. There is some question, for example, whether people voluntarily expose themselves to risks from smoking. But if they do, the person who is sitting next to them and breathing the same smoke is undergoing an involuntary risk which is part of being a member of this society. The same thing may be said of automobiles. People cheerfully accept the risk, the large risks involved, on the basis that this is voluntary. But many, many pedestrians are killed every year involuntarily, and I would like to say that, in a city like Oak Ridge with no public transportation system, driving an automobile is not an altogether voluntary activity.

So we have risks. What do we wish to do with them. We want to quantify them so that we can deal with them consciously, if not voluntarily—so we can handle these risks and factor them into our risk-benefit estimates. There are two hazards, which we can identify, which are associated with the compounds that we have been dealing with in this conference. These are carcinogenic risks and mutagenic risks. Of those two, carcinogenesis seems to be, the way we read it now, the most important. For qualitative investigations of these kinds of hazards, we have available to us a battery of submammalian tests which really are good for alerting us to problems. But they don't tell us anything about risks. And they don't really tell us, in many cases, whether that particular compound is really a hazard to humans. To determine that we have to have other kinds of information, other kinds of tests.

There are really only two kinds of information which give us any kind of quantitative information—information that might be useful ultimately for risk assessment. These two are extrapolation of data obtained from studies on intact mammals and, secondly, epidemiology. Where it can be applied successfully, epidemiology provides the best information. This is information that is directly applicable to human populations. But the problem is that very frequently there are problems which limit the use of epidemiology. There are confounding factors which make the interpretation impossible or very difficult. So, while epidemiology from some points of view can provide information, good information, frequently it is not available to us. Epidemiology is not likely, for example, to be useful in assessing mutagenic risks to the human population. I do not see at any time in the future when it will be a useful technique for mutagenicity. But it is or could be useful, could be crucially valuable, in certain cases for evaluating carcinogenicity. For dealing with mutagenicity, all that we have is quantitative extrapolation from mammalian test models, and this also provides a useful technique in evaluating carcinogenic problems.

Joseph E. Draley, Argonne National Laboratory. I am glad I wasn't last because I kept thinking I might have to change what I wanted to say because somebody else said it in front of me. What I wanted to do was to address the "chlorine problem." To address anything that broad it will be necessary to restrict what you want to say. I have put three headings down and I'm going to spend a few minutes going into a little detail under these. One is to identify this problem. The second one is to determine what to do. And the third one is to regulate and develop standards or to apply standards.

To know what the problem is, to add some detail to the problem so you know if there is a problem or if there is a significant problem or a bad one or a very small one, you need to know some things—to be able to answer some questions. First of all, how necessary or important is it to use chlorine of the type that would lead to any discharge? Secondly, what is released? You can measure releases and you can calculate releases. I suggest to you that it is not now possible to measure everything you release, and I expect it will not become possible to measure everything you release. Some of this problem is related to low concentration, and some of it is related to the almost myriad composition of your discharge. So there's going to have to be some uncertainty, and some calculation estimation involved. The third question is what harmful effects do these have or will they have, as we build more plants of whatever kind of plants that we're talking about? That problem has to do with the effect on the environment and the effect directly on man. Again, you can measure some of them and you are going to have to estimate some. I think that we've just heard some comment on what can be meausred in one area, and similar kinds of comments will turn out to be made in other areas.

Under those circumstances, I have, in the day or so that I've been aware I had to give this little talk, wondered what one ought to do about it. I must say that it is a credit to the gentleman who thought up the conference and also organized it, etc. In my own opinion, an interdisciplinary and interorganizational meeting is one of the very best things to do to lead to the development of appropriate activities. It turns out that the communication is often a major part of the problem of getting things done or taken care of. This kind of meeting, it seems to me, ought to go a long way over some period of time in helping to remedy a situation that I think started out with quite bad communication between the kinds of people that need to communicate. It still isn't great, but it's quite a bit better. I think this meeting helps it a lot.

Secondly, I would like to suggest for your consideration in the group here today, whether a panel could be constituted to make recommendations to appropriate government boards about what kinds of activities ought to be started or continued. I guess I mean funded—because I think that is really what it boils down to. What kinds of activities ought to be funded? That is not an easy question to answer—very difficult. You really are going to have to take all of the points of view that you see here, and that you hear—and you'll presumably hear more after we're through talking—to get a decent answer to that. I am not sure that a panel is of any more than a limited value. But I think we need something that will help, and a panel is the first thing that occurs to me that sounds like it might be beneficial. I will leave that.

Finally, with respect to regulations and standards, we cannot realistically expect that there will be no regulations or standards until we know exactly how they ought to be constituted and operated. I think we are going to have to make some kinds of regulations and standards earlier than that. As you know that has been done for some of the things. I think there will probably have to be other standards too. So, the activity of getting optimum regulation and standards needs to go on. What I mean by that is, although I think it is inevitable and not undesirable that there be standards before we know how to write them perfectly or near perfectly, we need to provide a mechanism to have a review system and improve on the standards as time goes on. As has been mentioned, the risk-benefit comparison has to be made. I don't think I need to pursue that. And, in establishment of standards you ought to have—and this is not a new thought at all, none of these are really—you ought to have input from the varied kinds of people who should have input for standards and regulations.

D. Heyward Hamilton, U.S. Energy Research and Development Administration. I have some skepticism about the possibility of someone in an administrative post, like myself, bringing appropriate perspective to the question that has been put to us, which is to try to suggest something that might be provocative in an interdisciplinary sense with respect to cooling problems and chlorination problems. I'd like to run through a couple of thoughts that I've had in advance of the meeting itself and in listening to the discussions that have gone on since I've been here. There is a good deal of work which has only very briefly been referred to.

First, let me back up and say somebody indicated this morning that they thought this would be a very difficult act to follow. The question has been brought up and I'm sure several people have talked about it, is there a need for another conference of this kind? It seems to me that it is probable that something like this again would be useful, in the not too distant future, when we have a little more information in hand about these problems in estuarine and in marine systems.

I do have some feeling that on the question of "chlorine problems" as in other similar situations, the industrial growth and the development that is going on is somewhat ahead of us. The number of power plants being sited and likely to be sited in estaurine and marine situations is growing and increasing rapidly, and we really don't have, as far as I can tell, a great deal of information about what to expect in these situations. For example, the importance of pH in controlling what kinds of residuals, if any, are going to be formed. Whether it is or is not advantageous to dilute the chlorinated effluent stream from a condenser system with nonchlorinated effluent streams. What kind of pH changes do you induce in that?

There is at least one person here that feels very strongly that we should not use chlorine gas or hypochlorite in estuaries. There is a modest amount of work, that I am aware of at this time, going on in marine environments. There is a program at Woods Hole under Ryther and Goldman and there are two representatives of that program here. That is a toxicity study. We've heard from Dr. Huggett at VIMS and the potentially frightening issues that he's raised with respect to the impact on oyster spat settlement, and the problems related to that in measuring residuals and understanding what is present in an estuarine environment. Oysters are an important crop in many locations. Dr. Will Davis, Bears Bluff Field Station; Dr. Carpenter, University of Miami; Dr. Helz, University of Maryland; Dr. Tom Thatcher at Battelle-Northwest; Dr. Donald Hoss at Beaufort, North Carolina—those are some of the people, some supported by EPA, some by NSF, some by us (ERDA) that I am aware of who are presently carrying out work related to the potential for "chlorine problems" in marine and estuarine environments. It doesn't amount to a very great deal, and I think most of it is in waters of full ocean salinity rather than in estuaries. So it seems to me, there is somewhat of a gap there that we are going to need to be working on.

For those of you who may be particularly interested in the marine aspect, let me—you may not be aware of it—let me come back to something I started to mention before, and then I'm going to stop. The current work at the food chain research group at Scripps includes three or four papers on the impact of the San Onofre plant on that coastal environment, and also on their efforts to screen some marine organisms for the presence of chlorinated hydrocarbons or halogenated hydrocarbons, rather, that may be produced in the chlorination of that plant. Some of these have been submitted, some are in press, and I think I've actually sent copies to a number of people recently that I thought might be interested. They are finding some very interesting things. They are also doing some interesting experimentation. For example, they are looking at the question of whether or not the toxic residuals formed on chlorination of sea water may interfere with the deposition of silicon in diatoms. This kind of second- and third-order problem that may exist, I think, is illustrative of the sort of thing we need to worry about in the marine and estuarine environment.

J. Carrell Morris, Harvard University. It's easy to get a reputation, I think, for being wise sometimes, by repeating obvious things that most everyone knows in a very firm voice often enough. And so much of what I have to say this morning will fall into that category of repeating for you the multiplication tables.

It seems to me the information we have been dealing with here comes on two levels. These are not distinct and one can find instances of considerable overlapping and ambiguity, but nonetheless, I do think we ought to think in terms of information at the pragmatic level and information at the fundamental level. We do have to have, in order to meet immediate situations and to deal with regulation and control of immediate factors, a lot of pragmatic research that simply says when you put so much chlorine into it you get thus and thus toxic effect. The difficulty is that information like this is restricted more or less to the particular situation. For example, you cannot move from the Chesapeake Bay to the Pacific Ocean with this kind of information and translate it directly. In order to do that, in order to universalize the applicability of your data, in order to understand the factors that are operating, you do have to have investigations at the fundamental level.

I would make a plea here that in attempting to meet the problems of the immediate situation we not overlook the need also for long-term information that leads to long-term predictability and understanding of the factors that are going on. Otherwise, we will be faced with a continual need for repeated investigation each time we come up with a new situation. We will be faced with these continuing inconsistencies in the data.

I don't think that we ought to blame the biologists because the chemists have been unable to come up with adequate analytical techniques for distinguishing exactly the forms and species of chlorine that we have available, in some of these situations, particularly the estuarine and marine situations. I think this is a challenge to the analytical and the physical chemists to find out a little more fundamental information about these kinds of situations, so that they can inform the biologists as to what is really going on, so that the biologists are able to do their fundamental work more adequately.

I do think that in some instances the biologists have been a little bit careless informing themselves of the extent of information that has been available, and in setting up their experiments so that they can know the kinds of things that they are dealing with. It does not, in many instances, take much adaptation of the experimental regime in order to have a clear-cut rather than an ambiguous situation.

At the same time I also think that the toxicological and biological people should always be aware and should look with some suspicion on the claims of analytical methods or even of the claims of physical chemists, although they are a noble breed, because they do carry out their work in relatively clean systems. When Don Johnson tells you that he is able to differentiate between thus and so, and thus and so, he has picked a clean system in which to do this. And while he has tried to imagine all sorts of interferences that might occur, and to look at these, when one gets into a real situation there are likely to be unsuspected interferences. So one must not always take the claims, particularly claims of differentiation, at their face value when they are applied to these real situations. There should be some investigation of the reliability of analytical methods under the situations, either by spiking techniques or things like that.

George C. White, Consulting Engineer. The question that was put to me about this discussion was, "What was the message that you got after being here for three days?" And since I'm a practitioner of chlorination, the message comes through to me like this, and it's loud and clear. We have fouled our own nest, and it's about time we recognize it and we should clean it up. I'm going to take this in three steps: potable water, wastewater and cooling water.

But first I'm going to say, doesn't it seem ridiculous when you look at the bathtub and I'll site a local bathtub, which is San Francisco Bay—when you can just reach down anywhere in San Francisco Bay, near the ferry building, and grab a gallon jug of water and go run a couple of amperometric residuals, do some chlorine demands, and the biggest chlorine demand you can get is only a part and a half per million. That's free chlorine. Yet you take a jug out of Passaic River and you've got to put in 16 mg/l to get a proper water. To me this is ridiculous. So, let's take them one at a time.

Cooling water—the problem that has been presented with the residual in cooling water—I cannot get myself excited about. And dechlorination, to me, is utter desperation. It seems to me that with the engineering brains we have that cooling waters can be properly diluted and also the systems can be optimized. One of the good things that has come out of all this scare on the chloro-organics is that people are looking at what they are really doing with chlorine. As you know, human nature is so "want to do" that if a little bit is good, a whole lot more is much better, and that's what we've been doing all along. I can't get excited about the poor oysters, if we've got to trade off something. Because we in San Francisco lost our whole oyster crop, which was considerable, simply because we had labor problems. We couldn't get people out there to build the necessary devices to grow the oysters. You can call that an involuntary risk but that is part of the whole problem. Also I can't get too excited about the effluent of the San Onofre plant, when the anchovies in the ocean out off of Los Angeles coast put in more waste than do the outfalls from L.A. County and Orange County Sanitation Districts So, we've got all these things to look at.

But I am concerned about potable water and wastewater. If you recall my opening remarks, I said that a good clean water can be disinfected with 1 mg/l and that the range of chlorination dosage in potable waters in the USA was 1 to 16. Now, there we have a factor of 16 to 1 that we can reduce our chlorine dosages if we clean up our environment. You've got to first start cleaning up the environment with the wastewater. Now 16 mg/l in wastewater, and this is the part that is ridiculous, can achieve good disinfection—even for a primary effluent, if you do it right. So just think of the chlorine we could save if we go to better wastewater treatment.

I don't know how we're going to get it, or whether it's too late or not, but with the new clean water act the cities can now bear down on the industries and get the industries to clean up and to conserve. I think we are going in the right direction. We are just like the race cars at Indianapolis—they only turn left. We can't turn right. We are still going left and we are going to stay going that way.

To get back to just one other thing that bothers me. I'll let somebody else worry about cleaning up the environment, but you can take a good secondary—let me just back up a little bit. I am a great believer in letting nature do the work. So I am all in favor of biological systems for cleaning things up. You can take a good secondary effluent, nitrify it, get a free chlorine residual with very low dosages, and you can get virus kills. That is really cleaning up the environment. I don't know whether we need to kill viruses in sewage effluents, but if we clean up the wastewater then we're going to clean up our raw potable water. Maybe this requires storage, but I think it is absolutely ridiculous to subject the people in Passaic Valley to that kind of water. That's like direct water reuse, which is something aesthetically not acceptable.

Now I'll just close by saying that I agree with Joe Draley that communication is necessary. To me, it is very frustrating to hear a paper at one of these conferences and you know that if it is an excellent paper, you have to wait a year before it is in print. There ought to be a clearinghouse for all these presentations so that we don't have to wait a year for them. I am impressed with this conference because I never realized so many people could gather in one place and only talk about chlorine. That is all I want to say. That ought to start something.

AUDIENCE PARTICIPATION IN DISCUSSION

C. Sengupta, Public Service Electric and Gas Company. What I am going to say is a comment. My choices are between either chlorinating and running the plant or not chlorinating and not running the plant. So until someone gives me an alternative to chlorine, it doesn't matter whether the fish are dying or not. I can go only down to a certain level.

We will try our best to go down to that, but beyond that point we need an alternate. We have heard so many people doing research here, so many millions of dollars being spent, but very little on alternates. What about alternatives to chlorination?

Draley. I made a very brief reference to the subject by making my number one item in the "chlorine problem" to identify how necessary or important it is to use chlorine. It becomes less necessary as you provide alternatives, which is what you're really asking. There are some things that are being done at the moment, but White's statement was that he didn't think there was any alternative now. Is that right?

White. I just cannot get worried about the power plant problem because with all the things that we have listened to here, the message is coming through clear to me that the people that are designing the plants now—we may have some problems with some existing plants that you cannot change—but with the new ones that are coming on the lines, I am sure that there is some way that dilution will allow you to operate below that acute lethal line.

Brungs. I agree entirely that in many new plants they can do things. In terms of a lot of the older ones, I don't know how broadly this is being done. But at least in the state of Michigan, the state in cooperation with Region 5, EPA, they have been doing a lot of work on what they call minimization programs at generating stations. I will pick the most extreme one, a plant on the Chicago ship channel or canal, which was chlorinating fantastically, because it's principally a sanitary waste effluent. So they had a tremendous slime problem, etc. They dosed a long time. So they kept cutting back and cutting back on their chlorine use until they started having back pressure problems, I believe. I'm a biologist, not an engineer. And then they had to kick it back up a little bit. But they ended up with, I think, a single dose for 10 minutes a day from something that had been multiple hours per day. Ignoring the savings in chlorine, I imagine they saved a few aquatic organisms. That is something that can be done with older plants. It may or may not meet the regulations or the criteria or standards. I'm quite sure, also that some plants may never be able to do that in existing situations where they are at low flow for 100% of every flow. I think somebody will have to turn their face on those until they are phased out.

Sengupta. If the power investor does go back to the board and minimizes the amount of chlorine that is necessary, is that going to be acceptable? Or do we still have to look and see where the fish are dying? As I saw it, the level at which we don't have any chronic effect is 0.0015 ppm and I don't think that we are going to go to that level.

Brungs. The 0.0015 ppm is a chronic value that gets out beyond 96 hours, so we don't have that concern there. I think, as I mentioned, sometimes somebody in a regulatory agency is going to turn their face and ignore existing problems because nothing can be done.

J. Donald Johnson, University of North Carolina at Chapel Hill. I wanted to ask a question of the panel as to what they thought of the recent proposed elimination of the coliform standards for waters which are being polluted by wastewater, but which are not subject to direct reuse or swimming? I wonder if we are throwing out the baby with the bath water with that kind of approach. Perhaps we should go back and better control our wastewater disinfection process, as we are suggesting that we should do for our cooling waters—to minimize the amount of chlorine being used.

White. I don't want to waste time going into philosophy of disinfection of wastewater. It's here. I'm sure it's here to stay. But the important thing about the disinfection of wastewater is that you clean up the operation. We're going to have to have some patience. There is going to be some risk in the mean time. But as we go to secondary effluents, and as the various states set these requirements for the receiving waters, we are going to stick with the coliform.

Johnson. They are not. They are proposing to throw it out.

White. That's not a good idea. We have got to stick with the coliform count for two very good reasons. It serves us well in identifying a safe potable water and, if you will recall in my opening remarks at the beginning of this session, there is an implication in the coliform regulation of the type of effluent that you should have to meet the regulation. Now, a 2.2 effluent implies tertiary treatment. A 23.2 implies a very good consistent secondary effluent, and this is why we have got to keep the coliform count in there, because it is directly related. Now, you can have all the other things in there you want, you can have BOD, you can have suspended solids—that's great—and grease and so on. That keeps the operator on his toes. It's something the operator can see. It's something we can work from. And you've got to give him the coliform in order to maintain that quality.

Johnson. We are not talking about California.

White. No.

Johnson. These standards they are proposing to throw out are 200 per 100 ml.

White. That is pretty sad.

Johnson. They are talking about eliminating that altogether so there will not be any coliform standard for nondirect use.

White. That would be a very bad mistake.

Guy R. Nelson, U.S. Environmental Protection Agency. I'd like to direct my question to Bill Brungs. I want to do it very carefully because I am an engineer asking as best I can a biological question to a biologist. In most of the criteria for chlorine levels and the effects on various animals and plants, there are two parameters: time and maximum concentration. Is that the best way that we can present them? For example, I am thinking of one of your famous criteria, that is, 0.2 mg/l for two hours—that being a maximum concentration. Now, to me it is more lethal to any organism to have 0.2 mg/l throughout that two-hour span versus having it maximum maybe for two minutes during that two hours.

Brungs. The criteria referred to, I assume, would include those such as Chuck Coutant showed today for Jack Mattice. Those to me are criteria. The point you bring up is somewhat of an academic one in the sense that the data that one has available to him when he is working up a criterion does not give you the precision that would let you make a comment on that. The assumption inherent in the criteria is that theoretically somebody can meet them. If you say, "the residual shall not exceed 0.2 ppm for two hours, somebody can do it for two hours and maintain a constant 0.2 ppm. That's sort of the assumption one makes. It can be done, and therefore the implication is that when you come up with a number it should be protective.

I like what Jack has worked up on the time-concentration relationship. I've mentioned, and other people have mentioned in the past quite frequently, that this is the ultimate for all kinds of criteria, not just chlorine. It is time-concentration dependent. I think a system like this will be extremely useful. I know Jack didn't do it intentionally, but if you take some of the older criteria from a few years ago and compare it to his line, the similarity for that one point, at least in relation to that line, is unusually good considering that a biologist did the playing around with the data. A long answer that didn't provide you with much information.

Johnson. Along that same line, I wanted to ask a question about transients and spikes of concentration that might be missed looking at toxicity data. I look at the data that is coming out of the studies and I see median values and relations of the toxicity to those median values. But when you look back at the data, you find some very large spikes within those medians. Might not the toxicity really be due to the sudden surge that you don't see in the medians.

Brungs. The more recent studies, that really have not even produced manuscripts yet, are utilizing curves such as Guy showed this morning on the cooling tower blowdown, where they actually are trying to design their intermittent exposures dose systems so that you get a rapid rise and then a gradual decline. The difficulty that all of these people have had is there is no presently acceptable biological concept, like LC50 or TLM or anything like that, that provides an answer to a variable exposure condition. In one of Jim Truchan's papers, he has introduced a term called intermittent lethal concentration, and you have to put in the number of exposures, the maximum concentration and all these. You can no longer get a median value. It's meaningless. You cannot just give the extremes either. You almost have to plot a curve and say that these are the experimental conditions and these are my results. Ideally, we will come up with a term that will be so many milligrams per liter per minute—a chlorine-minute concept.

Robert C. Paladino, Edison Electric Institute. If utilities minimize their use of chlorination and, in the face of no alternatives, must chlorinate to protect their plants, on what basis should regulations be written? Protection of fish? Protection of plant? Or both?

White. I thought that when I came here that we were worried about the chloro-organics in our drinking water. Now it seems that we are more worried about what we are doing to fish with the cooling waters. Maybe I have missed a point, but along that line I would like to interject this when you start talking about regulations. I think one of the regulations that the government ought to look at very closely, and I'm thinking of drinking water, that there should be a limit placed on the amount of chlorine required in a raw water to produce—just to toss a number out—to produce a 1.0 mg/l free chlorine at the end of 30 minutes. That would stop difficulties with various kinds of lousy water. Now there are a lot of ways to reduce the chlorine demand of water. You can do as they do in England. They store water for long periods of time. It can be aerated. The chlorine demand can be reduced drastically with a little bit of natural pretreatment. I think that anything that we can do to get some natural pretreatment we should look at. I am more concerned about the potable water than I am about the cooling water. I think that is a solvable problem.

Draley. Did that answer your question?

Paladino. No.

Draley. My answer would be that you can write regulations and standards with times of applicability; that is, it will become applicable to plants at different times. And some plants can be exempted for the lifetime of the plant. When you make those exemptions, you have to do it with your eyes open. That is, you can make your choice. You can either make an exemption and take the consequences or you can shut the plant down, if it's bad enough, which nobody likes to do. That's the kind of thing you have to face. You have no alternatives to facing it that way. As you know, a lot of the regulations that are being developed do have times on them, and you can find an occasional one that has exemptions too.

William J. Ross, Nuclear Regulatory Commission. I came to the conference to get a status report. Dr. White may have already given me my report, because we're more interested in the cooling water, of course, than the actual drinking water. But I am going back with three or four very serious questions. I would like to throw them out and then give you a very brief reason for them. One, are we in the power plant business over-regulating right now? Because we do have regulations that are based on the EPA regulations and I'm not infringing upon their prerogatives. Second, speaking as one who has been an analytical chemist for many more years than a regulator, I was very concerned yesterday to hear that in our special methods for measuring chlorine, we are not quite sure what we are measuring. We are now talking about a meter reading. There was quite a bit of debate about the limit of detection, and yet the biologists continue to talk about very low concentrations of chlorine. That question is, "When will this situation be improved?" Third, can we (NRC) or EPA do something by making our regulations more site-specific for the near term? I am very interested in this because we are beginning now to talk more and more of marine siting on both the Pacific and the Atlantic. We run into problems immediately—on the scope, on the chemistry, on everything. But I guess my main problem is we are trying to protect the public—and I don't want to get into our banquet speech again—but are you able to give us what we need right now? Dr. Draley's comment a while ago was apropos. Can we do something to speed it up?

White. Those are good questions.

Draley. Your questions are duly recorded.

Johnson. I wish to speak to the question. I think the first question is, "What kind of analytical procedures should we use?" First of all, I can tell you what not to use. This doesn't help much, does it? But yet you find it in many regulations still today—that is the acid orthotolidine test. Let's throw those out of the regulations. This method is coming out of *Standard Methods*, and it sure ought to come out of the regulations, state or otherwise. Don't use residuals of acid orthotolidine to regulate with. The next question is, "What can we use today?" My answer is that the amperometric titration is a good laboratory method. It is the method of choice for analyzing the chlorine species. I realize it is not a very good field method, and it also lacks some sensitivity. There are modifications you can make in the amperometric titration to make it sensitive enough, but you don't find those in the commercial units. So there is a problem of sensitivity and selectivity. I don't want to brag but I talked about an electrode we have been working on and it is commercially available, if anybody wants to buy it. I don't get a nickel from every unit sold so I'm not selling it. But that is one method. But is is not sensitive enough. So it is a good method for controlling in the plant, but it is not a good method for the field. How do we determine Bill Brungs' levels? Two ppb? There is a new method on the horizon, the flux monitor. That is the only method I know of that is able to measure that kind of level of chlorine out in the environment. So that is my answer.

White. I would like to make a comment on that. First of all, you have to accept some kind of an intermediate philosophy and take off from there. I am not a zero discharge man by any means. This is a sort of dilemma we are in now. These marine biologists keep talking about 0.000 something or other and they have modified titration procedures that you have to walk on your tiptoes around—you have to ground the jar with aluminum foil to keep the noise out and all this sort of thing. First, if you have a good control system, which very few of these power plants have, you can control those plumes going out of the condenser at some very low chlorine residuals. If a regulatory agency would come along and say 0.1 mg/l in the plume going into the receiving water is acceptable under all conditions, then equipment manufacturers would sit down and they would make control equipment that could do that. But right now—let's take California for example—they have a 0.02 or 0.05 chlorine residual in the wastewater effluent. They do not say how you measure it. I can calibrate a chlorine residual analyzer to put out a 0.05 orthotolodine residual and yet it will be putting out a 1.5 amperometric residual. That is the status of the situation right now. So what do we do? We say the best way is to go to zero residual and so we feed forward all of our dechlorination equipment. We are going to excess SO_2 which fortunately, so far, is not indictable for fouling the environment. So we go to about a 0.5 SO_2 residual, but there's no way to measure SO_2 residual practically. You can take probe, and you can fiddle around in the lab, and you can sort of guess at it, but you can't automate equipment with it. So that is all we have, feed forward control.

418 WATER CHLORINATION

You have got to have a primary meter that is controlling the chlorinator. You have got to have an analyzer controlling the chlorinator. You have got to have an analyzer at the end of the contact time that is controlling the sulfonator. You feed a closely coupled meter signal with the residual signal into the sulfonator, and then you just hope that everything is all right. Then you monitor it intermittently. The trouble with monitoring it intermittently is that a chlorine residual analyzer will go out of calibration when it doesn't see the halogen oxidant over a long period of time. So we are in trouble. But if somebody would say we will allow 0.1 mg/l residual, then the equipment manufacturers could put in closed-loop control on dechlorination. I want to get that across because I think that is very important.

Carol H. Tate, James M. Montgomery Consulting Engineers, Inc. I would like to go back to the drinking water question and throw the question to the panel as to whether they foresee, or whether they would advocate, standards for any of these 200 organic compounds in drinking water?

White. I can't answer the question, but I would make the comment that Russia has standards on 400. Did you mention that? I was talking when you started, I don't know whether you mentioned that or not. It makes us look pretty sad. What do we have, thirty-five or something like that? Thirty-two?

Tate. We have a CCE.

White. No. I mean the total number of standards for all constituents of water. It is very low. There is much talk about other standards.

Tate. How about 5-chlorouracil?

Cumming. I think that the compounds that we know about are present in such small levels that standards are not really meaningful. I presume that one could set standards. But at this point we know so little about the distribution of these compounds in surface waters that it is not clear to me what would be achieved by attempting to set standards at this time. I think more information is necessary.

Albert Dietz, Jones Chemical Company. I wanted to follow up on the question of alternatives for chlorine. It has been suggested that chlorination may be eliminated in some instances of sanitation in favor of an alternative sanitizing chemical. What are the alternative chemicals proposed for disinfecting raw potable water, municipal wastewater, cooling water, etc., and what is the availability of these alternative chemicals?

White. I don't think there is any alternative to chlorine. I know that is a cop-out. I am just a believer in cleaning up the environment, optimizing these systems, cutting the dosage down to 1 to 2 mg/l, straightening out our thinking a little on disinfecting wastewater, and cleaning up the industrial discharge. This will do more to turn this situation around than anything else. We are just going to have to live with some of these risks. We just can't make a 180-degree turn now with the commitment, the investment, the education. Do you realize that even in some of the most sophisticated plants you can't get operators to really appreciate the difference in the species of chlorine residual. It gets so frustrating, you will have an "umpteen" million dollar plant and here you see a guy running around with an orthotolodine kit. You just sort of want to throw up your hands. Not only that, they will get in trouble with the regulatory agency, and they will say you are not disinfecting and that you have 2300/100 coliform. So what does the operator do? He cranks up the chlorinator. He goes out to the end of the outfall. He takes an orthotolodine residual. He has a 0 to 10 mg/l disc and he looks at it and says it's 10 mg/l. But if you put a titrator on it, it might be 20. He can't measure orthotolodine higher than 10 mg/l, and he won't take the trouble to go through a dilution. So these are the things that we are up against. If we are going to turn around 180 degrees and do something else, we just could never educate enough people.

Morris. I think George White is an extremist. There are a number of possible alternatives to chlorine for disinfection purposes. George has pointed out that we use chlorine for a variety of purposes—not just for disinfection alone. Moreover, as an introduction here let me also say that in thinking of alternatives, one should think not only of alternative chemical substances, but also of alternative ways of applying chlorine. We have done very little investigation of possible effects on some of these undesirable side reactions—of what might happen for, say, a very short-time high dosage of chlorine for disinfection followed by almost immediate dechlorination before any of the chemical reactions occur. We have done very little in terms of seeing what might happen with small multiple doses of chlorine at different points. None of this has been dealt with systematically and might enable us to minimize a great many of these problems.

But besides this I think that one should give some credit to the fact that ozone is able to do a good many of the things that chlorine can do. If, as in George White's case, one is interested in having a water with just a little bit of chlorine demand when you add chlorine to it, one of the ways to clean up the water to this point is to make use of ozone as a pretreatment. One does not really even need chlorine to go out into the distribution system. It is possible, as they are doing in Zurich, to use chlorine dioxide as a substance going out into the distribution system. While these substances have many of the properties of chlorine, one of the things that we do know is that they do not cause substitution of chlorine into the organic molecule in the same way that chlorine itself does. So that one must keep these in one's eye and possibly the other new things that still may be coming along and give them a reasonable chance and not, simply because of our long experience with chlorine, just say, "Oh, it's the traditional thing. We can't do anything else."

White. Yes. If you recall in my opening remarks this is one of the things that I want to see funded. I think we have glossed over the benefits of chlorine dioxide. We have not investigated ozone. We don't know enough about the chemistry of the combinations. I totally agree with Dr. Morris that we need to do this investigation. I was trying to emphasize cleaning up the environment first, because we can do that at the same time. In other words, the wastewater people can be working with that part of the environment while we are working with these different chemicals. We need to look into water storage before treatment. We need to look into natural means of pretreatment. We need to look into ozone first, then chlorine, dechlorination, aeration, rechlorination, and also preforming of chloramines. There are a whole lot of things that we haven't done anything about. No real serious chlorination chemistry on combinations has been done—not to speak of—ever since Dr. Morris did it in the 40's and 50's at Harvard. So we have just sort of been limping along.

Draley. I think we are a little unfair to your question. We have not addressed it in an "honest-to-God" serious way about practical alternatives. The first thing I would say is that it is site-specific. There is not any across-the-board generality about whether you need to use chlorine or not. For example, the Point Beach Nuclear Plant has never done anything to defoul, and they've been running for some three years. They do not have evidence of any problem. Incidentally, the people's thought about that is, primarily, that there is just enough of the right kind of fine silt in the water (the plant is located on Lake Michigan) that the scrubbing is helping them as well as the low contamination level. That's one thing to say. There are other things. Most everybody here that has something to do with the power industry knows about the Ampertap system. We have not mentioned it. I think we ought to say that there are mechanical devices that will clean condenser tubes. There are places where those things have done quite a satisfactory job for condenser tubes. I am not sure that it meets the needs of the power plants, because there are other things than condenser tubes. My answer is that a mixture of things that include nonchemicals is also possible.

Tate. Is Point Beach once-through? And will the regulators reconsider the regulation with respect to once-through for site-specific cases?

Draley. It is once-through. The answer of the question about the regulators is going to depend partly upon what kind of pressure is put on the regulators. I don't know the answer, but I think that they ought to be site-specific or an effort ought to be made to have them site-specific. I think we are doing a public disservice if we insist that all regulations have to be the same for everybody.

Gehrs. I think perhaps that this is a good time to bring this together. I believe this latter point was, in my opinion, the most important facet with respect to regulation of chlorination at power plants. Is the real need for the regulations to be flexible and viable, as you will, with respect to specific sites?

I want to thank each of you for your participation. I personally think it was a very valuable conference bringing together scientists of diverse backgrounds. The interactions started here will go on through verbal and written interactions or future meetings.

. . .

The following represent the editor's summation of two statements too lengthy for inclusion in this report:

David H. Rosenblatt, U.S. Army Medical Bioengineering Research and Development Laboratory. Dr. Rosenblatt summarized the complex equilibria involved in the simple case of the hydrolysis of low concentrations of chlorine in distilled water. He pointed out the simplifying assumptions necessary for development of computer programs for calculation of equilibrium concentrations of the chemical species of interest. The chemical model and program developed will be adapted to other more complex equilibrium and dynamic situations more closely approximating the real world.

Dr. David D. Woodbridge, Florida Institute of Technology. Dr. Woodbridge summarized his experimental data on the use of irradiation for disinfection and sewage treatment as an alternative to chlorine. Experimental results were shown which indicated significant reductions in concentration of chlorine and chloro-organics in water and sewage effluents within time periods of several minutes using upwards to 80-kilorad doses.

The following are written comments or questions received after the discussion had ended:

Jerome McKersie, Wisconsin Division of Natural Resources. The proposed secondary treatment standards, published in the *Federal Register* August 15, 1975, deletes disinfection requirements for municipal wastewater treatment plants. Instead, disinfection requirements would be based on state water quality standards. Will the EPA develop criteria or guidance on where chlorination, chlorination with disinfection, no disinfection or alternate methods of disinfection should be applied? This type of information is vital to application of water quality standards.

John R. J. Sorenson, QUAD Corporation. It is probably true that, as Mr. George C. White has put it, "we have fouled our own nest." Chlorination of waters has resulted in the introduction of too many unneeded chlorinated compounds. This has been the result of our misuse of chlorine, which was only intended as a water disinfectant but has been excessively used to the point that excess quantities of hypochlorous acid are available for the unwanted reaction processes leading to the production of undesirable chlorinated compounds. Since the rate of disinfection is fast compared to the rate of formation of chlorine-containing compounds in the water chlorination process, efficient mixing of water with chlorine should allow the reduction of the amount of chlorine used for its intended purpose—disinfection. The use of chlorine should not be allowed for the purpose of oxidation in the treatment of wastewaters just to speed up the wastewater treatment. Wastewater treatment should be done as it was in the past with natural processes employing aeration and microbial decomposition of the components of wastewater. Water cannot be continually chlorinated on intake and discharge as it passes from city to city along the waterway. It should be clear that additions to this water by each city will become part of the water environment of the next city. After a number of cities have made their additions to this water it is no wonder that cities at the end of the waterway have a grossly adulterated water supply.

Consistent with this is the need for the use of *pure* chlorine in the water disinfection process. The chlorine used today is the chlorine that no one else will buy. This chlorine is the dregs of the chlorine production process. Chemicals users of chlorine specify that it be nearly 100% pure chlorine. This is done because they require that their chemical syntheses be run as cleanly as possible to obtain good yields of the desired products. To furnish this high-quality chlorine the chlorine produced in the electrolytic process is purified by allowing the impure chlorine to pass through a fractionating tower which brings about the removal of chlorinated compounds produced as a result of the reaction of chlorine with lubricants and the graphite electrodes used in the electrolytic production of chlorine. As I understand it, these impurities are removed from the equipment along with the chlorine that is used for sanitation purposes. Since these impurities are composed of chloroform, methylene chloride and carbon tetrachloride in addition to chlorinated aromatic and aliphatic hydrocarbons, this chlorine is contaminated with materials which are suspected as being responsible for liver cancers seen in humans. Even if these impurities are added in small quantities, if they are being added by each city along a waterway, those individuals in cities along and at the end of that waterway may be exposed to larger and larger quantities of chlorinated compounds which may represent a substantial health risk.

At the moment and in the foreseeable future there will be no alternative to disinfection by chlorination. Chlorination is a remarkably safe and effective method of disinfection and its applicability can be sustained with preliminary natural treatment such as settling, filtration and microbial removal of wastewater components prior to chlorination, by the efficient mixing of waters to be disinfected with pure chlorine.

Some have suggested that ozonization is an alternative to chlorination. However, ozonides and epoxides, which are known to be the active species of the carcinogenic aromatic hydrocarbons, represent a clearly defined cancer risk and should not be quickly adopted as an alternative. No method of disinfection should be seriously considered until the research has been done to show that the method produces water which has been shown to be safer or less toxic than that produced by chlorination.

Research should continue with regard to the better utilization of chlorine and the effects of compounds produced in this process. Research on alternative methods should be done to establish that the alternative method represents a lesser health risk.

Stephen A. Hubbs, Louisville Water Company. Drawing an analogy between the concern of organic and heavy metal pollutants of ten years ago and the thrust toward watershed and river basin planning today, I wondered if it might be beneficial to compile an inventory of chlorine discharged into receiving streams on an entire river basin level. Then, as the kinetics and chemical mechanisms of chlorine chemistry unravel, the aquatic effects of water chlorination may be immediately evaluated based on known exposures to flora and fauna for extended time periods. In essence, we would be gathering raw field data for an experiment which has not yet been designed.

Chronologically, the situation for such data accumulation could not be better. In Kentucky, all waste discharges of any quantity and quality must provide information for **NPDES** permits (as in other states). Those discharges utilizing chlorine have the capability to measure the amount of chlorine in their effluent. Likewise, the total amount of chlorine used in the process should be easily obtainable. These data, along with a note concerning the method of chlorine analysis used, may provide adequate information to evaluate bioaccumulation and bottom sediment accumulation of chlorinated compounds. Thus, by requiring this information on **NPDES** permits, a data base could be established that may greatly aid the evaluation of long-term effects of water chlorination.

T. F. Craft, Georgia Institute of Technology. In spite of the work that has already been carried out, it is apparent that we have only nibbled at the edges of the problem. Our lack of knowledge of aqueous chlorine reactions, their products, yields and significance has been brought into sharp focus by this meeting. A need for periodic reviews of progress is obvious, and I hope that subsequent conferences at appropriate intervals can be held.

INDEX

abietic acid 73
Acartia tonsi 289
acclimation 292
acetaldehyde 80
acetanilide 68,313
acetone 80,93-96
acetyl compounds 88
acetylenes 70
acetyl moiety 99
acid catalysis 22,29
acid orthotolodine 51
activated carbon 14
   filtration 103
activation 27
activation-inactivation process 205
addition reaction 109,116
   to double bonds 28
adenosine triphosphate 287,288
aeration 14
aflatoxins 186
aldrin 220
alevins 297
algae 292,302
   Also see freshwater algae
algicides 17
Allen Steam Plant 123-125
Alosa aestivalis (blueback herring) 295
alum coagulation 84
alum sedimentation 84
amides 26,112
amines 26,27
aminating agents 109
amino acids 112,137
ammonia 4,26,27,43,48,96,109, 128,129,297,310
ammoniation 14,99

amperometric
   measurements 365
   membrane electrode 55
   method 168
   monitors 50
   procedure 353
   titration 16,50,52,281,302,417
amphipods 291
analytical methods 167,168,189
anemones 290,292
anion exchange chromatography 118
   Also see chromatography
angiosarcoma 185,249
anisole 68,313,316
annelid worm 291
anoxia 301
antagonism 203,335,339
antifoulants 123
   Also see biocide
aqueducts 5,7
aqueous chemistry of chlorine 19-35
   Also see aqueous chlorination, aqueous chlorination reactions and chlorine
aqueous chlorination 67,131
   chlorine species 23
   reactions 108-116
aromatic compounds 81,155
   hydrocarbons 315
   organic acids 115,127
   ring 27,28
aresenic 221,223
arsenite blank 54
asbestos 184,221,249,251,253
   in drinking water 222

ATP 287,288
avoidance 295,296,297,299

bacteria
   filamentous 5
   regrowth of 16
   thymine-requiring 234
bactericidal free chlorine 40
   Also see free chlorine
Balanus sp. 289,292
barnacles 290,291,292
base-catalyzed reaction 29
base-pair substitutions 204
basicity 26
bathing waters 10
behavioral response 307
1,2-benzanthracene 221
benzene 220
benzidine 189,219,221,251
benzo(a)pyrene 220
beryllium 223
bioaccumulation 119,258,285,311-328
   Also see bioconcentration factor
bioassay 119,200,293,297,298, 299,308,316,322,402
biocentration factor 316,321-323
   Also see bioaccumulation
biocide 262,380
biofouling 12
   Also see fouling
biological oxygen demand (BOD) 12,68
Bimaria franciscana 290
biomedical effects of chloro-organics 193-209
biota 285
biotransformation 204
biphenyl 68
1,2-*bis*-chloroethxyethane 205
bis (2-chloroethyl) ether 204,220
bis (2-chloroisopropyl) ether 204,218
bis-chloromethyl ether (BCME) 189,218,219
blocking effect 293
blowdown 13,364
   flow rate 347
   streams 347

Brantford, Ontario 6
breakpoint 7
   chlorination 44
   reaction 31,43,45,48
Brevoortia tyrannus (menhaden) 294,298
British Isles 7
bromamines 34,50,166,285,286
bromate 166,176,285,286
bromide 32,34,46,78,80,165,285, 286
   catalysts 47
   effects 31,32
   in sea water 35,285
brominated hydrocarbons 33
bromination 285,302
bromine 14,16,17,50,57,285,308
bromine chloride 14,18
bromobenzenes 204
bromochloroiodomethane 155
bromodichloromethane 78,79,204
bromoform 78,79,103,197,204, 205,219,286
   Also see haloforms and trihalomethane formation
bromo-organics 131
5-bromouracil 234-236,241

calcium, hardness effect 282
cadmium 221
calibration factors 60
California 301
California State Department of Health 6,10
California Water Plan 6
California Water Resources Quality Control Board 6
cancer
   age at first exposure 250
   mortality studies 252-255
   Also see carcinogens, carcinogenesis, carcinogenic
carbohydrate 111
α-carbon 29
carbon, organic 129,310
carbon tetrachloride 15,31,78,102, 213,219,220,223,332
carbonyl groups 29
   multiple 75

carcinogenesis 194,200,212,231
　confounding factors 250
　latent period 250
　magnitude of risk 250
carcinogenic
　activity, evaluation of 218,219
　chemicals in drinking water 211-228
　potency 217
　properties 196
　risk 230,405
carcinogens 205
　bioassay 201
　classification of 215-217
　human sensitivity to 252
　in water, assessment of 218-221
　occupational 215
　potential 221
　thresholds 212
　water-borne 243-258
carcinoma 202
carp eggs 298
Catastomus commersonni (white suckers) 270
cause-effect relationships 223,253
chemical characterization 201
　Also see chlorinated organic compounds
chemical communication systems 131
　Also see pheromone systems
chemical composition 285
　prediction 347-352
chemical industry, chlorine use in 2
chemical transformations 302
chemicals in drinking water 211-228
chemicals, regulation of 181-191
Chesapeake Bay 165,284,308
Chicago, Illinois 5,6
Chicago Ship Canal 6
Chlamydomonas sp. 288
chloralkyl acetate 151
chloramination 26
chloramine 4,6,97,165,269,271, 286,290,301,368,380
　concentrations in fish 263
　Also see dichloramine and monochloramine
chlordane 205,220

chloride 129
chlorinated acetone derivatives 151
chlorinated benzenes 206
chlorinated benzyl alcohol 151
chlorinated cooling waters 127
chlorinated effluent injector water 158
chlorinated effluents 284
chlorinated hydrocarbons 33,102, 222,408
　insecticides 206
chlorinated organic compounds 110,151,206,232,249,268, 286,308,313
　analysis of 118,139-159
　effects of 329-342
　extraction and analysis of 142
　in cooling waters and process effluents 105-135
　in disinfected municipal effluent 317
　in drinking water 77-104
　in wastewater effluent 311-328, 381
　Also see chlorinated hydrocarbons, chloro-organics and halogenated organics
chlorinated sewage effluent 234, 288,289,293,294,301,318,392
　secondary 120,127
　Also see superchlorinated wastewater
chlorinated surface waters 380
chlorinated waters 288,289,294
chlorinating agents 109
chlorination 284,291,299
chlorination by-products 284,302
chlorination cycle 360,361
chlorination/dechlorination practices in water treatment 1-18
chlorination effects 285
chlorination-induced oxidant residual 287
chlorination of organic compounds 313
chlorination, point of 99
chlorination practice 3
chlorination procedure 83
chlorination processes 283,285
chlorination programs 348-352,360

chlorination time 385
chlorination yield 121,123,125,128
chlorine
    analytical methods 50-60
    annual production 1,2
    attributes of 14
    chemistry 368-377
    control 11
    degradation model 285
    discharge 284
    hydrolysis 22
    impurities 16
    injection 288
    measurement 51,307,308
    persistence of 49,285
    stock solution preparation 81
    Also see aqueous chemistry of chlorine, chlorine residual, combined available chlorine, decay rates, degradation pathways, halogen, residual chlorine, and total residual chlorine
chlorine-containing constituents in chlorinated effluents 313
chlorine-containing species 108,330
chlorine demand 3,13,44,110,123, 170,308,347,359,396
    reactions 48
chlorine dioxide 4,14,420
chlorine dosage 4,44,119,129,198
    dose-time 387-390
chlorine exposure 389,390
Chlorine Flux Monitor 53
chlorine gas 40,118,285
chlorine radical 66
chlorine residual 5,12,129,293
    false 39
    measurement and persistence of 37-63
chlorine substitution reactions 113-116
chlorine toxicity 382
    Also see toxicity
chlorine use 3,422
    inventory data base 423, 424
    Also see chlorine
chlorite 176
chloroalkyl acetate 314

6-chloro-2-aminopurine 127
    Also see 6-chloroguanine
chloroanisoles 320
chlorobenzene 151,204,206
chlorobenzoic acid 127
3-chlorobenzoic acid 127,313
4-chlorobenzoic acid 127,313
chlorobromination 315
chlorobromobenzenes 204
cis-chlordane 317
trans-chlordane 317
8-chlorocaffeine 127,313
chlorocumene 155,314
chlorocyclohexane 151,155,314
5-chlorodeoxyuridine 236
chlorodibromomethane 204
chloroethane 197
chloroether 203
chloroethylbenzene 155,314
chloroform 15,28,29,31,34,78,79, 102,103,128,138,155,189,197, 204,213,219,220,223,254,255, 314,315
    concentration in drinking water 197
    formation of 26
    Also see haloform and trihalomethane formation
N-chloroglycine 63
6-chloroguanine 313
    Also see 6-chloro-2-aminopurine
chlorohydrin 70,116
3-chloro-4-hydroxybenzoic acid 127,313
4-chloromandelic acid 127,313
chloro-α-methylbenzyl alcohol 155,314
3-chloro-2-methylbut-1-ene 155,314
chloromethyl ethyl ether 221
chloromethyl methyl ether (CCME) 219
4-chloro-3-methylphenol 127,313
m-chloronitrobenzene 205
chloronium ion 28,66
chloro-organics 8,16,70,106,131, 392
    biomedical effects 193-209
    compounds 45,123,125,126,380
    dosages 258

production of 130,131
reaction yields 119
volatile 127
Also see chlorinated organic compounds
chlorophenol 68,69,127
p-chlorophenol 27
2-chlorophenol 127,313
3-chlorophenol 127,313
4-chlorophenol 127,313
4-chlorophenylacetic acid 127,313
chlorophyll *a* 273
N-chloropiperidine 47
4-chlororesorcinol 126,127,263, 268,273,295,298,313,330,332
5-chlorosalicylic acid 127,313
N-chlorosuccinimide 47
chlorotoluene 151
N-chloro-p-toluenesulfonamide 47
5-chlorouracil 119,126,127,234, 235,237,238,263,268,273, 296,298,313,330,332,418
5-chlorouridine 127,313
8-chloroxanthine 127,313
cholera 244-249
mortality rates 245
chromatographic profile 121,125, 126
chromatography 118,143,149-155
Also see gas chromatography/ mass spectrometry
chromium 59
hexavalent 221
chromosomal effects 233
chromosome breakage 237
chronic diseases 244
epidemiology 249-256
Cincinnati, Ohio 200,210
coagulation 11
Also see alum coagulation
cocarcinogens 217,221
Also see carcinogens
coelenterata 292
coliform 10
concentration, fecal 11
concentration, total 11
effluent 10
kill 41
standards 413
color removal 4

combined available chlorine (CAC) 43-46
Also see chlorine, chlorine residual, combined residual chlorine, and residual chlorine
combined residual chlorine 6,98, 359,403
stability of 48-50
concentration isopleth 385
condenser tubes 12
contact time 10,11
continuous amperometric cells 52
continuous exposure 380
continuous treatment 292
control 284
cooling systems 287,292
notation 363
power plant 287,367-378
Also see cooling tower systems and cooling water
cooling tower systems 345-365
cooling water 12,13,105-138
chlorinated 121-125
circuits 3
Also see cooling systems
copepods 291
copper 292
copper catalysts 47
cost-benefit analysis 182,183
critical thermal maxima (CTM) 292
cyanide 46
wastes 9
cyanogen chloride 268
cyclohexanedione-1,3 30
Cyclops bicuspidatus thomasi 272
Cynoscion regalis (grey seatrout) 294,298
Cyprinus carpio (carp) 263,295
cyst disinfection 41

Daphnia magna 271,273
data analysis 300
DDE 220,317
DDT 220,317
decane 221
decay rates 47,48
constants 47
for free chlorine 48
dechlorination 4,14,16,315,365, 417

degradation pathways 285
degradation processes 284,286
dehydroabietic acid 71,73
demographic variables 256
Denton, Texas 147,155
detachment 290,292
1,2-dibromoethane 220
dibromoiodomethane 155
dichloramine 43-45,108,129,130, 269
   organic 108
   Also see chloramine
dichloroacetate derivative 314
dichloroaniline derivative 314
dichloroaromatic derivative 314
dichlorobenzene 315,317,318
dichloro-*bis* (ethoxy) benzene 314
o-dichlorobenzene 155,314
p-dichlorobenzene 155,314
dichlorobiphenyl 317
dichlorobutane 314
1,1-dichloroethane 220
1,2-dichloroethane 78,205-219
1,1-dichloroethylene 205
dichloroethylenebenzene 314
dichloroiodomethane 155
dibromochloromethane 78,79,155, 219,314
dichloro-α-methyl benzyl alcohol 314
dichloromethoxytoluene 314
dichlorotoluene 155,314
dichromate 50
dieldrin 220
diethylparaphenylenediamine (DPD) 51
m-dihydroxy aromatic compounds 81
m-dihydroxybenzene 28
dilution 396
   rates 385
5,5-dimethylcyclohexanedione-1,3 30
diphenyl 322
diphenylether 322
disease incidence 256
disinfectants 96
disinfection 3,40,57,68,119,196, 312,421
   efficiency 41,43,347

objectives of 10
pH effect 41,42
proof of 9
temperature effect 42
displacement reaction 26
dissociation constant 22
dissolved oxygen 293,308
distress symptoms 298
diversity 294
diversity index 293
DNA 234,236,241,330
dodecane 221
dosage regimen 199
dose estimates 256
dose-response relationship 185,203, 337
dose-time estimation 398
drinking water 77-104,107,211-228
   contaminants 244
   sample extraction 143
   Also see tap water
Duluth, Minnesota 253

echiuroid 291
E. coli 42,234,236
ecosystems 284
   estuarine 285
   marine 285
effluents 284
   process 105-138
   secondary 11,47
eggs 291,295,297
E. histolytica 42
electric power generating plants 12,121,123,284
   Also see power plants and thermal electric power plants
electrode fouling 60
electrode interference 60
electrode sensitivity 62
electrophilic
   agents 25
   aromatic substitution 67
   attack 28,29,31
   chlorine 26
   properties 24
electrolytic conductivity detection 82
electropositive atom 25

Eliminius modestus 289
embryos 295,296,300
endrin 220
energy of permeation 59
enolate carbanion 29
Entermorpha sp. 292
entrainment 291
environmental chemical carcinogens 186
environmental stressors 212
EPA requirements 11
epidemiological approach 232
  to evaluation of water-borne carcinogens 243-258
epidemiologic studies 223,255
epizootics 223
epoxides 70,205
equilibrium concentrations 129
equilibrium constants 371,372
estuarine fish 292,294-296
estuarine water 48
ethanol 80
ethylene thiourea 221
Eubranchus sp. 292
exposure criteria 403,404
exposure, duration 300
extrapolation factors 224

fate of compounds 285,302
Federal Water Pollution Control Act 9
ferrous ion 44
  Also see iron
fertility 291
fertilization 287,289
filtration 10,11,86,103
fingerlings 297
fish analysis 325
fish egg hatching studies 332,335-337
fish contamination 321
fish kills 298
Florida Current 165
fluoride 80,165
5-fluorouracil 241
food-chain mechanisms 320
food packaging process water 3
fouling organisms 291
  Also see biofouling

frameshift mutations 204
free available chlorine 39,40
  Also see free chlorine
free chlorine 40-43,45,60,265,269, 380,403
  measurement 53
  residual 5,6,12,41
  stability of 46-48
  Also see free residual
free radical 33
  mechanisms 33
  processes 71
  reaction mechanisms 25
free residual 4
  chlorine 359
  Also see free chlorine
freshwater algae 273-274,382
  Also see algae
freshwater fish 263-270,294,382
  avoidance to chlorine 264
  larvae 263
  reproductive behavior 263
  size 263
  Also see estuarine fish and marine fish
freshwater invertebrates 270-273, 382
  Also see invertebrates and marine invertebrates
fulvic acid 81
  Also see humic substances
Fundulus heteroclitus (mummichog) 295,297

Gambusia affinis (mosquitofish) 322
gametes 287,289,291
gas chromatography (GC) 143
  Also see chromatography
gas chromatography/mass spectrometry (GC/MS) 202, 314,315,317,326
  Also see chromatography
Gasterosteus aculeatus (three-spine sticklebacks) 263
genetic damage 230
genetic risk 237,238
  Also see risk-benefit calculations
genome 236
germicidal activity 47
  Also see biocide

gill damage 296,297,299
glycine 45
Grand River (Canada) 4
Grandville, Michigan 315
granular activated carbon (GAC) 4
   and adsorption 84
Great Britain 292
growth rate 287,288

halobenzenes 203,204
haloethers 218,223
haloforms 203,204
   formation 32
   reaction 28-30,80,116,117
   Also see bromoforms, chloroforms, and trihalomethanes
halogen
   chemistry of 161-179
   electrode 55,58,59
   total organic 140
   Also see bromine, chlorine and iodine
halogenated compounds 285
halogenated hydrocarbons 408
halogenated organics 203,284,286
   in tap water 195-209
   Also see chlorinated organic compounds
halomethanes 197
   Also see trihalomethane formation
hatch, reduced 295,296
health effects assessment 189,190
hemoglobin 301
hepatocarcinogenic compounds 223
heptachlor 220
hexachloroacetone 155,314
hexachlorobenzene 15,322
hexachloro-1,2-butadiene 205
hexachloroethane 15
hexane extraction 325
high-pressure liquid chromatography (HPLC) 118
   Also see chromatography
histidine auxotrophs 233
Homarus americanus 290
humic acid 87-89,91-93
   Also see humic substances
humic substances 81,99,107,111,128

hydrocarbons 206,222
   chlorinated 15,102
   chlorinated and brominated 33
α-hydrogens 28
hydroids 291
hydrolysis 22,163
hypobromite 78,165,286
hypobromous acid (HOBr) 31,34, 165
hypochlorite 22,78,118,129,163, 286,291,346,368
   interference 59
   solutions, decomposition of 164
hypochlorite ion (OCl^-) 38,40,108, 285
hypochlorous acid (HOCl) 22,24, 34,37,38,40,108,109,129,163, 285,368
   as electrophile 25,26
   membrane electrode 55-60
   reactivity with inorganic anions 25

impact 283-310
indanedione 30
indoles 112
industrial wastes 12
inhibition 289
insecticides 206
interference 54,144,365
   error 53
interhalogen complexes 167
intermittent discharge of chlorine 265
intermittent doses 13
intermittent exposure 380,415
intermittent treatment 292
intoxication, chronic 200
invertebrates 287,292,302
iodate 165
iodide 80,165
iodine 14,17,46,50,57
   Also see halogen
iodometric method 168
ion exchange chromatography, high pressure 118,149-151
   Also see chromatography
iron 6,46,286,310
   removal process 7

irradiation 421
isopropylbenzene 205

James River 298,299

Kansas City, Missouri 5
kepone 299
Keratella cochlearis 272
kinetic data 369-377
Kingston Steam Plant 121,122

lagoons 293
Lake Michigan 5,6,273
larvae 291,295,296,300
LC_{50} values 292,297,299,307,308
lead 223
Leistomus xanthuras 294,298
Lepomis macrochirus (bluegills) 266
leuco crystal violet 51
lime softening 91
Limnocalanus macrurus 272
lipophilic compounds 320
liquid scintillation counting 118
Little Patuxent River 293
liver cancer 186
London cholera study 244
lung cancer 254

main sterilization 7
mammalian cellular transformation assay 202
manganese 6,46,286,310
manganese dioxide 50,60
marine chemistry of chlorination 402
marine ecosystems 283,285
marine fish 292-301
marine invertebrates 287,289-292
marine phytoplankton 287,288
Maryland 284,293
mass spectrometry-gas chromatography 143
maximum tolerable concentration 346
MBAS 297
measurement and persistence of chlorine residuals 37-63
mechanical draft cooling tower 347
mechanistic organic chemistry 66,71

Melita nitida 289
Memphis, Tennessee 123
Menidia beryllina (tidewater silversides) 295
Menidia menidia (Atlantic silversides) 295,297
mesothelioma 185,249,254
methemoglobin 301
methemoglobinemia 270
methylation 316
methylbromide 205
methylchloride 205
3-methylcholanthrene 221
methylene chloride 15,315
methyl ketones 28,80
N-methyl-trichloroaniline 155,314
Miami, Florida 200,201
microbial degradation 338
microcoulometric analysis 141
migration 290,293
minimization programs 412
minimum safe residual 41
minnows 262-264,270
mirex 221
Mississippi River 5,123-125,128, 368
Missouri River 5
mixing of chlorine with wastewater 11
model analysis 347-352
model expressions 255-259
modeling composition of chlorinated water from power plant cooling system 367-378
modeling residual chlorine levels 345-365
model notation 362
mollusk 292
monochloramine (NH_2Cl) 38,44, 45,47,49,53,63,108,129,170
   as an oxidizing agent 48
   decay rates 50
   organic 108
   persistence 48
   Also see chloramine
monochloroglycine 45
morone sp. 294,295
morone americana (white perch) 297
morone saxatilis (striped bass) 263,295,299

434 WATER CHLORINATION

mortality 289-291,293-297
  chlorine-caused 289-291,293-296,391
  estimates 392
  prediction 381
  threshold 382,385
  threshold, chronic 384
  Also see toxicity
most probable number (MPN) of coliform organisms 10
motility 289,291
municipal treatment plant 284
municipal waste effluent 315
municipal wastewater chlorination 139-159
mussels 290,292
mustard gas 251
mutagenesis 200,201
  assay 202
mutagenic
  activity 204,236
  compounds 205
  risk 229-241,405
  risk assessment 233-235
mutagenicity 231
  in vitro tests for 232
mutation
  dominant lethal 237
  frequency 239
  spontaneous frequency 238,239
Mytilus edulus 290

B-naphthylamine 251
nauplii 291
National Organics Reconnaissance Survey (NORS) 78,79,214,254
natural draft cooling tower 347
natural waters 107
neonatal rat 202
neoplasia 217,223
New Orleans, Louisiana 189,200,201,223
nickel 222
nitrate 310
nitrification 10,11
nitrile 286
nitrite 25,44,310
nitroanisole 205

nitrogen-containing constituents in sewage effluents 112
nitrogen, organic 129,310
nitrogen, proteinaceous 310
nitrogen, total 310
nitrogen trichloride (NCl_3) 108
cis-nonachlor 317
trans-nonachlor 317
nonaqueous solvents 24
Notropis volucellus (mimic shiners) 264
nucleophilicity 26
nucleosides 127

Oak Ridge, Tennessee 118
octadecane 221
Ohio River 82
olefins 70
oleic acid 70
once-through systems 13
onchorynchus gorbuscha (pink salmon) 293-297
Onchorynchus kitusch (coho salmon) 263
Onchorynchus nerka (sockeye salmon) 293-297
Onchorynchus tshawytscha (chinook salmon) 266
optimum design 11
Orcanectes virilis (crayfish) 271
organic acids 114
organic carbon 129
organic chemicals, classification in drinking water 214-215
organic chloramines 46,55
organic compounds, occurrence in tap water 198
organic concentrates from tap water 201
organic constituents 107,108
organic contaminants 78,204
organic content of finished water 241
organic decomposition 293
organochemical implications of water chlorination 65-75
organochlorine compounds 281,312,315
  Also see chlorinated organic compounds

organohalides 155,156
organophosphate 206
Ostrea edulis 289
osyter larvae 291
Ottuma, Iowa 200,201
Overton-Meyer theory 320
oxidants 285,291
oxidation 109,110
oxidative pyrolysis 142
oxidizing agents 39,47,50
oxidizing reactions 24
oxidizing species 22
oxygen electrode 59
ozonation 222,315
    of sea water 176
ozone 4,8,14,50,419,420

Palaemonetes pugio 290
Paleomynetes (decapod) 308
Pandalus danae 292
particulates 84-87
partition coefficient 317,322
    water-solvent 324
Passaic River 50
pathogenesis 203
pathologic lesions 203
Patuxent River 6,291
    Also see Little and Upper
        Patuxent Rivers
Pennsylvania 293
pentachloroacetone 155,314
pentachloroanisole 316,317,319
pentachlorobiphenyl 317
pentachlorophenol 317
Perca flavescens (yellow perch)
    266
peroxide 50
persistence 46-50,285
pesticides 315
pH 308
pH effect 68,285
phenol 67,69,127,310,313
phenolic compounds 27,114
pheromone systems 130
Philadelphia, Pennsylvania
    6,200,201
photochemical generation 25
photolysis 285,338
photosynthetic activity 287,288

photosynthetic rate 273
Phragmatopoma californica 289
phytoplankton 273,287,288
    Also see marine phytoplankton
phytoplankton productivity 287
Pimephales promelas (fathead
    minnows) 263,268,270,311
Pimephales vigilax (bullhead
    minnows) 264
plankton populations 397
Pleuronectes platessa (plaice) 295,
    297
plume
    chlorine concentration estimates
        392
    discharge 386
    entrainment 385
    mortality 388
polar solvents 28
polychlorinated aromatics
    70
polychlorinated biphenyls (PCB)
    315
polycyclic aromatic hydrocarbons
    222
polymers 128
polynuclear aromatic hydrocarbons
    206
polynuclear hydrocarbons 223
polyurethane 221,228
polyzoa 292
Pomatomus saltratrix (bluefish)
    294,298
population genome 204
population size 294
postammoniation 6
posttreatment 4
potable water 3-9,410
potassium permanganate 3
Potomac River 6
power plants 367-378,381
    effluent 35
    Also see electric power-
        generating plants
prechlorination 4
precursor compounds 80,99,100
    Also see acetone, acetyl com-
        pounds, carbonyl groups,
        fluvic acid and humic acid
precursor removal 84,99

prediction and composition of chlorinated cooling waters 347-352,369,370
prediction of toxic effects of chlorinated effluents 379-398
  Also see toxicity
predictive capability 369
productivity 287,288
prokaryotic systems 234
proteinaceous materials 110,138
proton equilibrium in sea water 163
public health problems 231
pulp and paper industry 3
purines 112,116,127
pyridine 28,112
pyrimidine 112,115,127
pyrolysis 142
pyrolysis/coulometric titration 144
pyrrole 28

radiation 251
range-finding 201
  assay 200
rate constants 369,372,373
rate-determining step 29
reaction mechanism 27,32
reaction
  pathway 23,285-286
  pattern 32
  rate 23
reactivity 23,24,32
rechlorination 14
recirculation systems 13,347
redox catalysts 47
reducing agents 44
regulation 284
report data, significant 302,308,310
reservoirs 7
residence time 385
residual chlorination 299-301
residual chlorine 55,168,264,269, 289,290,294-296,403
  Also see chlorine and combined residual chlorine
residual data feedback 351,352, 360
residual effective disinfectant 60

residual halogen concentration, measurement of 59
residual oxidizing components 174
resorcinol 28
respiration 273
reverse osmosis 199
Rhinchthys atratulus (blacknose dace) 264,265,269
risk-benefit calculations 404
RNA 241
Runge-Kutte technique 370

Sacramento, California 6
safety evaluation 197-207
saline waters 286
salinity 285,310
Salmo clarki (cutthroat trout) 268
Salmo gairdneri (rainbow trout) 264,265,268,321
Salmon
  chinook 266,297
  coho 263,267,269
  pink 266,297
  sockeye 297
Salmonella typhimurium 200,204 233
salmonids 262
Salvelinus fontinales (brook trout) 267
San Francisco 4,5
San Joaquin, California 6
Santa Monica Bay 301
screening tests 202
sea urchin, *Strongylocentrotus purpuratus* 287
sea water 18,161,179,283-310
Seattle, Washington 200,201
secondary treatment standards 422
sediments 308
sedimentation 338
Seine River (France) 4
sewage treatment plant effluents 118-121,234,293
  chlorination 116-125
  domestic 111
  secondary 128,129
shellfish 10
shrimp
  coonstripe 292
  grass 291

site-specific regulations 416,421
Skeletonema costatum 287,288
slime control 246
smoking 251
Snow, John 244
sodium hypochlorite 285,288, 294
Solea solea (Dover sole) 295,297
species-dependent response 274
species diversity index 293
species shifts 130
species specificity 217
specific-locus test 238
sperm 289,291
split stream chlorination 352,360
St. Croix River 273
St. Louis, Missouri 5
stabilized neutral orthotolodine (SNORT) 52
standards 302
stereochemistry 70
Strasburg, Pennsylvania 8
strobane 221
Strongylocentrotus purpuratus 289
sublethal effects 384
substitution reaction 109
sulfide 44
sulfite 25
sulfur dioxide 4
sunlight 285,308
sunlight as a catalyst 47
superchlorinated wastewater 140, 145,149,155
suspended solids 12,365
swimming pools 3,17
synergism 203,301,339
synergistic effects 268,269
syringaldazine 52

tap water 156,195-207
 Also see drinking water
target organs 217
taste and odor control 4,7
taste threshold 321
temperature 285,297,301,308
 acclimation and shock 267
 effects on trihalomethane formation 96

temperature dependence electrode 59
temperature-dependent expressions 370
temperature-dependent response 300
Tennessee Valley Authority 121
teratogenesis 200
α-terpineol 70,72
test species 206
tetrachloroacetone 155,314
tetrachloroanisole 317
tetrachlorobiphenyl 317
tetrachlorodimethoxybenzene 155, 314
tetrachloroethane 220
1,1,2,2-tetrachloroethylene 315
tetrachloroethylene 220
tetrachloroethylstyrene 155,314
tetrachloromethyoxytoluene 314
tetrachlorophenol 155,314,317
tetrachlorophthalate 155
 derivative 314
tetrachloropropylmethoxynaph-thalene 317
Thames River 245
thermal electric power plants 347
thermal stress 291,296,300
thermodynamic data 369-377
thymine-requiring bacteria 234
Tilapia aurea 266
time-concentration relationship 414
Tl$_m$ 297
toluene 68,313
total organic carbon (TOC) 82
 nonvolatile (NVTOC) 82
total organic chlorine determinations 141-143
total organic halogen (TOCl) 140
 method, evaluation of 144-149
total residual chlorine 287,359,382
 Also see chlorine
toxic effects of chlorinated effluents, assessment and prediction 379-398
toxicity 190,194-209,235, 284,287
 acute 199,383,387
 estuarine data 292-302

exposure in freshwater organisms 261-282
exposure in marine organisms 287-303
exposure time in fish 265,266
exposure level for invertebrates 271,287
marine data 182,287-302, 402
mechanisms of action 269,270
pH effect on fish 267,268
screening methods 324
species-related factors 262-265
temperature effect in fish 266, 267,272
thresholds 337,381
water quality and chemistry in fish 267
water quality effect on invertebrates 272,273
toxicokinetics 203,204
toxicologic screens 201
toxicological evaluation 199
toxicology program on halogenated organics 203
toxic potential of organic compounds 199
trace metal contaminants 222
transient species 22
transport 285
tribromoanisole 317
trichlorobenzene 314,315,317
trichlorobiphenyl 317
trichlorocumene 155,314
trichlorodimethoxybenzene 155
1,1,1-trichloroethane 315
1,1,2-trichloroethane 220
trichloroethylbenzene 155,314
trichloroethylene 220
1,1,2-trichloroethylene 315
trichloro-α-methyl benzyl alcohol 314
trichloromethylstyrene 155,314
trichlorophenol 155,317
trichlorophthalate 155,314
  derivative 314
trichloropropylmethoxynapthalene 317
trickling filter effluents 48

trihalomethane formation 16,78,79, 137,214
  analysis 82
  effect of disinfectant 96-99
  effect of filtration of humic acid substances 87-89
  effect of particulates 84-87
  effect of temperature 96
  health significance 80
  pH effect 89-96
  reaction rate of humic acid 91-93
  Also see bromoform and chloroform
Trinectes maculatus (hog choker) 297
tumor induction 213
turbidity 310

ultraviolet catalysis 75
ultraviolet light 46,285
Upper Patuxent River 293
Urechis caupo 289

vinyl chloride 185,197,205,220, 249,251,253
vinylidine chloride 197
Virginia 293
virus inactivation 41
virus removal 11
volatile fraction 197
volatile organohalides 155,156
  Also see haloforms and trihalomethane formation

Washington Suburban Sanitary Commission 6
waste treatment effluents 311-328
wastewater 46,107,287,410
  chlorination 139-159
  extract 152-154
  treatment 3,9-12
  Also see sewage treatment plant effluents, municipal waste effluent, chlorinated sewage effluent, and effluents

water analysis   368
water contact sports   10
water hardness   282
water quality   3
water treatment   21,214
Watts Bar Lake   121-123,368
white catfish   293
white perch   293
Winnipeg, Canada   7
Wyoming, Michigan   315

XAD-2-resins   141,144,151, 159,325

York River   49

zooplankton mortality data   334
zooplankton studies   331-335

DATE DUE

~~July 25/98~~	MAY 0 4 1995
3660561	
ILL 5438992	due 10/25/98 COF

DEMCO, INC. 38-2931